ELECTRIC POWER SYSTEMS

ELECTRIC POWER SYSTEMS

A CONCEPTUAL INTRODUCTION

Alexandra von Meier

IEEE PRESS

A JOHN WILEY & SONS, INC., PUBLICATION

Copyright © 2006 by John Wiley & Sons, Inc. All rights reserved

Published by John Wiley & Sons, Inc., Hoboken, New Jersey
Published simultaneously in Canada

No part of this publication may be reproduced, stored in a retrieval system or transmitted in any form or by any means, electronic, mechanical, photocopying, recording, scanning or otherwise, except as permitted under Sections 107 or 108 of the 1976 United States Copyright Act, without either the prior written permission of the Publisher, or authorization through payment of the appropriate per-copy fee to the Copyright Clearance Center, 222 Rosewood Drive, Danvers, MA 01923, (978) 750-8400, fax (978) 750-4470. Requests to the Publisher for permission should be addressed to the Permissions Department, John Wiley & Sons, Inc., 111 River Street, Hoboken, NJ 07030, (201) 748-6011, fax (201) 748-6008.

Limit of Liability/Disclaimer of Warranty: While the publisher and author have used their best efforts in preparing this book, they make no representations or warranties with respect to the accuracy or completeness of the contents of this book and specifically disclaim any implied warranties of merchantability or fitness for a particular purpose. No warranty may be created or extended by sales representatives or written sales materials. The advice and strategies contained herein may not be suitable for your situation. You should consult with a professional where appropriate. Neither the publisher nor author shall be liable for any loss of profit or any other commercial damages, including but not limited to special, incidental, consequential, or other damages.

For general information on our other products and services or for technical support, please contact our Customer Care Department within the United States at (800) 762-2974, outside the United States at (317) 572-3993 or fax (317) 572-4002.

Wiley also publishes its books in a variety of electronic formats. Some content that appears in print may not be available in electronic formats. For more information about Wiley products, visit our Web site at www.wiley.com.

Library of Congress Cataloging-in-Publication Data:

Meier, Alexandra von.
  Electric power systems: a conceptual introduction/by Alexandra von Meier.
    p. cm.
  "A Wiley-Interscience publication."
  Includes bibliographical references and index.
  ISBN-13: 978-0-471-17859-0
  ISBN-10: 0-471-17859-4
    1. Electric power systems. I. Title

TK1005.M37 2006
621.31–dc22
                                                            2005056773

Printed in the United States of America

15 14 13 12

*To my late grandfather
Karl Wilhelm Clauberg
who introduced me to
The Joy of Explaining Things*

CONTENTS

Preface — xiii

1. The Physics of Electricity — 1

  1.1 Basic Quantities — 1
    1.1.1 Introduction — 1
    1.1.2 Charge — 2
    1.1.3 Potential or Voltage — 3
    1.1.4 Ground — 5
    1.1.5 Conductivity — 5
    1.1.6 Current — 6
  1.2 Ohm's law — 8
    1.2.1 Resistance — 9
    1.2.2 Conductance — 10
    1.2.3 Insulation — 11
  1.3 Circuit Fundamentals — 11
    1.3.1 Static Charge — 11
    1.3.2 Electric Circuits — 12
    1.3.3 Voltage Drop — 13
    1.3.4 Electric Shock — 13
  1.4 Resistive Heating — 14
    1.4.1 Calculating Resistive Heating — 15
    1.4.2 Transmission Voltage and Resistive Losses — 17
  1.5 Electric and Magnetic Fields — 18
    1.5.1 The Field as a Concept — 18
    1.5.2 Electric Fields — 19
    1.5.3 Magnetic Fields — 21
    1.5.4 Electromagnetic Induction — 24
    1.5.5 Electromagnetic Fields and Health Effects — 25
    1.5.6 Electromagnetic Radiation — 26

2. Basic Circuit Analysis — 30

  2.1 Modeling Circuits — 30

2.2	Series and Parallel Circuits		31
	2.2.1	Resistance in Series	32
	2.2.2	Resistance in Parallel	33
	2.2.3	Network Reduction	35
	2.2.4	Practical Aspects	36
2.3	Kirchhoff's Laws		37
	2.3.1	Kirchhoff's Voltage Law	38
	2.3.2	Kirchhoff's Current Law	39
	2.3.3	Application to Simple Circuits	40
	2.3.4	The Superposition Principle	41
2.4	Magnetic Circuits		44

3. AC Power 49

3.1	Alternating Current and Voltage		49
	3.1.1	Historical Notes	49
	3.1.2	Mathematical Description	50
	3.1.3	The rms Value	53
3.2	Reactance		55
	3.2.1	Inductance	55
	3.2.2	Capacitance	58
	3.2.3	Impedance	64
	3.2.4	Admittance	64
3.3	Power		66
	3.3.1	Definition of Electric Power	66
	3.3.2	Complex Power	68
	3.3.3	The Significance of Reactive Power	73
3.4	Phasor Notation		75
	3.4.1	Phasors as Graphics	75
	3.4.2	Phasors as Exponentials	78
	3.4.3	Operations with Phasors	80

4. Generators 85

4.1	The Simple Generator		86
4.2	The Synchronous Generator		92
	4.2.1	Basic Components and Functioning	92
	4.2.2	Other Design Aspects	97
4.3	Operational Control of Synchronous Generators		100
	4.3.1	Single Generator: Real Power	100
	4.3.2	Single Generator: Reactive Power	101
	4.3.3	Multiple Generators: Real Power	107
	4.3.4	Multiple Generators: Reactive Power	112

4.4	Operating Limits		115
4.5	The Induction Generator		118
	4.5.1	General Characteristics	118
	4.5.2	Electromagnetic Characteristics	120
4.6	Inverters		123

5. Loads   127

5.1	Resistive Loads		128
5.2	Motors		131
5.3	Electronic Devices		134
5.4	Load from the System Perspective		136
	5.4.1	Coincident and Noncoincident Demand	137
	5.4.2	Load Profiles and Load Duration Curve	138
5.5	Single- and Multiphase Connections		140

6. Transmission and Distribution   144

6.1	System Structure		144
	6.1.1	Historical Notes	144
	6.1.2	Structural Features	147
	6.1.3	Sample Diagram	149
	6.1.4	Topology	150
	6.1.5	Loop Flow	153
	6.1.6	Stations and Substations	156
	6.1.7	Reconfiguring the System	158
6.2	Three-Phase Transmission		159
	6.2.1	Rationale for Three Phases	160
	6.2.2	Balancing Loads	163
	6.2.3	Delta and Wye Connections	164
	6.2.4	Per-Phase Analysis	166
	6.2.5	Three-Phase Power	166
	6.2.6	D.C. Transmission	167
6.3	Transformers		168
	6.3.1	General Properties	168
	6.3.2	Transformer Heating	170
	6.3.3	Delta and Wye Transformers	172
6.4	Characteristics of Power Lines		175
	6.4.1	Conductors	175
	6.4.2	Towers, Insulators, and Other Components	179
6.5	Loading		182
	6.5.1	Thermal Limits	182
	6.5.2	Stability Limit	183

6.6	Voltage Control	184
6.7	Protection	188
	6.7.1 Basics of Protection and Protective Devices	188
	6.7.2 Protection Coordination	192

7. Power Flow Analysis 195

7.1	Introduction	195
7.2	The Power Flow Problem	197
	7.2.1 Network Representation	197
	7.2.2 Choice of Variables	198
	7.2.3 Types of Buses	201
	7.2.4 Variables for Balancing Real Power	201
	7.2.5 Variables for Balancing Reactive Power	202
	7.2.6 The Slack Bus	204
	7.2.7 Summary of Variables	205
7.3	Example with Interpretation of Results	206
	7.3.1 Six-Bus Example	206
	7.3.2 Tweaking the Case	210
	7.3.3 Conceptualizing Power Flow	211
7.4	Power Flow Equations and Solution Methods	214
	7.4.1 Derivation of Power Flow Equations	214
	7.4.2 Solution Methods	217
	7.4.3 Decoupled Power Flow	224
7.5	Applications and Optimal Power Flow	226

8. System Performance 229

8.1	Reliability	229
	8.1.1 Measures of Reliability	229
	8.1.2 Valuation of Reliability	231
8.2	Security	233
8.3	Stability	234
	8.3.1 The Concept of Stability	234
	8.3.2 Steady-State Stability	236
	8.3.3 Dynamic Stability	240
	8.3.4 Voltage Stability	249
8.4	Power Quality	250
	8.4.1 Voltage	251
	8.4.2 Frequency	253
	8.4.3 Waveform	255

9. System Operation, Management, and New Technology — 259

- 9.1 Operation and Control on Different Time Scales — 260
  - 9.1.1 The Scale of a Cycle — 261
  - 9.1.2 The Scale of Real-Time Operation — 262
  - 9.1.3 The Scale of Scheduling — 264
  - 9.1.4 The Planning Scale — 267
- 9.2 New Technology — 268
  - 9.2.1 Storage — 268
  - 9.2.2 Distributed Generation — 271
  - 9.2.3 Automation — 278
  - 9.2.4 FACTS — 280
- 9.3 Human Factors — 281
  - 9.3.1 Operators and Engineers — 281
  - 9.3.2 Cognitive Representations of Power Systems — 282
  - 9.3.3 Operational Criteria — 285
  - 9.3.4 Implications for Technological Innovation — 291
- 9.4 Implications for Restructuring — 292

Appendix: Symbols, Units, Abbreviations, and Acronyms — 298

Index — 302

PREFACE

This book is intended to bridge the gap between formal engineering texts and more popularly accessible descriptions of electric power technology. I discovered this gap as a graduate student struggling to understand power systems—especially transmission and distribution systems—which had always fascinated me but which now invited serious study in the context of research on implementing solar energy. Although I had studied physics as an undergraduate, I found the subject of power systems difficult and intimidating.

The available literature seemed to fall into two categories: easy-to-read, qualitative descriptions of the electric grid for the layperson, on the one hand, and highly technical books and papers, on the other hand, written for professionals and electrical engineering majors. The second category had the information I needed, but was guarded by a layer of impenetrable phasor diagrams and other symbolism that obviously required a special sort of initiation.

I was extremely fortunate to have access to some of the most highly respected scholars in the field at the University of California, Berkeley, who were also generous, patient, and gifted teachers. Thus I survived Leon Chua's formidable course on circuit analysis, followed by two semesters of power engineering with Felix Wu. This curriculum hardly made me an expert, but it did enable me to decipher the language of the academic and professional literature and identify the issues relevant to my work.

I enjoyed another marvelous learning opportunity through a research project beginning in 1989 at several large nuclear and fossil-fueled steam generation plants, where our team interviewed the staff as part of a study of "High-Reliability Organizations." My own subsequent research on power distribution took me into the field with five U.S. utilities and one in Germany. Aside from the many intriguing things we learned about the operating culture in these settings, I discovered how clearly the power plant staff could often explain technical concepts about their working systems. Their language was characteristically plain and direct, and was always guided by practical considerations, such as what this dial tells you, or what happens when you push that button.

In hindsight, the defining moment for inspiring this book occurred in the Pittsburg control room when I revealed my ignorance about reactive power (just after having boasted about my physics degree, to the operators' benign amusement). They generously supplied me with a copy of the plant operating manual, which turned out to contain the single most lucid and comprehensible explanation of electric generators, including reactive power, I had seen. That manual proved to me that it is possible to write about electric power systems in a way that is accessible to audiences who have not undergone the initiation rites of electrical engineering, but who nevertheless want to get the real story. This experience suggested there might be other people much like myself—outside the power industry, but vitally concerned with it—who could benefit from such a practical approach.

After finishing my dissertation in 1995, I decided to give it a try: My goal was to write the book that I would have liked to read as a student six or seven years earlier. Considering that it has taken almost a decade to achieve, this turned out to be a much more ambitious undertaking than I imagined at the outset. A guiding principle throughout my writing process was to assume a minimum of prior knowledge on the part of the readers while trying to relate as much as possible to their direct experience, thus building a conceptual and intuitive understanding from the ground up. I hope the book will serve as a useful reference, and perhaps even as a source of further inspiration for others to study the rich and complex subject of electric power.

I envision two main audiences for this book. The first consists of students and researchers who are learning about electric circuits and power system engineering in an academic setting, and who feel that their understanding would be enhanced by a qualitative, conceptual emphasis to complement the quantitative methods stressed in technical courses. This audience might include students of diverse backgrounds or differing levels of preparation, perhaps transferring into an engineering program from other disciplines. Such students often need to solidify their understanding of basic information presumed to be second nature for advanced undergraduates in technical fields. As a supplement to standard engineering texts, this volume aims to provide a clear and accessible review of units, definitions, and fundamental physical principles; to explain in words some of the ideas conventionally shown by equations; to contextualize information, showing connections among different topics and pointing out their relevance; and to offer a glimpse into the practical world of the electric power industry.

The second major audience consists of professionals working in and around the power industry whose educational background may not be in electrical engineering, but who wish to become more familiar with the technical details and the theoretical underpinnings of the system they deal with. This group might include analysts and administrators and managers coming from the fields of business, economics, law, or public policy, as well as individuals with technical or multidisciplinary training in areas other than power engineering. In view of the scope and importance of contemporary policy decisions about electricity supply and delivery, both in the United States and abroad—from the siting of power generation and transmission facilities to market regulation and restructuring—a real need appears for a coherent, general education on the subject of power systems. My hope is that this volume can make a meaningful contribution.

ACKNOWLEDGMENTS

Many individuals and organizations have made the writing of this text possible. I am deeply grateful to my teachers for their mentorship, inspiration, and clarity, most especially Gene Rochlin, Felix Wu, and Oscar Ichazo. I am also indebted to the many professionals who took the time to show me around power systems in the field and teach me about their work. The project was supported directly and indirectly by a University of California President's Postdoctoral Fellowship, the University of California Energy Institute (UCEI), the California Energy Commission's Public Interest Energy Research (PIER) program, and the California Institute for Energy Efficiency (CIEE). I especially thank Carl Blumstein, Severin Borenstein, Ron Hofmann, Laurie ten Hope, and Gaymond Yee for their encouragement at various stages in the process of writing this book.

I also thank all my colleagues who graciously read, discussed, and helped improve portions of the draft manuscript. They include Raquel Blanco, Alex Farrell, Hannah Friedman, John Galloway, Chris Greacen, Sean Greenwalt, Dianne Hawk, Nicole Hopper, Merrill Jones, Chris Marnay, Andrew McAllister, Alex McEachern, Steve Shoptaugh, Kurt von Meier, and Jim Williams. My gratitude also extends to all others who participated in the development of the text—particularly my students who never cease to ask insightful and challenging questions—and to my friends and family who offered encouragement and support. Darcy McCormick, Thomas Harris, and Steve Shoptaugh helped prepare illustrations, Cary Berkley organized the manuscript, and Trumbull Rogers improved it by careful editing. Of course, I am solely responsible for any errors. As an endeavor that has not, to my knowledge, been attempted before, this text is necessarily a work in progress. Suggestions from readers for improving its accuracy and clarity will be warmly welcomed.

ALEXANDRA VON MEIER
Sebastopol, California
August 2005

CHAPTER 1

The Physics of Electricity

1.1 BASIC QUANTITIES

1.1.1 Introduction

This chapter describes the quantities that are essential to our understanding of electricity: charge, voltage, current, resistance, and electric and magnetic fields. Most students of science and engineering find it very hard to gain an intuitive appreciation of these quantities, since they are not part of the way we normally see and make sense of the world around us. Electrical phenomena have a certain mystique that derives from the difficulty of associating them with our direct experience, but also from the knowledge that they embody a potent, fundamental force of nature.

Electric charge is one of the basic dimensions of physical measurement, along with mass, distance, time and temperature. All other units in physics can be expressed as some combination of these five terms. Unlike the other four, however, charge is more remote from our sensory perception. While we can easily visualize the size of an object, imagine its weight, or anticipate the duration of a process, it is difficult to conceive of "charge" as a tangible phenomenon.

To be sure, electrical processes are vital to our bodies, from cell metabolism to nervous impulses, but we do not usually conceptualize these in terms of electrical quantities or forces. Our most direct and obvious experience of electricity is to receive an electric shock. Here the presence of charge sends such a strong wave of nervous impulses through our body that it produces a distinct and unique sensation. Other firsthand encounters with electricity include hair that defiantly stands on end, a zap from a door knob, and static cling in the laundry. Yet these experiences hardly translate into the context of electric power, where we can witness the *effects* of electricity, such as a glowing light bulb or a rotating motor, while the essential happenings take place silently and concealed within pieces of metal. For the most part, then, electricity remains an abstraction to us, and we rely on numerical and geometric representations—aided by liberal analogies from other areas of the physical world—to form concepts and develop an intuition about it.

Electric Power Systems: A Conceptual Introduction, by Alexandra von Meier
Copyright © 2006 John Wiley & Sons, Inc.

2 THE PHYSICS OF ELECTRICITY

1.1.2 Charge

It was a major scientific accomplishment to integrate an understanding of electricity with fundamental concepts about the microscopic nature of matter. Observations of static electricity like those mentioned earlier were elegantly explained by Benjamin Franklin in the late 1700s as follows: There exist in nature two types of a property called *charge*, arbitrarily labeled "positive" and "negative." Opposite charges attract each other, while like charges repel. When certain materials rub together, one type of charge can be transferred by friction and "charge up" objects that subsequently repel objects of the same kind (hair), or attract objects of a different kind (polyester and cotton, for instance).

Through a host of ingenious experiments,[1] scientists arrived at a model of the atom as being composed of smaller individual particles with opposite charges, held together by their electrical attraction. Specifically, the nucleus of an atom, which constitutes the vast majority of its mass, contains *protons* with a positive charge, and is enshrouded by *electrons* with a negative charge. The nucleus also contains neutrons, which resemble protons, except they have no charge. The electric attraction between protons and electrons just balances the electrons' natural tendency to escape, which results from both their rapid movement, or kinetic energy, and their mutual electric repulsion. (The repulsion among protons in the nucleus is overcome by another type of force called the *strong nuclear interaction*, which only acts over very short distances.)

This model explains both why most materials exhibit no obvious electrical properties, and how they can become "charged" under certain circumstances: The opposite charges carried by electrons and protons are equivalent in magnitude, and when electrons and protons are present in equal numbers (as they are in a normal atom), these charges "cancel" each other in terms of their effect on their environment. Thus, from the outside, the entire atom appears as if it had no charge whatsoever; it is *electrically neutral*.

Yet individual electrons can sometimes escape from their atoms and travel elsewhere. Friction, for instance, can cause electrons to be transferred from one material into another. As a result, the material with excess electrons becomes *negatively charged*, and the material with a deficit of electrons becomes *positively charged* (since the positive charge of its protons is no longer compensated). The ability of electrons to travel also explains the phenomenon of *electric current*, as we will see shortly.

Some atoms or groups of atoms (molecules) naturally occur with a net charge because they contain an imbalanced number of protons and electrons; they are called *ions*. The propensity of an atom or molecule to become an ion—namely, to release electrons or accept additional ones—results from peculiarities in the geometric pattern by which electrons occupy the space around the nuclei. Even electrically neutral molecules can have a local appearance of charge that results from

---

[1]Almost any introductory physics text will provide examples. For an explanation of the basic concepts of electricity, I recommend Paul Hewitt, *Conceptual Physics*, Tenth Edition (Menlo Park, CA: Addison Wesley, 2006).

imbalances in the spatial distribution of electrons—that is, electrons favoring one side over the other side of the molecule. These electrical phenomena within molecules determine most of the physical and chemical properties of all the substances we know.[2]

While on the microscopic level, one deals with fundamental units of charge (that of a single electron or proton), the practical unit of charge in the context of electric power is the *coulomb* (C). One coulomb corresponds to the charge of 6.25×10^{18} protons. Stated the other way around, one proton has a charge of 1.6×10^{-19} C. One electron has a negative charge of the same magnitude, -1.6×10^{-19} C. In equations, charge is conventionally denoted by the symbol Q or q.

1.1.3 Potential or Voltage

Because like charges repel and opposite charges attract, charge has a natural tendency to "spread out." A local accumulation or deficit of electrons causes a certain "discomfort" or "tension":[3] unless physically restricted, these charges will tend to move in such a way as to relieve the local imbalance. In rigorous physical terms, the discomfort level is expressed as a level of *energy*. This energy (strictly, electrical potential energy), said to be "held" or "possessed" by a charge, is analogous to the mechanical potential energy possessed by a massive object when it is elevated above the ground: we might say that, by virtue of its height, the object has an inherent potential to fall down. A state of lower energy—closer to the ground, or farther away from like charges—represents a more "comfortable" state, with a smaller potential fall.

The potential energy held by an object or charge in a particular location can be specified in two ways that are physically equivalent: first, it is the *work*[4] that would be required in order to move the object or charge *to* that location. For example, it takes work to lift an object; it also takes work to bring an electron near an accumulation of more electrons. Alternatively, the potential energy is the work the object or charge would do in order to move *from* that location, through interacting with the objects in its way. For example, a weight suspended by a rubber band will stretch the rubber band in order to move downward with the pull of gravity (from higher to lower gravitational potential). A charge moving toward a more comfortable location might do work by producing heat in the wire through which it flows.

---

[2] For example, water owes its amazing liquidity and density at room temperature to the electrical attraction among its neutral molecules that results from each molecule being polarized: casually speaking, the electrons prefer to hang out near the oxygen atom as opposed to the hydrogen atoms of H_2O; a chemist would say that oxygen has a greater *electronegativity* than hydrogen. The resulting attraction between these polarized ends of molecules is called a *hydrogen bond*, which is essential to all aspects of our physical life.

[3] The term *tension* is actually synonymous with voltage or potential, mainly in British usage.

[4] In physics, work is equivalent to and measured in the same units as energy, with the implied sense of exerting a force to "push" or "pull" something over some distance (Work = Force × Distance).

This notion of work is crucial because, as we will see later, it represents the physical basis of transferring and utilizing electrical energy. In order to make this "work" a useful and unambiguous measure, some proper definitions are necessary. The first is to explicitly distinguish the contributions of charge and potential to the total amount of work or energy transferred. Clearly, the amount of work in either direction (higher or lower potential) depends on the amount of mass or charge involved. For example, a heavy weight would stretch a rubber band farther, or even break it. Similarly, a greater charge will do more work in order to move to a lower potential.

On the other hand, we also wish to characterize the location proper, independent of the object or charge there. Thus, we establish the rigorous definition of the electric *potential*, which is synonymous with *voltage* (but more formal). The electric potential is the potential energy possessed by a charge at the location in question, relative to a reference location, divided by the amount of its charge. Casually speaking, we might say that the potential represents a measure of how comfortable or uncomfortable it would be for any charge to reside at that location. A potential or voltage can be positive or negative. A positive voltage implies that a positive charge would be repelled, whereas a negative charge would be attracted to the location; a negative voltage implies the opposite.

Furthermore, we must be careful to specify the "reference" location: namely, the place where the object or charge was moved from or to. In the mechanical context, we specify the height *above ground level*. In electricity, we refer to an electrically neutral place, real or abstract, with *zero* or *ground potential*. Theoretically, one might imagine a place where no other charges are present to exert any forces; in practice, ground potential is any place where positive and negative charges are balanced and their influences cancel. When describing the potential at a single location, it is implicitly the potential *difference* between this and the neutral location. However, potential can also be specified as a difference between two locations of which neither is neutral, like a difference in height.

Because electric potential or voltage equals energy per charge, the units of voltage are equivalent to units of energy divided by units of charge. These units are *volts* (V). One volt is equivalent to one joule per coulomb, where the joule is a standard unit of work or energy.[5]

Note how the notion of a difference always remains implicit in the measurement of volts. A statement like "this wire is at a voltage of 100 volts" means "this wire is at a voltage of 100 volts *relative to ground*," or "the voltage *difference between the wire and the ground* is 100 volts." By contrast, if we say "the battery has a voltage of 1.5 volts," we mean that "the voltage *difference between the two terminals* of the battery is 1.5 volts." Note that the latter statement does not tell us the potential of either terminal in relation to ground, which depends on the type of battery and whether it is connected to other batteries.

---

[5] A joule can be expressed as a watt-second; 1 kilowatt-hour = 3.6×10^6 joules, as there are 3600 seconds in an hour.

In equations, voltage is conventionally denoted by *E*, *e*, *V*, or *v* (in a rare and inelegant instance of using the same letter for both the symbol of the quantity and its unit of measurement).

1.1.4 Ground

The term *ground* has a very important and specific meaning in the context of electric circuits: it is an electrically neutral place, meaning that it has zero voltage or potential, which moreover has the ability to absorb excesses of either positive or negative charge and disperse them so as to *remain* neutral regardless of what might be electrically connected to it. The literal ground outdoors has this ability because the Earth as a whole acts as a vast reservoir of charge and is electrically neutral, and because most soils are sufficiently conductive to allow charge to move away from any local accumulation. The term *earth* is synonymous with ground, especially in British usage. A circuit "ground" is constructed simply by creating a pathway for charge into the earth. In the home, this is often done by attaching a wire to metal water pipes. In power systems, ground wires, capable of carrying large currents if necessary, are specifically dug into the earth.

1.1.5 Conductivity

To understand conductivity, we must return to the microscopic view of matter. In most materials, electrons are bound to their atoms or molecules by the attraction to the protons in the nuclei. We have mentioned how special conditions such as friction can cause electrons to escape. In certain materials, some number of electrons are always free to travel. As a result, the material is able to *conduct* electricity. When a charge (i.e., an excess or deficit of electrons) is applied to one side of such a conducting material, the electrons throughout will realign themselves, spreading out by virtue of their mutual repulsion, and thus conduct the charge to the other side.

For this to happen, an individual electron need not travel very far. We can imagine each electron moving a little to the side, giving its neighbor a repulsive "shove," and this shove propagating through the conducting material like a wave of falling dominoes.

The most important conducting materials in our context are metals. The microscopic structure of metals is such that some electrons are always free to travel throughout a fixed lattice of positive ions (the atomic nuclei surrounded by the remaining, tightly bound electrons).[6] While all metals conduct, their *conductivity* varies quantitatively depending on the ease with which electrons can travel, or the extent to which their movement tends to be hampered by microscopic forces and collisions inside the material.

---

[6]This property can be understood through the periodic table of the elements, which identifies metals as being those types of atoms with one or a few electrons dwelling alone in more distant locations from the nucleus (orbitals), from where they are easily removed (ionized) so as to become free electrons.

Besides metals, there are other types of material that conduct electricity. One is water, or any other fluid, with dissolved ions (such as salt or minerals). In this case, it is not electrons but entire charged molecules that travel through the fluid to carry a current. Only small concentrations of ions are needed to make water conductive; while pure distilled water does not conduct electricity, normal tap water and rain water conduct all too well.[7]

Some materials, including air, can also become temporarily conductive through *ionization*. In the presence of a very strong potential gradient (defined as an *electric field* in Section 1.5), or intense heat, some electrons are stripped from their molecules and become free to travel. A gas in this state is called a *plasma*. Plasmas exist inside stars, nuclear fusion reactors, or fluorescent lights. More often, though, ionization tends to be local and transient: it occurs along a distinct trail, since ionized molecules incite their neighbors to do the same, and charge flows along this trail until the potential difference (charge imbalance) is neutralized. This is precisely what happens in an electric spark across an air gap, an arc between power lines, or a lightning bolt.[8]

In many engineering situations, it is important to predict just when ionization might occur; namely, how great a potential difference over how short a distance will cause "arcing." For air, this varies according to temperature and especially humidity, as well as the presence of other substances like salt suspended in the air. Exact figures for the *ionizing potential* can be found in engineering tables. For units of conductivity and the relationship to resistance, see Section 1.2.

Finally, some materials can become *superconducting*, generally at very low temperatures. Here, electrons undergo a peculiar energetic transition that allows them to travel with extreme ease, unimpeded by any obstructive forces or collisions. Thus, electrons in the superconducting state do no work on anything in their path, and therefore lose no energy. Some ceramic materials attain superconductivity at a temperature easily sustained by cooling with liquid nitrogen (at minus 319°F).[9] While liquid nitrogen is quite cheap in a research setting, large-scale refrigeration systems aimed at taking advantage of superconductivity in electric power applications are generally considered too formidable in cost to be justified by the savings in electric losses (see Section 1.3). Another conceivable application of superconductivity in power systems is superconducting magnetic energy storage (SMES).

1.1.6 Current

When charge travels through a material, an *electric current* is said to flow. The current is quantified in terms of the number of electrons (or equivalent charge, in

---

[7]In fact, conductivity is used as an indicator of water purity. Of course, it says nothing about the *kind* of ions present, only the amount.

[8]The ionization trail is visible because, as the electrons return to their normal state, the balance of their energy is released in the form of light.

[9]The first such material to be discovered was yttrium-barium-copper oxide, $YBa_2Cu_3O_7$.

the case of ions) moving past a given point in the material in a certain period of time. In other words, current is a *flow rate* of charge. In this way, electric current is analogous to a flow rate of water (say, in gallons per minute) or natural gas (cubic feet per second).

These analogies are also helpful in remembering the distinction between current and voltage. Voltage would be analogous to a height difference (say, between a water reservoir and the downhill end of a pipe), or to a pressure difference (between two ends of a gas pipeline). Intuitively, voltage is a measure of "how badly the stuff wants to get there," and current is a measure of "how much stuff is actually going."

Current is conventionally denoted by the symbol I or i and is measured in units of *amperes* (A), often called "amps." Since current represents a flow rate of charge, the units of current are equivalent to units of charge divided by units of time. Thus, one ampere equals one coulomb per second.

A subject that often causes confusion is the "direction" in which current flows, though in practice, having an accurate picture of this is not all that important. Most often, the reasons one is concerned with current have to do with the amount of power transferred or the amount of heating of the wires, neither of which depend on direction.

When in doubt, we can always refer back to the fact that opposite charges attract and like charges repel. Thus, a positive charge will be attracted by a negative potential, and hence flow toward it, and vice versa: electrons, which have negative charge, flow toward a positive potential or voltage. In a mathematical sense, negative charge flowing in one direction is equivalent to positive charge flowing the opposite way. Indeed, our practical representation of electric current does not distinguish between these two physical phenomena. For example, the current flowing through a lead–acid battery at various times consists of negative electrons in the terminals and wires, and positive ions in the battery fluid; yet these flows are thought of as the same current.

In circuit analysis, it often becomes necessary to define a direction of current flow, so as to know when to add and when to subtract currents that meet on a section of the circuit. The general convention is to label a current flow as "positive" in the direction from positive toward negative potential (as if a *positive* charge were flowing). Once this labeling has been chosen, all currents in the circuit will be computed as positive or negative so as to be consistent with that requirement (positive currents will always point toward lower potential). However, the convention is arbitrary in that one can define the currents throughout an entire circuit "backwards," and obtain just as "correct" a result. In other words, for purposes of calculation, the quantity "current" need not indicate the actual physical direction of traveling charge.

In the power systems context, the notion of directionality is more complicated (and less revealing) because the physical direction of current flow actually alternates (see "Alternating Current" in Section 3.1). Instead, to capture the relationship between two currents (whether they add or subtract), the concept of *phase*, or relative timing, is used.

8   THE PHYSICS OF ELECTRICITY

As for the speed at which current propagates, it is often said that current travels at the speed of light (186,000 miles per second). While this is not quite accurate (just as the speed of light actually varies in different materials), it is usually sufficient to know that current travels *very fast*. Conceptually, it is important to recognize that what is traveling at this high speed is the *pulse* or *signal* of the current, not the individual electrons. For the current to flow, it is also not necessary for all the electrons to physically depart at one end and arrive at the other end of the conductor. Rather, the electrons inside a metal conductor continually move in a more or less random way, wiggling around in different directions at a speed related to the temperature of the material. They then receive a "shove" in one direction by the *electric field* (see Section 1.5.2). We can imagine this shove propagating by way of the electrical repulsion among electrons: each electron need not travel a long distance, just enough to push its neighbor over a bit, which in turn pushes *its* neighbor, and so on. This chain reaction creates a more orderly motion of charge, as opposed to the usual random motion, and is observed macroscopically as the current. It is the signal to "move over" that propagates at essentially the speed of light.[10]

The question of the propagation speed of electric current only becomes relevant when the distance to be covered is so large that the time it takes for a current pulse to travel from one point to another is significant compared to other timing parameters of the circuit. This can be the case for electric transmission lines that extend over many hundreds of miles.[11] However, we will not deal with this problem explicitly (see Chapter 7, "Power Flow Analysis," for how we treat the concept of time in power systems). A circuit that is sufficiently small so that the speed of current is not an issue is called a *lumped circuit*. Circuits are treated as lumped circuits unless otherwise stated.

1.2 OHM'S LAW

It is intuitive that voltage and current would be somehow related. For example, if the potential difference between two ends of a wire is increased, we would expect a greater current to flow, just like the flow rate of gas through a pipeline increases when a greater pressure difference is applied. For most materials, including metallic conductors, this relationship between voltage and current is linear: as the potential difference between the two ends of the conductor increases, the current through the conductor increases proportionally. This statement is expressed in Ohm's law,

$$V = IR$$

---

[10] We can draw an analogy with an ocean wave: the water itself moves essentially up and down, and it is the "signal" to move up and down that propagates across the surface, at a speed much faster than the bulk motion of water.

[11] For example, traveling down a 500-mile transmission line at the speed of light takes 2.7 milliseconds. Compared to the rate at which alternating current changes direction (60 times per second, or every 16.7 milliseconds), this corresponds to one-sixth of a cycle, which is not negligible.

where V is the voltage, I is the current, and R is the proportionality constant called the *resistance*.

1.2.1 Resistance

To say that Ohm's law is true for a particular conductor is to say that the resistance of this conductor is, in fact, constant with respect to current and voltage. Certain materials and electronic devices exhibit a nonlinear relationship between current and voltage, that is, their resistance varies depending on the voltage applied. The relationship $V = IR$ will still hold at any given time, but the value of R will be a different one for different values of V and I. These nonlinear devices have specialized applications and will not be discussed in this chapter. Resistance also tends to vary with temperature, though a conductor can still obey Ohm's law at any one temperature.[12] For example, the resistance of a copper wire increases as it heats up. In most operating regimes, these variations are negligible. Generally, in any situation where changes in resistance are significant, this is explicitly mentioned. Thus, whenever one encounters the term "resistance" without further elaboration, it is safe to assume that within the given context, this resistance is a fixed, unchanging property of the object in question.

Resistance depends on an object's material composition as well as its shape. For a wire, resistance increases with length, and decreases with cross-sectional area. Again, the analogy to a gas or water pipe is handy: we know that a pipe will allow a higher flow rate for the same pressure difference if it has a greater diameter, while the flow rate will decrease with the length of the pipe. This is due to *friction* in the pipe, and in fact, an analogous "friction" occurs when an electric current travels through a material.

This friction can be explained by referring to the microscopic movement of electrons or ions, and noting that they interact or collide with other particles in the material as they go. The resulting forces tend to impede the movement of the charge carriers and in effect limit the rate at which they pass. These forces vary for different materials because of the different spatial arrangements of electrons and nuclei, and they determine the material's ability to conduct.

This intrinsic material property, independent of size or shape, is called *resistivity* and is denoted by ρ (the Greek lowercase rho). The actual resistance of an object is given by the resistivity multiplied by the length of the object (l) and divided by its cross-sectional area (A):

$$R = \frac{\rho l}{A}$$

The units of resistance are *ohms*, abbreviated Ω (Greek capital omega). By rearranging Ohm's law, we see that resistance equals voltage divided by current. Units of

---

[12] If we graph V versus I, Ohm's law requires that the graph be a straight line. With temperature, the slope of this line may change.

resistance are thus equivalent to units of voltage divided by units of current. By definition, one ohm equals one volt per ampere ($\Omega = V/A$).

The units of resistivity are ohm-meters (Ω-m), which can be reconstructed through the preceding formula: when ohm-meters are multiplied by meters (for l) and divided by square meters (for A), the result is simply ohms. Resistivity, which is an intrinsic property of a material, is not to be confused with the *resistance per unit length* (usually of a wire), quoted in units of ohms *per* meter (Ω/m). The latter measure already takes into account the wire diameter; it represents, in effect, the quantity ρ/A. The resistivities of different materials in Ω-m can be found in engineering tables.

1.2.2 Conductance

It is sometimes convenient to refer to the resistive property of a material or object in the inverse, as *conductivity* or *conductance*. Conductivity is the inverse of resistivity and is denoted by σ (Greek lowercase sigma): $\sigma = 1/\rho$. For the case of a simple resistor, conductance is the reciprocal of resistance and is usually denoted by G (sometimes g), where $G = 1/R$.[13] Not without humor, the units of conductance are called *mhos*, and 1 mho $= 1/\Omega$. Another name for the mho is the siemens (S); they are identical units. The conductance is related to the conductivity by

$$G = \frac{\sigma A}{l}$$

and the units of σ are thus mhos/m.

For the special case of an insulator, the conductance is zero and the resistance is infinite. For the special case of a superconductor, the resistance is zero and the conductance is, theoretically, infinite (a truly infinite conductance would imply an infinitely large current, which does not actually occur since its magnitude is eventually constrained by the number of electrons available).

Example

Consider two power extension cords, one with twice the wire diameter of the other. If the cords are of the same length and same material, how do their resistances compare?

Since resistance is inversely proportional to area, the smaller wire will have *four times* the resistance. We can see this through the formula $R = \rho l/A$, where ρ and l are the same for both. Thus, using the subscripts 1 and 2 to refer to the two cords, we can write $R_1/R_2 = A_2/A_1$. The areas are given by the familiar geometry formula, $A = \pi(d/2)^2$ (where $\pi = 3.1415...$), which includes the *square* of the diameter or radius. If the length of either cord were doubled, its resistance would also double.

---

[13]See Section 3.2.4 for the complex case that includes both resistance and reactance.

To put some numbers to this example, consider a typical 25-ft, 16-gauge extension cord, made of a copper conductor. The cross-sectional area of 16-gauge wire is 1.31 mm² (or 1.31×10^{-8} m²) and the resistivity of copper is $\rho = 1.76 \times 10^{-8}$ Ω-m. The resistance per unit length of 16-gauge copper wire is 0.0134 Ω/m, and a 25-ft length of it has a resistance of 0.102 Ω. By contrast, a 10-gauge copper wire of the same length, which has about twice the diameter, has a resistance of only 0.025 Ω.

Suppose the current in the 25-ft, 16-gauge cord is 5 A. What is the voltage difference between the two ends of each conductor?
The voltage drop in the wire is given by Ohm's law, $V = IR$. Thus, $V = 5$ A $\cdot 0.102$ Ω $= 0.51$ V. Because the voltage drop applies to each of the two conductors in the cord, this means that the line voltage, or difference between the two sides of the electrical outlet, will be diminished by about one volt (say, from 120 to 119 V), as seen by the appliance at the end.

1.2.3 Insulation

Insulating materials are used in electric devices to keep current from flowing where it is not desired. They are simply materials with a sufficiently high resistance (or sufficiently low conductance), also known as *dielectric* materials. Typically, plastics or ceramics are used. When an insulator is functional, its resistance is infinite, or the conductance zero, so that zero current flows through it.

Any insulator has a specific voltage regime within which it can be expected to perform. If the voltage difference between two sides of the insulator becomes too large, its insulating properties may break down due to microscopic changes in the material, where it actually becomes conducting. Generally, the thicker the insulator, the higher the voltage difference it can sustain. However, temperature can also be important; for example, plastic wire insulation may melt if the wire becomes too hot.

The insulators often seen on high-voltage equipment consist of strings of ceramic bells, holding the energized wires away from other components (e.g., transmission towers or transformers). The shape of these bells serves to inhibit the formation of arcs along their surface. The number of bells is roughly proportional to the voltage level, though it also depends on climate. For example, the presence of salt water droplets in coastal air encourages ionization and therefore requires more insulation to prevent arcing.

1.3 CIRCUIT FUNDAMENTALS

1.3.1 Static Charge

A current can only flow as long as a potential difference is sustained; in other words, the flowing charge must be replenished. Therefore, some currents have a very short duration. For example, a lightning bolt lasts only a fraction of a second, until the charge imbalance between the clouds and the ground is neutralized.

When charge accumulates in one place, it is called *static charge*, because it is not moving. The reason charge remains static is that it lacks a conducting pathway that enables it to flow toward its opposite charge. When we receive a shock from static electricity—for example, by touching a doorknob—our body is providing just such a pathway. In this example, our body is charged through friction, often on a synthetic carpet, and this charge returns to the ground via the doorknob (the carpet only gives off electrons by rubbing, but does not allow them to flow back). As our fingers approach the doorknob, the air in between is actually ionized momentarily, producing a tiny arc that causes the painful sensation.[14] Static electricity occurs mostly in dry weather, since moisture on the surface of objects makes them sufficiently conductive to prevent accumulations of charge.

However startling and uncomfortable, static electricity encountered in everyday situations is harmless because the amount of charge available is so small,[15] and it is not being replenished. This is true despite the fact that very high voltages can be involved (recall that voltage is energy *per* charge), but these voltages drop instantaneously as soon as the contact is made.

1.3.2 Electric Circuits

In order to produce a sustained flow of current, the potential difference must be maintained. This is achieved by providing a pathway to "recycle" charge to its origin, and a mechanism (called an *electromotive force*, or *emf*[16]) that compels the charge to return to the less "comfortable" potential. Such a setup constitutes an *electric circuit*.

A simple example is a battery connected with two wires to a light bulb. The chemical forces inside the battery do work on the charge to move it to the terminals, where an electric potential is produced and sustained. Specifically, electrons are moved to the negative terminal, and positive ions are moved to the positive electrode, where they produce a deficit of electrons in the positive battery terminal. The wires then provide a path for electrons to flow from the negative to the positive terminal. Because the positive potential is so attractive, these electrons even do work by flowing through the resistive light bulb, causing it to heat up and glow. As soon as the electrons arrive at the positive terminal, they are "lifted" again to the negative potential, allowing the current to continue flowing. In analogy with flowing water, the wires are like pipes that carry water downhill, and the battery is like a pump that returns the water to the uphill end of the circuit.

---

[14]Charge will accumulate more densely in the point, being attracted to the opposite charge across the gap. The charge density in turn affects the gradient of the electric potential across the gap, which is what causes the ionization. Therefore, approaching the doorknob with a flat hand can prevent the formation of an arc, and charge will simply flow (unnoticeably) after the contact has been made. This is also why lightning arresters work: a particularly pointed object like a metal rod will "attract" an electric arc toward its high charge density. By the same token, lightning tends to strike tall trees and transmission towers.

[15]The same is *not* true of electrical equipment that has been specifically designed to hold a very large amount of static charge!

[16]Unrelated to the EMF that stands for "Electromagnetic Fields."

When the wires are connected to form a complete loop, they make a *closed circuit*. If the wire were cut, this would create an *open circuit*, and the current would cease to flow. In practice, circuits are opened and closed by means of switches that make and break electrical contacts.

1.3.3 Voltage Drop

In describing circuits, it is often desirable to specify the voltage at particular points along the way. The difference in voltage between two points in a circuit is referred to as the *voltage drop* across the wire or other component in between. As in Ohm's law, $V = IR$, this voltage drop is proportional to the current flowing through the component, multiplied by its resistance.

As in the analogy of water pipes running downhill, the voltage drops continuously throughout a circuit, from one terminal of the emf to the other. However, just like the slope of the pipes may change, the voltage does not necessarily drop at a steady rate. Rather, depending on the resistance of a given circuit component, the voltage drop across it will be more or less: a component with high resistance will sustain a greater voltage drop, whereas a component with low resistance such as a conducting wire will have a smaller voltage drop across it, perhaps so small as to be negligible in a given context. For small circuits, it is often reasonable to assume that the wire's resistance is zero, and that therefore the voltage is the same all the way along the wire. In power systems, however, where transmission and distribution lines cover long distances, the voltage drop across them is significant and indeed accounts for some important aspects of how these systems function.

Importantly (and in contrast to the water analogy), the magnitude of the current also determines the voltage drop (along with resistance). For example, at times of high electric demand and thus high current flow, the voltage drop along transmission and distribution lines is greater; that is, the voltage drops more rapidly with distance. If this condition cannot be compensated for by other adjustments in the system (see Section 6.7), customers experience lower voltage levels associated with dimmer lights and impaired equipment performance, known as "brownouts." Similarly, if a piece of heavy power equipment is connected through a long extension cord with too high a resistance, the voltage drop along this cord can result in damage to the motor from excessively low voltage at the far end.

1.3.4 Electric Shock

Any situation where a high voltage is sustained by an electromotive force (or a very large accumulation of charge) constitutes a shock hazard. Our bodies are not noticeably affected by being "charged up," or raised to a potential above ground, just as birds can sit on a single power line. Rather, harm is done when a current flows *through* our body. A current as small as a few milliamperes across the human heart can be lethal.[17] For current to flow through an object, there must be a

---

[17] A saying goes, "It's the volts that jolts, and the mils that kills."

voltage drop across it. In other words, our body must be simultaneously in contact with two sources of different potential—for example, a power line and the ground.

Though it is the current that causes biological damage, Ohm's law indicates that shock hazard is roughly proportional to the voltage encountered. However, the resistance is also important. On an electrical path through the human body, the greatest resistance is on the surface of the skin and clothing, while our interior conducts very well. Thus, the severity of a shock received from a particular voltage can vary, depending on how sweaty one's palms are, or what type of shoes one is wearing.

The physical principles of electric current can be applied to suggest a number of practical precautions for reducing electric-shock hazards. For example, when touching an object at a single high voltage, we are safe as long as we are insulated from the ground. A wooden ladder might serve this purpose at home, while utility linemen often work on "hot" equipment out of raised plastic "buckets." Linemen can also insulate themselves from the high-voltage source by wearing special rubber gloves, which are commonly used for work on up to 12 kilovolts. The important thing is to know the capability of the insulator in relation to the voltage encountered.

A different safety measure often used by electricians when touching a questionable component (such as a wire that might be energized) is to make contact with ground potential with the same hand, for example by touching the little finger to the wall. In this way, a path of low resistance is created through the hand, which will greatly reduce the current flowing through the rest of the body and especially across the heart. Though the hand might be injured (improbable at household voltage), such a shock is far less likely to be lethal.

Around high-voltage equipment, in order to avoid the possibility of touching two objects at different potentials with a current pathway across the heart, a common practice is to "keep one hand in your pockets." Near very high potentials, where the concern is not just about touching equipment, but even drawing an arc across the air, the advice is to "keep both hands in your pockets" so as to avoid creating a point with high charge density to attract an arc.

Finally, another factor to consider is the muscular contraction that often occurs in response to an electric shock. Thus, a potentially energized wire is better touched with the back of the hand, so as to prevent involuntary closing of the hand around it.

If a person is in contact with an energized source, similar precautions should be exercised in removing them, lest there be additional casualties. If available, a device like a wooden stick would be ideal; in the worst case, kicking is preferable to grabbing.

1.4 RESISTIVE HEATING

Whenever an electric current flows through a material that has some resistance (i.e., anything but a superconductor), it creates heat. This *resistive heating* is the result of "friction," as created by microscopic phenomena such as retarding forces and collisions involving the charge carriers (usually electrons); in formal terminology, the heat corresponds to the work done by the charge carriers in order to travel to a lower potential. This heat generation may be intended by design, as in any heating appliance (for example, a toaster, an electric space heater, or an electric blanket). Such an appliance essentially consists of a conductor whose resistance is chosen so as to produce the desired amount of resistive heating. In other cases, resistive heating may be undesirable. Power lines are a classic example. For one, their purpose is to transmit energy, not to dissipate it; the energy converted to heat along the way is, in effect, lost (thus the term *resistive losses*). Furthermore, resistive heating of transmission and distribution lines is undesirable, since it causes thermal expansion of the conductors, making them sag. In extreme cases such as fault conditions, resistive heating can literally melt the wires.

1.4.1 Calculating Resistive Heating

There are two simple formulas for calculating the amount of heat dissipated in a resistor (i.e., any object with some resistance). This heat is measured in terms of *power*, which corresponds to energy per unit time. Thus, we are calculating a *rate* at which energy is being converted into heat inside a conductor. The first formula is

$$P = IV$$

where P is the power, I is the current through the resistor, and V is the voltage drop across the resistor.

Power is measured in units of *watts* (W), which correspond to amperes × volts. Thus, a current of one ampere flowing through a resistor across a voltage drop of one volt produces one watt of heat. Units of watts can also be expressed as joules per second. To conceptualize the magnitude of a watt, it helps to consider the heat created by a 100-watt light bulb, or a 1000-watt space heater.

The relationship $P = IV$ makes sense if we recall that voltage is a measure of energy per unit charge, while the current is the flow rate of charge. The product of current and voltage therefore tells us how many electrons are "passing through," multiplied by the amount of energy each electron loses in the form of heat as it goes, giving an overall rate of heat production. We can write this as

$$\frac{Charge}{Time} \cdot \frac{Energy}{Charge} = \frac{Energy}{Time}$$

and see that, with the charge canceling out, units of current multiplied by units of voltage indeed give us units of power.

The second formula for calculating resistive heating is

$$P = I^2 R$$

where P is the power, I is the current, and R is the resistance. This equation could be derived from the first one by substituting $I \cdot R$ for V (according to Ohm's law). As we discuss in Section 3.1, this second formula is more frequently used in practice to calculate resistive heating, whereas the first formula has other, more general applications.

As we might infer from the equation, the units of watts also correspond to amperes$^2 \cdot$ ohms (A$^2 \cdot \Omega$). Thus, a current of one ampere flowing through a wire with one ohm resistance would heat this wire at a rate of one watt. Because the current is squared in the equation, two amperes through the same wire would heat it at a rate of 4 watts, and so on.

Example

A toaster oven draws a current of 6 A at a voltage of 120 V. It dissipates 720 W in the form of heat. We can see this in two ways: First, using $P = IV$, 120 V $\cdot$ 6 A = 720 W. Alternatively, we could use the resistance, which is 20 Ω (20 $\Omega \cdot$ 6 A = 120 V), and write $P = I^2R$: (6 A)$^2 \cdot$ 20 Ω = 720 W.

It is important to distinguish carefully how power depends on resistance, current, and voltage, since these are all interdependent. Obviously, the power dissipated will increase with increasing voltage and with increasing current. From the formula $P = I^2R$, we might also expect power to increase with increasing resistance, assuming that the current remains constant. However, it may be incorrect to assume that we can vary resistance without varying the current.

Specifically, in many situations it is the voltage that remains (approximately) constant. For example, the voltage at a customer's wall outlet ideally remains at 120 V, regardless of how much power is consumed.[18] The resistance is determined by the physical properties of the appliance: its intrinsic design, and, if applicable, a power setting (such as "high" or "low"). Given the standard voltage, then, the resistance determines the amount of current "drawn" by the appliance according to Ohm's law: higher resistance means lower current, and vice versa. In fact, resistance and current are inversely proportional in this case: if one doubles, the other is halved.

What, then, is the effect of resistance on power consumption? The key here is that resistive heating depends on the *square* of the current, meaning that the power is more sensitive to changes in current than resistance. Therefore, at constant voltage, the effect of a change in current outweighs the effect of the corresponding change in resistance. For example, decreasing the resistance (which, in and of itself, would tend to decrease resistive heating) causes the current to increase, which increases resistive heating by a greater factor. Thus, at constant voltage, the net effect of decreasing resistance is to increase power consumption. An appliance that draws more power has a lower internal resistance.

For an intuitive example, consider the extreme case of a *short circuit*, caused by an effectively zero resistance (usually unintentional). Suppose a thick metal bar were

---

[18]This is generally true because (a) changes in power consumption from an individual appliance are small compared to the total power supplied to the area by the utility, and (b) the utility takes active steps to regulate the voltage (see Section 6.6). Dramatic changes in demand do cause changes in voltage, but for the present discussion, it is more instructive to ignore these phenomena and treat voltage as a fixed quantity.

placed across the terminals of a car battery. A very large current would flow, the metal would become very hot, and the battery would be drawn down very rapidly. If a similar experiment were performed on a wall outlet by sticking, say, a fork into it, the high current would hopefully be interrupted by the circuit breaker before either the fork or the wires melted (DO NOT actually try this!). The other extreme case is simply an open circuit, where the two terminals are separate and the resistance of the air between them is infinite: here the current and the power consumption are obviously zero.

Example

Consider two incandescent light bulbs, with resistances of 240 Ω and 480 Ω. How much power do they each draw when connected to a 120 V outlet?

First we must compute the current through each bulb, using Ohm's law: Substituting $V = 120$ V and $R_1 = 240\ \Omega$ into $V = IR$, we obtain $I_1 = 0.5$ A. For $R_2 = 480\ \Omega$, we get $I_2 = 0.25$ A.

Now we can use these values for I and R in the power formula, $P = I^2 R$, which yields $P_1 = (0.5\ \text{A})^2 \cdot 240\ \Omega = 60$ W and $P_2 = (0.25\ \text{A})^2 \cdot 480\ \Omega = 30$ W.

We see that at constant voltage, the bulb with twice the resistance draws half the power.

There are other situations, however, where the current rather than the voltage is constant. Transmission and distribution lines are an important case. Here, the reasoning suggested earlier does in fact apply, and resistive heating is directly proportional to resistance. The important difference between power lines and appliances is that for power lines, the current is unaffected by the resistance of the line itself, being determined instead by the load or power consumption at the end of the line (this is because the resistance of the line itself is very small and insignificant compared to that of the appliances at the end, so that any reasonable change in the resistance of the line will have a negligible effect on the overall resistance, and thus the current flowing through it). However, the voltage drop along the line (i.e., the difference in voltage between its endpoints, not to be confused with the line voltage relative to ground) is unconstrained and varies depending on current and the line's resistance. Thus, Ohm's law still holds, but it is now I that is fixed and V and R that vary. Applying the formula $P = I^2 R$ for resistive heating with the current held constant, we see that doubling the resistance of the power line will double resistive losses. Since in practice it is desirable to minimize resistive losses on power transmission and distribution lines, these conductors are chosen with the minimal resistance that is practically and economically feasible.

1.4.2 Transmission Voltage and Resistive Losses

Resistive losses are the reason why increasingly high voltage levels are chosen for power transmission lines. Recalling the relationship $P = IV$, the amount of power transmitted by a line is given by the product of the current flowing through it and its voltage level (as measured either with respect to ground or between two lines or phases of one circuit). Given that a certain quantity of power is demanded, there is a choice as to what combination of I and V will constitute this power. A higher voltage level implies that in order to transmit the same amount of power, less current needs to flow. Since resistive heating is related to the square of the current, it is highly beneficial from the standpoint of line losses to reduce the current by increasing the voltage.

Before power transformers were available, transmission voltages were limited to levels that were considered safe for customers. Thus, high currents were required, causing so much resistive heating that it posed a significant constraint to the expansion of power transmission. With increasing power carried at a given voltage, an increasing fraction of the total power is lost on the lines, making transmission uneconomical at some point. The increase in losses can be counteracted by reducing the resistance of the conductors, but only at the expense of making them thicker and heavier. A century ago, Thomas Edison found the practical limit for transmitting electricity at the level of a few hundred volts to be only a few miles.

With the help of transformers that allow essentially arbitrary voltage conversion (see Section 6.3), transmission voltage levels have grown steadily in conjunction with the geographic expansion of electric power systems, up to about 1000 kilovolts (kV), and with the most common voltages around 100–500 kV. The main factor offsetting the economic benefits of very high voltage is the increased cost and engineering challenge of safe and effective insulation.

1.5 ELECTRIC AND MAGNETIC FIELDS

1.5.1 The Field as a Concept

The notion of a *field* is an abstraction initially developed in physics to explain how tangible objects exert forces on each other at a distance, by invisible means. Articulating and quantifying a "field" particularly helps to analyze situations where an object experiences forces of various strengths and directions, depending on its location. Rather than referring to other objects associated with "causing" such forces, it is usually more convenient to just map their hypothetical effects across space. Such a map is then considered to describe properties of the space, even in the absence of an actual object placed within it to experience the results, and this map represents the field.

For example, consider gravity. We know that our body is experiencing a force downward because of the gravitational attraction between it and the Earth. This gravitational force depends on the respective masses of our bodies and the Earth, but it also depends on our location: astronauts traveling into space feel less and less of a pull toward the Earth as they get farther away. Indeed, though the effect is small, we are even slightly "lighter" on a tall mountain or in an airplane at high altitude. If we were interested in extremely accurate measurements of gravity (for example, to calculate the exact flight path of a ballistic missile), we could construct a map of a "gravitational field" encompassing the entire atmosphere, which would indicate the strength of gravity at any point. This field is caused by the Earth, but does not explicitly refer to the Earth as a mass; rather, it represents in abstract terms the effect of the Earth's presence. The field also does not refer to any object (such as an astronaut) that it may influence, though such an object's mass would need to be taken into account in order to calculate the actual force on it. Thus, the gravitational field is a way of mapping the influence of the Earth's gravity throughout a region of space.

An alternative interpretation is to consider the field as a physical entity in its own right, even though it has no substance of its own. Here we would call gravity a property of *the space itself*, rather than a map telling us about objects such as the Earth *in* space. Indeed, the field itself can be considered a "thing" rather than a map, because it represents potential energy distributed over space. We know of the presence of this potential energy because it does physical *work* on objects: for example, a massive object within the field is accelerated, and in that moment, the energy becomes observable. With this in mind, we can understand the field as the answer to the question, Where does the potential energy reside while we are not observing it?

This notion of the field as a physical entity is a fairly recent one. Whereas classical physics relied on the notion of action-at-a-distance, in which only tangible objects figured as "actors," the study of very large and very small things in the 20th century has forced us to give up referring to entities that we can touch or readily visualize when talking about how the world works. Instead, modern physics has cultivated more ambiguity and caution in declaring the "reality" of physical phenomena, recognizing that what is accessible to our human perception is perhaps not a definitive standard for what "exists." Even what once seemed like the most absolute, immutable entities—mass, distance, and time—were proved ultimately changeable and intractable to our intuition by relativity theory and quantum mechanics.

Based on these insights, we might conclude that any quantities we choose to define and measure are in some sense arbitrary patterns superimposed on the vast web of energy and movement that constitutes reality, for the purpose of helping us apprehend this reality with our thoughts. In this sense, we are no more justified in considering a planet a "thing that really exists" than we are a gravitational field. What we really care about as scientists, though, is how useful such a conceptual pattern might be for describing the world in concise terms and making predictions about how things will behave. By this standard, the notion of a "field" does wonders. Physicists and engineers are therefore accustomed to regarding fields, however devoid of substance, as real, manipulable, and legitimate physical entities just like tangible objects. In any case, the reader should rest assured that it is quite all right to simply accept the "field" as a strange instrument of analysis that grows more palatable with familiarity.

1.5.2 Electric Fields

In Section 1.1, we characterized the electric potential as a property of the location at which a charge might find itself. A map of the electric potential would indicate how much potential energy would be possessed by a charge located at any given point. The *electric field* is a similar map, but rather of the electric *force* (such as attraction or repulsion) that would be experienced by that charge at any location. This force is the result of potential differences between locations: the more dramatically the potential varies from one point to the next, the greater the force would be on an electric charge in between these points. In formal terms, the electric field represents the *potential gradient*.

Consider the electric field created by a single positive charge, just sitting in space. Another positive charge in its vicinity would experience a repulsive force. This repulsive force would increase as the two charges were positioned closer together, or decrease as they moved father apart; specifically, the electric force drops off at a rate proportional to the square of the distance. This situation can be represented graphically by drawing straight arrows radially outward from the first charge, as in Figure 1.1a. Such arrows are referred to as *field lines*. Their direction indicates the direction that a "test charge," such as the hypothetical second charge that was introduced, would be pushed or pulled (in this case, straight away). The strength

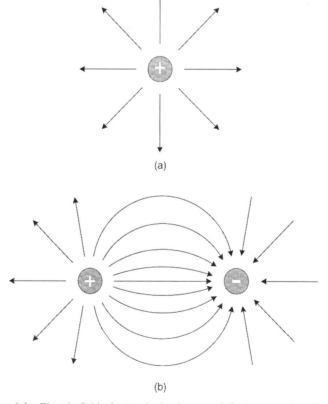

Figure 1.1 Electric field of (a) a single charge and (b) two opposite charges.

of the force is indicated by the proximity of field lines: the force is stronger where the lines are closer together.

This field also indicates what would happen to a negative charge: At any point, it would experience a force of equal strength (assuming equal magnitude of charge), but opposite direction as the positive test charge, since it would be attracted rather than repelled. Thus, a negative test charge would also move along the field lines, only backwards. By convention, the direction of the electric field lines is drawn so as to represent the movement of a *positive* test charge.

For a slightly more complex situation, consider the electric field created by a positive and a negative charge, sitting at a fixed distance from each other. We can map the field conceptually by asking, for any location, "What force would be acting on a (positive) test charge if it were placed here?" Each time, the net force on the test charge would be a combination of one attractive force and one repulsive force, in different directions and at different strengths depending on the distance from the respective fixed charges. Graphically, we can construct an image of the field by drawing an arrow in the direction that the charge would be pulled. The arrows for points along the charge's hypothetical path then combine into continuous field lines. Again, these field lines will be spaced more closely where the force is stronger. This exercise generates the picture in Figure 1.1b.

1.5.3 Magnetic Fields

The pattern of the electric field in Figure 1.1 may be reminiscent to some readers of the pattern that many of us produced once upon a time in science class by sprinkling iron filings on a sheet of paper over a bar magnet. The two phenomena, electric and magnetic forces, are indeed closely linked manifestations of a common underlying physics.

As we know from direct tactile experience, magnets exert force on each other: opposite poles attract, and like poles repel. This is somewhat analogous to the fact that opposite electric charges attract and like charges repel. But, unlike a positive or negative electric charge, a magnetic pole cannot travel individually. There is no such thing as an individual north or south pole (a "monopole" in scientific terms, which has never been found). Every magnet has a north and a south pole. Thus, unlike electric field lines that indicate the direction of movement of an individual test charge, magnetic field lines indicate the *orientation* of a test magnet. The iron filings in the familiar experiment—which become little test magnets since they are magnetized in the presence of the bar magnet—do not move toward one pole or the other, but rotate and align themselves with the direction of the field lines.

It is important to emphasize that, despite the similar shape of field lines, magnetic poles are not analogous to single electric charges sitting in space. Rather than thinking of magnetism as existing in the form of "stuff" like electric charge (which could conceivably be decomposed into its "north" and "south" constituents), it is more appropriate to think of magnetism as an expression of *directionality*, where north is meaningless without south. If you cut a magnet in half, you get two smaller magnets that still each have a north and a south pole.

If we pursued such a division of magnets again and again, down to the level of the smallest particles, we would find that even individual electrons or protons appear as tiny magnets. In ordinary materials, the orientation of all these microscopic magnets varies randomly throughout space, and they therefore do not produce observable magnetic properties at the macroscopic level. It is only in *magnetized* materials that the direction of these myriad tiny magnets becomes aligned, allowing their magnetic fields to combine to become externally noticeable. This alignment stems from the force magnets exert on each other, and their resulting tendency to position themselves with their north poles all pointing in the same direction. Some substances like *magnetite* occur naturally with a permanent alignment, making the familiar magnets that adhere to refrigerators and other things. Other materials like iron and steel can be temporarily magnetized in the presence of a sufficiently strong external magnetic field (this is what happens to the refrigerator door underneath the magnet), with the particles returning to their disordered state after the external field is withdrawn.

The magnetic property of microscopic particles is due to their electric charge and their intrinsic motion, which brings us to the fundamental connection between electricity and magnetism. Indeed, we can think of magnetism as nothing but a manifestation of directionality associated with electric charge in motion, whereby moving charges always exert a specific directional force on other moving charges. At the level of individual electrons, their motion consists of both an orbital movement around the atom's nucleus and an intrinsic *spin*, which we can visualize as if the particle were spinning like a top.[19] Both of these rotational motions combine to form what is referred to as a *magnetic moment*. Similarly, the protons inside atomic nuclei possess a magnetic moment due to their intrinsic spin.[20]

Knowing this, it would stand to reason that a large amount of moving charge such as a measurable electric current should produce a magnetic field as well. This phenomenon was in fact discovered in 1820, when Hans Christian Oersted observed that a compass needle was deflected by an electric current through a nearby wire. The magnetic field produced by an electric current points at a right angle to the flow of charge, in a direction specified by the "right-hand rule" illustrated in Figure 1.2. If the thumb of one's right hand is pointing in the direction of the current, then the curled fingers of the same hand indicate the direction of the magnetic field. Thus, the magnetic field lines surround the wire in a circular manner.

In order to make practical use of this phenomenon, we can alter the shape of the current-carrying wire by winding it into a coil, which brings many turns of wire closely together so that their magnetic fields will add to form a "straight" field in the center of the coil that is comparable to that of a bar magnet (an illustration of such a coil and its magnetic field is shown in Figure 3.4). This arrangement can be thought of as "concentrating" the magnetic field in space.

---

[19]Such a mechanistic representation is not quite accurate in terms of quantum mechanics, but it is nonetheless useful for constructing some intuitive picture.

[20]This effect is quite subtle and not important in our context, but is exploited in such technologies as Magnetic Resonance Imaging (MRI) for medical diagnostic purposes, which discriminates among tissues of different water content by way of the magnetic properties of the hydrogen nucleus.

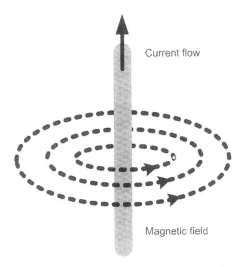

Figure 1.2 Magnetic field around a current-carrying wire.

Magnetic effects are essential for the generation and conversion of electric power. In order to successfully navigate the literature on these applications, it is important to be aware of a distinction between two types of quantities: one is called the *magnetic field* and the other *magnetic flux*. Despite the earlier caution, it is at times helpful (and indeed consistent with the Latin translation) to think of the flux as the directional "flow" of something, however immaterial, created in turn by the flow of electric current. Conceptually as well as mathematically, the flux is a very convenient quantity for analyzing electrical machines, while the magnetic field is particularly useful for describing the basic principles of electromagnetic induction in simplified settings.

Conventionally, the magnetic field is denoted by the symbol B and measured in units of *tesla* (T) or *gauss* (G). One tesla, which equals 10,000 or 10^4 gauss, corresponds to one newton (N) (a measure of force) per ampere (current) per meter: 1 T = 1 N/A-m. Magnetic flux is denoted by ϕ (the Greek phi) and is measured in units of weber (Wb). One tesla equals one weber per square meter.

From this relationship between the units of flux and field, we can see that the magnetic field corresponds to the density or concentration in space of the magnetic flux. The magnetic field represents magnetic *flux per unit area*. Stated in reverse, magnetic flux represents a measure of the magnetic field multiplied by the area that it intersects.

Unless "concentrated" by a coil, the magnetic field associated with typical currents is not very strong. For example, a current of 1 ampere produces a magnetic field of 2×10^{-7} T or 0.002 G (2 milligauss) at a distance of 1 meter. By comparison, the strength of the Earth's magnetic field is on the order of half a gauss.[21]

---

[21] The exact value of the Earth's magnetic field depends on geographic location, and it is less if only the horizontal component (to which a compass needle responds) is measured.

1.5.4 Electromagnetic Induction

While electric current creates a magnetic field, the reverse effect also exists: magnetic fields, in turn, can influence electric charges and cause electric currents to flow. However, there is an important twist: the magnetic field must be *changing* in order to have any effect. A static magnetic field, such as a bar magnet, will not cause any motion of nearby charge. Yet if there is any *relative* motion between the charge and the magnetic field—for example, because either the magnet or the wire is being moved, or because the strength of the magnet itself is changing—then a force will be exerted on the charge, causing it to move. This force is called an *electromotive force* (*emf*) which, just like an ordinary electric field, is distinguished by its property of accelerating electric charges.

The most elementary case of the electromotive force involves a single charged particle traveling through a magnetic field, at a right angle to the field lines (the direction along which iron filings would line up). This charge experiences a force again at right angles to both the field and its velocity, the direction of which (up or down) depends on the sign of the charge (positive or negative) and can be specified in terms of another right-hand rule, as illustrated in Figure 1.3.

This effect can be expressed concisely in mathematical terms of a *cross product* of vector quantities (i.e., quantities with a directionality in space, represented in boldface), in what is known as the Lorentz equation,

$$\mathbf{F} = q\mathbf{v} \times \mathbf{B}$$

where **F** denotes the force, q the particle's charge, **v** its velocity, and **B** the magnetic field. In the case where the angle between **v** and **B** is 90° (i.e., the charge travels at right angles to the direction of the field) the magnitude or numerical result for **F** is simply the arithmetic product of the three quantities. This is the maximum force

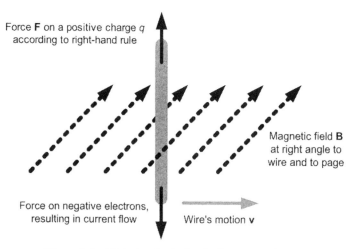

Figure 1.3 Right-hand rule for the force on a charge.

possible: as the term cross product suggests, the charge has to move *across* the field in order to experience the effect. The more **v** and **B** are at right angles to each other, the greater the force; the more closely aligned **v** and **B** are, the smaller the force. If **v** and **B** are parallel—that is, the charge is traveling along the magnetic field lines rather than across them—the force on the charge is zero.

Figure 1.3 illustrates a typical application of this relationship. The charges q reside inside a wire, being moved as a whole so that each of the microscopic charges inside has a velocity **v** in the direction of the wire's motion. If we align our right hand with that direction **v** and then curl our fingers in the direction of the magnetic field **B** (shown in the illustration as pointing straight back into the page), our thumb will point in the direction of the force **F** on a positive test charge. Because in practice the positive charges in a metal cannot move but the negatively charged electrons can, we observe a flow of electrons in the negative or opposite direction of **F**. Because only the *relative* motion between the charge and the magnetic field matters, the same effect results if the charge is stationary in space and the magnetic field is moved (e.g., by physically moving a bar magnet), or even if both the magnet and the wire are stationary but the magnetic field is somehow made to become stronger or weaker over time. As we will see in Chapter 4, a combination of these effects—movement through space of wires and magnets, as well as changing magnetic field strength—is employed in the production of electric power by generators.

The phenomenon of *electromagnetic induction* occurs when this electromagnetic force acts on the electrons inside a wire, accelerating them in one direction along the wire and thus causing a current to flow. The current resulting from such a changing magnetic field is referred to as an *induced current*. This is the fundamental process by which electricity is generated, which will be applied over and over within the many elaborate geometric arrangements of wires and magnetic fields inside actual generators.

1.5.5 Electromagnetic Fields and Health Effects

A current flowing through a wire, alternating at 60 cycles per second (60 Hz), produces around it a magnetic field that changes direction at the same frequency. Thus, whenever in the vicinity of electric equipment carrying any currents, we are exposed to magnetic fields. Such fields are sometimes referred to as *EMF*, for *electromagnetic fields*, or more precisely as *ELF*, for *extremely low-frequency fields*, since 60 Hz is extremely low compared to other electromagnetic radiation such as radio waves (which is in the megahertz, or million hertz range).

There is some concern in the scientific community that even fields produced by household appliances or electric transmission and distribution lines may present human health hazards. While such fields may be small in magnitude compared to the Earth's magnetic field, the fact that they are oscillating at a particular frequency may have important biological implications that are as yet poorly understood.

Research on the health effects of EMFs or ELFs continues. Some results to date seem to indicate a small but statistically significant correlation of exposure to ELFs from electric power with certain forms of cancer, particularly childhood leukemia, while other studies have found no effects.[22] In any case, the health effects of ELFs on adults appear to be either sufficiently mild or sufficiently rare that no obvious disease clusters have been noted among workers who are routinely exposed—and have been over decades—to vastly stronger fields than are commonly experienced by the general population.

From a purely physical standpoint, the following observations are relevant: First, the intensity of the magnetic field associated with a current in a wire is directly proportional to the current; second, the intensity of this field decreases at a rate proportional to the inverse square of the distance from the wire, so that doubling the distance reduces the field by a factor of about 4. The effect of distance thus tends to outweigh that of current magnitude, especially at close range where a doubling may equate to mere inches. It stands to reason, therefore, that sleeping with an electric blanket or even an electric alarm clock on the bedside table would typically lead to much higher exposure than living near high-voltage transmission lines. Measured ELF data are published by many sources.

1.5.6 Electromagnetic Radiation

Although not vital in the context of electric power, another manifestation of electromagnetic interactions deserves at least brief discussion: namely, electromagnetic waves or *radiation*, including what we experience as *light*. Visible light in fact represents a small portion of an entire spectrum of electromagnetic radiation, which is differentiated by *frequency* or *wavelength*. The nonvisible (to us) regions of this spectrum include infrared and ultraviolet radiation, microwaves, radio waves and others used in telecommunications (such as cellular phones), X rays, as well as gamma rays from radioactive decay. Physically, all these types of radiation are of the same basic nature.

The concept of a wave is familiar to most of us as the periodic movement of some material or medium: the surf on a Hawaiian beach, a vibrating guitar string, or the coordinated motion of sports fans in the bleachers. What is actually "waving" when electromagnetic radiation travels through space is much less tangible and challenging to the imagination; we can describe it only as a pulse of rapidly increasing and decreasing electric and magnetic fields, themselves completely insubstantial and yet measurably affecting their environment. These electric and magnetic fields are at right angles to each other, and at right angles again to the direction of propagation of the wave, as illustrated in Figure 1.4.

The *frequency* of the electromagnetic wave refers to the rate at which either the electric or magnetic field at any one point oscillates (i.e., changes direction back and

---

[22]Technical information and summaries of research have been published by the World Health Organization, http://www.who.int/mediacentre/factsheets/ (accessed November 2004), and by the National Institutes of Health, http://www.niehs.nih.gov/emfrapid/home.htm (accessed November 2004). See, for example, C.J. Portier and M.S. Wolfe (eds.), *Assessment of Health Effects from Exposure to Power-Line Frequency Electric and Magnetic Fields*, NIEHS Working Group Report (Research Triangle Park, NC: National Institute of Environmental Health Sciences of the National Institutes of Health, 1998).

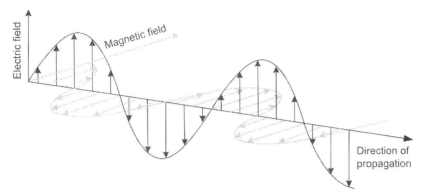

Figure 1.4 An electromagnetic wave.

forth). Frequency is measured in cycles per second, or *hertz* (Hz). The *wavelength* represents the distance in space from one wave crest to the next (analogously, the *period* of a wave represents the separation in time from one wave crest to the next). Depending on the range of the spectrum in question, wavelength may be measured in meters or any small fraction thereof. For example, the wavelength of a certain color of visible light might be quoted in microns (1 μm = 10^{-6} m), nanometers (1 nm = 10^{-9} m), or angstroms (1 Å = 10^{-10} m).

Wavelength is inversely proportional to the frequency; higher frequency implies shorter wavelength, and vice versa. This is because wavelength and frequency multiplied together yield the speed of propagation of the wave, which is fixed: the speed of light. Aside from the detail that the speed of an electromagnetic wave actually varies slightly depending on the medium through which it is traveling, the constancy of the speed of light is famously important.[23]

This constancy can be understood as a manifestation of the principle of the *conservation of energy*: It is at this speed and only at this speed of propagation, or rate of change of electric and magnetic fields, that they keep inducing each other at the same magnitude. Were the wave to propagate more slowly, the fields would decay (implying energy that mysteriously vanishes); were it to propagate any faster, the fields would continually increase (implying a limitless creation of energy). As we know from the first law of thermodynamics, energy can be neither created nor destroyed in any physical process. From this basic principle it is possible to derive the constant speed of light.

Electromagnetic radiation interacts with matter through charges—specifically, electrons—that are accelerated and moved by the field. Let us consider first the example of radio waves that are broadcast and received through conducting metal

---

[23]In the fine print of the physics textbook we learn that 3×10^8 meters per second, or 186,000 miles per hour, is the speed of light only *in a vacuum*. Light travels somewhat slower through various materials, and these small differences in the speed of light—both as a function of the medium and of wavelength—give rise to familiar optical phenomena like refraction in a lens, prism, or glass of water.

antennas, and then the more general case of photon absorption and emission in all types of materials.

As the music plays at the radio station, its specially encoded electronic signal travels in the form of a rapidly changing electric current into the station's large antenna, moving the electrons inside the metal up and down.[24] These moving electrons produce a pulse of a changing electric field that is "felt" in the region of space surrounding the antenna. This oscillating electric field induces a magnetic field, which in turn induces an electric field, and so forth, with the fields propagating away in the form of a wave—an electromagnetic wave of a very specific time signature. The wave becomes weaker with increasing distance from the antenna in that it spreads out through space, though the "pulse" itself is preserved as long as the wave is detectable.

Another antenna at a distance can now "receive" the wave because the electrons inside it will be accelerated by the changing electric field, in exactly the same fashion as the electrons responsible for "sending" it. We can see that an antenna needs to be conducting so as to allow the electrons to move freely according to the changing field. When this induced motion of electrons is decoded by the radio receiver, the electric signal travels through a wire and finally moves the magnet of a loudspeaker back and forth, the specific signature of the electromagnetic wave is translated back into sound.

This large-scale motion of electrons, as in radio antennas, is a special case of their interaction with electromagnetic radiation. More generally, electrons stay within their atomic orbitals, but they nevertheless undergo certain transitions that allow them to "send out" or "receive" electromagnetic radiation. These transitions are not readily represented in terms of physical motion, but can only be described in the language of quantum mechanics. Physicists say that an electron changes its *energy level*, and that the difference between the energy level before and after the transition corresponds to the energy carried by a "packet" of electromagnetic radiation. Such a packet is called a *photon*.[25] As an electron moves to a state of lower energy, it emits a photon, and conversely, as it absorbs a photon, it rises to a state of higher energy.

Although the photon itself has no mass and can hardly be conceived of as an "object," it is nonetheless transporting energy through space. The amount of energy is directly proportional to the frequency of the radiation: the higher the frequency, the greater the energy. Thus, we can think of the "waving" electric and

---

[24]There are two standard types of encoding: amplitude modulation (AM) and frequency modulation (FM). In each case, the sound signal (which itself is an electrical pulse of changing voltage and current that mimics the corresponding sound wave) is superimposed on a *carrier wave* of a given frequency (the broadcast frequency of that particular station, which, at many kilohertz or megahertz, is several orders of magnitude higher than the frequency of the signal). In AM, the amplitude of the carrier wave is continually changed (modulated) according to the signal; in FM, the frequency is changed by a small percentage.

[25]It was one of the most stunning discoveries in early 20th-century physics that radiation occurs in such packets, or *quanta*, that only interact with a single electron at a time; the crucial experiment that demonstrated this (the *photoelectric effect*) is what actually won the 1921 Nobel Prize in Physics for Einstein.

magnetic fields in space as a form of potential energy, whose presence does not become apparent until it interacts with matter.

The configuration of electrons within a given material, having a certain atomic and molecular structure, determines what energy transitions are available to electrons. They will interact with radiation only to the extent that the available transitions match precisely the energy of the photon, corresponding to its frequency (wavelength). This explains why materials interact differently with radiation of different frequencies, absorbing some and transmitting or reflecting others. A glass window, for example, transmits visible light but not ultraviolet. And we find ourselves—at this very moment—in a space full of radio waves, oblivious to their presence because the waves pass right through our bodies: the energy of their individual photons is insufficient to cause a transition inside our electrons.

In the context of electric power system operation, electromagnetic radiation does not play much of an explicit role. This is because the conventional frequency of alternating current at 50 or 60 hertz is so low that the corresponding radiation propagates with extremely little energy and is in practice unobservable. Stationary and alternating electric and magnetic fields, however, are central to the workings of all electric machinery.

CHAPTER 2

Basic Circuit Analysis

2.1 MODELING CIRCUITS

As a general definition, a *circuit* is an interconnection of electric *devices*, or physical objects that interact with electric voltages and currents in a particular manner. Typically, we would imagine the devices in a circuit to include a power *source* (such as a battery, a wall outlet, or a generator), *conductors* or wires through which the electric current can flow, and a *load* in which the electric power is being utilized (converted to mechanical or thermal energy). To analyze a circuit means to account for the properties of all the individual devices so as to predict the circuit's electrical behavior. By "behavior" we mean specifically what voltages and currents will occur at particular places in the circuit given some set of conditions, such as a voltage supplied by a power source. This behavior will depend on the nature of the devices in the circuit and on how they are connected.

For the purpose of circuit analysis, individual devices are represented as ideal objects or *circuit elements* that behave according to well-understood rules.[1] From the circuit perspective, the events inside these elements are irrelevant; rather, we focus on measurements at the elements' *terminals*, or points where the elements connect to others.

The scale of analysis can shift depending on the information that is of interest. For example, suppose we are analyzing a circuit in our house to see whether it might be overloaded. We find that several appliances are plugged into this circuit, including a radio. We would consider this radio as one of the loads and conceptualize it simply as a box with two terminals (the two prongs of its plug) that draws a particular amount of current when presented with 120 volts by the outlet. On the other hand, suppose we wish to understand how the radio can be tuned to different frequencies. Now we would draw a diagram that includes many of the radio's interior components. Still, we might not include every single little resistor or capacitor; rather, we would group some of the many electronic parts together and represent

---

[1] Most real devices match their simple idealized versions very closely in behavior (for example, a resistor that obeys Ohm's law). If not, there is always some way of combining a set of abstract elements so as to represent the behavior of the physical gadget to the desired accuracy.

Electric Power Systems: A Conceptual Introduction, by Alexandra von Meier
Copyright © 2006 John Wiley & Sons, Inc.

them as a single element so as to eliminate unnecessary detail. Our choice of scale for this grouping would be precisely such that what goes on *inside* the elements we have defined is not relevant to the question at hand.

Grouping components into functional elements on whatever the appropriate scale is a powerful technique in circuit analysis, and electrical engineers use it constantly without even thinking about it. As we will see, there are precise rules for "scaling up" or simplifying the representation of a circuit without altering the relevant properties.

The most basic circuit elements, and those mainly of interest in power systems, include resistors, capacitors, and inductors. These are also called *linear* circuit elements because they exhibit linear relationships between voltage and current or their rates of change. In electronics, nonlinear circuit elements such as transistors and diodes are also extremely important, but we do not discuss these here.[2] Finally, common circuit elements include d.c. and a.c. power sources, some of which may be broken down further for analysis in certain contexts (e.g., the internal modeling of electric generators).

Generally, the conducting wires that connect various elements are assumed to have a negligible or zero resistance. Thus, they are simply drawn as lines, of arbitrary length and shape, with no particular significance other than the endpoints they connect. For purposes of circuit analysis, two points connected by such a wire might as well be immediately adjacent; they are, in electrical terms, the same point.[3] Formally, we would say that the voltage difference between any two points connected by a zero-resistance conductor is nil; the points are at the same electric potential. A notable difference in the power systems context is that when one is dealing with transmission and distribution lines that extend over longer distances, the assumption of negligible resistance (or, more accurately, impedance) no longer holds. Thus, depending again on the scale of analysis, conductors in power systems may or may not be important as circuit elements in their own right.

2.2 SERIES AND PARALLEL CIRCUITS

When considering multiple devices in a circuit and their joint behavior, it is obviously important to consider how they are connected. There are two basic ways in which circuit devices can be connected together, referred to as *series* and *parallel*.

A series connection is one in which the electric current flows first through one element, then through the next. By necessity, all the current that goes through the first also goes through the second (and third, etc.); in other words, the current

---

[2] Nonlinear circuit elements can be made out of semiconducting materials, whose conductive properties are not fixed (as we assume for all other circuit elements), but change depending on ambient conditions, such as the voltage that is being applied. Resistors, capacitors, and inductors can also be nonlinear, but we will ignore those cases.

[3] Mathematically speaking, we are only modeling the *topology* of the connections.

through elements in series is equal. This requirement follows intuitively from the nature of current as a flow of charge. This charge is neither created nor destroyed at any point along the connection of interest; we say it is *conserved*. Therefore, what goes in one end must come out the other.

A parallel connection is one in which the devices present the current with two or more alternate paths: there is a branch point. Any individual charge will only go through one device, and thus the current flow divides. Conservation of charge in this case requires that the sum of currents through all the alternate paths remain constant; that is, they add together to equal the initial current that was divided up. This notion is formalized in Kirchhoff's current law (see Section 2.3).

Any network of circuit elements, no matter how intricate, can be decomposed into series and parallel combinations.

2.2.1 Resistance in Series

The simplest kind of combination of multiple circuit elements has resistors connected in series (Figure 2.1). The rule is easy: to find the resistance of a series combination of resistors, add their individual resistances. For example, if a 10-Ω resistor is connected in series with a 20-Ω resistor, their combined resistance is 30 Ω. This means that we could replace the two resistors with a single resistor of 30 Ω and make no difference whatsoever to the rest of the circuit. In fact, if the series resistors were enclosed in a box with only the terminal ends sticking out, there would be no way for us to tell by electrical testing on the terminals whether the box contained a single 30-Ω resistor or any series combination of two or more resistors whose resistances added up to 30 Ω. Thus, an arbitrary number of resistances can be added in series, and their order does not matter.

Intuitively, the addition rule makes sense because if we think of a resistor as posing an "obstacle" to the current, and note that the same current must travel through each element in the series, each obstacle adds to the previous ones. This notion can be formalized in terms of *voltage drop* (defined in Section 1.3.3). Across each resistor in a series combination, there will be a voltage drop proportional to its resistance. It is always true that, regardless of the nature of the elements (whether they are resistors or something else), the voltage drop across a set of elements connected in series equals the sum of voltage drops across the individual elements. This notion reappears in the context of Kirchhoff's Voltage Law (see Section 2.3.1).

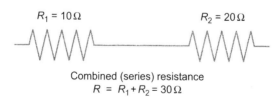

Figure 2.1 Resistors in series.

2.2.2 Resistance in Parallel

When resistors are combined in parallel, the effect is perhaps less obvious than for the series case: rather than adding resistance, we are in fact *decreasing* the overall resistance of the combination by providing alternative paths for the current. This is so because in the parallel case the individual charge is not required to travel through every element, only one branch, so that the presence of the parallel elements "alleviates" the current flow through each branch, and thereby makes it easier for the charge to traverse. It is convenient here to consider resistors in terms of the inverse property, *conductance* (Section 1.1.5). Thus, we think of the resistor added in parallel not as posing a further obstacle, but rather as providing an additional conducting option: after all, as far as the current is concerned, any resistor is still better than no path at all. Accordingly, the total resistance of a parallel combination will always be *less* than any of the individual resistances.

Using conductance ($G = 1/R$), the algebraic rule for combining any number of resistive elements in parallel is simply that the conductance of the parallel combination equals the sum of the individual conductances.

For example, suppose a 10-Ω and a 2.5-Ω resistor are connected in parallel, as in Figure 2.2. We know already that their combined (parallel) resistance must be less than 2.5 Ω. To do the math, it is convenient to first write each in terms of conductance: 0.1 mho and 0.4 mho. The combined conductance is then simply the sum of the two, 0.5 mho. Expressed in terms of resistance, this result equals 2 Ω. In equation form, we would write for resistors in parallel:

$$\frac{1}{R} = \frac{1}{R_1} + \frac{1}{R_2} + \cdots$$

where R is the combined resistance, and R_1, R_2, and so forth are the individual resistances.

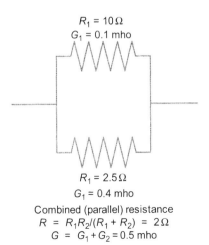

Figure 2.2 Resistors in parallel.

34 BASIC CIRCUIT ANALYSIS

In the case of only two resistances in parallel, it is often preferred to rearrange this equation to read[4]

$$R = \frac{R_1 R_2}{R_1 + R_2}$$

With more than two resistances, however, this notation becomes increasingly awkward. For example, in the case of three resistances in parallel it reads

$$R = \frac{R_1 R_2 R_3}{R_1 R_2 + R_1 R_3 + R_2 R_3}$$

Thus, when many resistances are combined in parallel, it is generally more convenient to express them in terms of conductance.

Note that the voltage drop across any number of elements in parallel is the same. This can easily be seen because all the elements share the same terminals: the points where they connect to the rest of the circuit are, in electrical terms, the same.

While elements connected in parallel thus have a common voltage drop across them, the current flowing through the various elements or branches[5] will typically differ. Intuitively, we might guess that more current will flow through a branch with a lower resistance, and less current through one with a higher resistance. This can be shown rigorously by applying Ohm's law for each of the parallel resistances: If V is the voltage drop common to all the parallel resistances, and R_1 is the individual resistance of one branch, then the current I_1 through this branch is given by V/R_1. Thus, the amount of current through each branch is inversely proportional to its resistance.

For example, in Figure 2.2, the current through the 2.5-Ω resistor will be four times greater than that through the 10-Ω resistor, whatever the applied voltage. If the voltage is, say, 10 V, then the currents will be 1 A and 4 A, respectively. Note that the sum of these currents is consistent with applying Ohm's law to the combined resistance: $10 \text{ V}/2 \text{ } \Omega = 5 \text{ A}$.

To summarize, there is a tidy correspondence between the series and parallel cases: In a series connection, the current through the various elements is the same, but the voltage drops across them vary (proportional to their resistance); in a parallel connection, the voltage drop across the various elements is the same, but the currents through them vary (inversely proportional to their resistance). This fundamental distinction has important practical implications, as we shall see in Section 2.2.4.

---

[4]When memorizing this formula, it is helpful to keep in mind that the units of both sides of the equation are resistance. Thus, the product term (units of resistance squared) must be in the numerator, and the sum term (units of resistance) in the denominator.

[5]The term "branch" for one path in a parallel connection is preferable because, in general, there could be more than one element (in series) along each parallel path; see Section 2.2.4.

2.2.3 Network Reduction

As stated earlier, any network of circuit elements is composed of some mixture of series and parallel combinations. To analyze the network, circuit branches are sequentially aggregated from the bottom up as series or parallel combinations, up to the desired scale. The point is best made through an illustrative example.

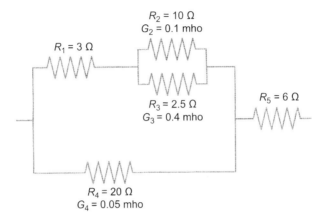

Figure 2.3 Network reduction.

Suppose we wish to determine the equivalent resistance of the network of five resistors shown in Figure 2.3, relying on the fact that any parallel and series combination of resistors can ultimately be reduced to a single resistance. Starting from the largest scale, we note that the total resistance will be the sum of R_5 and the combination of four resistors on the left-hand side. This combination, in turn, has a resistance corresponding to the parallel combination of R_4 and the branch on top with three resistances, and so on. For computation, we begin at the smallest scale, evaluating first the combination of R_1 and R_2 and working our way up from there. As demonstrated in the numerical example, it is convenient to switch back and forth between units of resistance and conductance so as to facilitate the arithmetic for evaluating the parallel combinations. Finally, by expressing the five individual resistances as one combined resistance, we have converted our initial model into a much simpler one with only a single circuit element, while its behavior in relation to anything outside it remains unchanged. Thus we have effectively scaled up our representation of the circuit. This kind of simplification process can be carried out repeatedly at various levels.

Example

What is the combined resistance of the five resistors in the network in Figure 2.3?
    We proceed by adding together conductances for parallel elements, and resistances for series elements, starting with the combined conductance of R_1 and R_2, which is $G_{12} = G_1 + G_2 = 0.5$ mho. We then invert G_{12} to express it as a resistance that can be added to R_3. Thus, $R_{12} = 1/G_{12} = 2\,\Omega$, and

$R_{123} = R_{12} + R_3 = 5\ \Omega$. We then invert to conductance, $G_{123} = 1/R_{123} = 0.2$ mho, and add it to G_4 to obtain $G_{1234} = G_{123} + G_4 = 0.25$ mho. Finally, after inverting again to resistance, $R_{12345} = R_{1234} + R_5 = 10\ \Omega$.

In analyzing power systems, it is often necessary to model the system at different scales, depending, for example, on whether the focus is on long-distance transmission or power distribution. The scaling process primarily involves aggregating individual loads within an area and representing them as a single block of load. This could be the load on a distribution transformer, a distribution feeder, or on an entire substation. In this case, however, the simplification is based on empirical measurement (e.g., of loads at the substation), as opposed to an algebraic procedure like the one just discussed, say, by combining the individual resistances or *impedances* (see Section 3.2.3).

In general, circuit analysis, and especially the technique of network reduction, gets more complicated when there is a large number of branches and circuit elements, and also when there are circuit elements of different types whose characteristics are not readily summarized in terms of resistance or impedance. For these situations, electrical engineers carry an arsenal of more sophisticated reduction techniques that are beyond the scope of this text.[6] Still, as we will see in Chapter 7, the complexity of power systems is such that elegant analytic procedures can be easily exhausted well before a system appears tractable by pencil and calculator. At this point, the task becomes assembling all the information neatly in tabular or *matrix* form, and then preferably passing it along to a computer to work out the numerical answers or approximations.

2.2.4 Practical Aspects

In real circuits for power delivery, we mostly think of circuit elements in terms of power sources or loads, such as appliances in the house or several homes on the distribution circuit that runs down the block. These loads are always connected in parallel, not in series. The reason for this is that a parallel connection essentially allows each load to be operated independently of the others, since each is supplied with the same standard voltage but can draw a current depending on its particular function (which determines the amount of power consumed; see Section 3.4).

Supplied by a constant voltage source, which we like to assume in the context of power systems (although it is only an approximation), resistive loads in parallel are essentially unaffected by each other. Interactive effects only occur as departures from the idealized situation; for example, when a particularly heavy load affects the local voltage, and thus indirectly the other loads.[7] By contrast, independent operation of loads in series would be impossible, since elements in series share the same

---

[6] The most important of these techniques are representing circuits as so-called Thévenin and Norton equivalents, where circuits containing multiple elements and power sources are modeled as a single voltage or current source with a series or parallel resistance, respectively.

[7] Dimming or flickering lights when a big motor switches on are a familiar example.

current. Thus, turning any one of them off will interrupt the current flow to all the others.[8] Even if all elements are operational, the amount of power consumed in each one cannot easily be adjusted, and the voltage across each represents only some fraction of the voltage across the combination.

The series connection is mostly relevant in power systems when elements are considered that represent successive steps between power generation and consumption. Thus, the loads are in series with a distribution line, a transmission line, and a generator. Actually, these elements may also be in parallel with many others, since the entire system forms a network with many links. Still, there is some minimum number of elements in series that constitute a path from power source to load. For example, if we are interested in the resistance of the conductors between a distribution substation and a customer (where there is usually only a single path due to the *radial* layout of the distribution system; see Section 6.1), we would have to *add* all the contributions to resistance along the way.

The important conceptual point here is that because of this series connection, there is no escaping the interdependence among the elements: there is literally no way around the other elements on a series path. This becomes important in the context of transmission constraints (see Sections 6.5, 8.2, 9.2) and excessive voltage drops due to high loads (see Section 6.6). While it may be perfectly obvious to an engineer that any devices through which the same current must travel are necessarily dependent on each other, this presents a very fundamental problem when legal and institutional arrangements concerning power systems have these devices under the auspices of different parties.

2.3 KIRCHHOFF'S LAWS

Anything we learn about the behavior of a circuit from the connections among its elements can be understood in terms of two constraints known as Kirchhoff's laws (after the 19th-century German physicist Gustav Robert Kirchhoff). Specifically, they are Kirchhoff's voltage law and Kirchhoff's current law. Their application in circuit analysis is ubiquitous, sometimes so obvious as to be done unconsciously, and sometimes surprisingly powerful. While Kirchhoff's laws are ultimately just concise statements about the basic physical properties of electricity discussed in Chapter 1, when applied to intricate circuits with many connections, they turn into sets of equations that organize our knowledge about the circuit in an extremely elegant and convenient fashion.[9]

---

[8]Christmas tree lights of older vintages are a classic example of this phenomenon, and of the painstaking process to identify the culprit element. Newer models have a contact that allows the current to bypass the light bulb filament when it is broken. The other lights on the string will only go out if a bulb is removed altogether.

[9]In mathematical terms, Kirchhoff's laws yield a number of linearly independent equations, often arranged in matrix form, that are just sufficient to determine the voltages and currents in every circuit branch, given information about all the circuit elements present.

2.3.1 Kirchhoff's Voltage Law

Kirchhoff's voltage law (often abbreviated KVL) states that the sum of voltages around any closed loop in a circuit must be zero. In essence, this law expresses the basic properties that are inherent in the definition of the term "voltage" or "electric potential." Specifically, it means that we can definitively associate a potential with a particular point that does not depend on the path by which a charge might get there. This also implies that if there are three points (A, B, and C) and we know the potential differences between two pairings (between A and B and between B and C), this determines the third relationship (between A and C). Without thinking in such abstract and general terms, we apply this principle when we move from one point to another along a circuit by *adding* the potential differences or voltages along the way, so as to express the *cumulative* voltage between the initial and final point. Finally, when we go all the way around a closed loop, the initial and final point are the same, and therefore must be at the same potential: a zero difference in all.

The analogy of flowing water comes in handy. Here, the voltage at any given point corresponds to the elevation. A closed loop of an electric circuit corresponds to a closed system like a water fountain. The voltage "rise" is a power source—say, a battery—that corresponds to the pump. From the top of the fountain, the water then flows down, maybe from one ledge to another, losing elevation along the way and ending up again at the bottom. Analogously, the electric current flows "down" in voltage, maybe across several distinct steps or resistors, to finish at the "bottom" end of the battery. This notion is illustrated by in the simple circuit in Figure 2.4 that includes one battery and two resistors. Note that it is irrelevant which point we choose to label as the "zero" potential: no matter what the starting point,

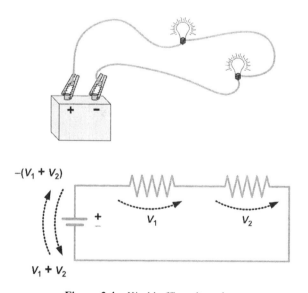

Figure 2.4 Kirchhoff's voltage law.

adding all the potential gains and drops encountered throughout the complete loop will give a zero net gain.

2.3.2 Kirchhoff's Current Law

Kirchhoff's current law (KCL) states that the currents entering and leaving any branch point or *node* in the circuit must add up to zero. This follows directly from the conservation property: electric charge is neither created nor destroyed, nor is it "stored" (in appreciable quantity) within our wires, so that all the charge that flows into any junction must also flow out. Thus, if three wires connect at one point, and we know the current in two of them, they determine the current in the third.

Again, the analogy of flowing water helps make this more obvious. At a point where three pipes are connected, the amount of water flowing in must equal the amount flowing out (unless there is a leak). For the purpose of computation, we assign positive or negative signs to currents flowing in and out of the node, respectively. It does not matter which way we call positive, as long as we remain consistent in our definition. Then, the sum of currents into (or out of) the node is zero. This is illustrated with the simple example in Figure 2.5, where KCL applied to the branch point proves that the current through the battery equals the sum of currents through the individual resistors.

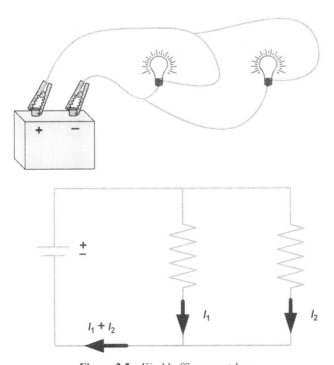

Figure 2.5 Kirchhoff's current law.

Despite their simple and intuitive nature, the fundamental importance of Kirchhoff's laws cannot be overemphasized. They lie at the heart of the interdependence of the different parts and branches of power systems: whenever two points are electrically connected, their voltages and the currents through them must obey KVL and KCL, whether this is operationally and economically desirable or not. For example, managing transmission constraints in power markets is complicated by the fact that the flow on any one line cannot be changed independently of others. Thus the engineer's response to the economist's lamentation of how hard it is to manage power transmission: "Blame Kirchhoff."

2.3.3 Application to Simple Circuits

When Kirchhoff's laws are combined with information about the characteristics of all the elements within a circuit (which specify the relationship between the voltage across and the current through each element), the voltages and currents at every location in a circuit can be specified regardless of the number of branch points, as long as all the circuit elements behave according to certain rules and the topology (connectedness) of the circuit meets certain criteria. Actually carrying out such a calculation, however, can be quite tedious, and we will only give a qualitative description of the process here.

The technique of choice is to write the relationships implied by Kirchhoff's laws as a list or table of equations. Such a table, when properly organized, is known as a *matrix*. This matrix shows which branch point or *node* of the circuit is connected to which other node, and also includes information about each branch, such as the conductance (or, as in Chapter 7, the *admittance*) between pairs of nodes. For large circuits, one writes this matrix in a systematic fashion by first labeling nodes and assigning reference directions for voltages between nodes and for the current through each branch (since it is important whether the voltage and current in any one direction is to be considered positive or negative). When combined according to the rules of linear algebra with current and voltage *vectors*, which are just properly ordered listings of all the branch currents and voltages, this formulation spells out Kirchhoff's laws. It is then a matter of standard linear algebra procedure to solve for any unknown current and voltage variables, given a sufficient number of "knowns." This type of procedure is detailed in all standard textbooks on circuit analysis.

Rather than analyzing complicated circuits, let us illustrate some simple, familiar applications of Kirchhoff's Laws through the following two examples.

Example

A string of Christmas lights that plugs into a 120 V outlet has 50 identical bulbs connected in series. What is the voltage across each bulb?

The voltage across each bulb is $120 \text{ V}/50 = 2.4 \text{ V}$. This seems obvious, but is a consequence of KVL, which requires that the voltage drops along the string add up to the same amount as the voltage drop from one to the other terminal of the wall outlet.

Example

A college student sets up a kitchen in his studio and decides to run several appliances from the same extension cord: a hot plate with an internal resistance of 20 Ω, a light with 60 Ω, and a toaster oven with 10 Ω. The extension cord has a resistance of 0.1 Ω on each of its two wires. What is the current through each device, when all are in use?

To a first approximation, we would say that the appliances all "see" a voltage of 120 V. The current through each is then given by Ohm's law: 120 V/20 Ω = 6 A, 120 V/60 Ω = 2 A, and 120 V/10 Ω = 12 A. KCL then tells us that the current through the cord must equal the sum of all three, or 6 A + 2 A + 12 A = 20 A. The complication, however, is that at the end of the extension cord, the voltage is not exactly 120 V. Rather, with 20 A, there would be a voltage drop according to Ohm's law of 20 A · 0.1 Ω = 2 V on each "leg" of the cord. From KVL, it follows that the actual voltage as "seen" by the appliances is only 120 V − (2 · 2 V) = 116 V.

To obtain an exact answer, the easiest approach is to write an expression for the total resistance in the circuit, which is

$$R = 0.1 + \frac{1}{\frac{1}{20} + \frac{1}{60} + \frac{1}{10}} + 0.1 = 6.2 \, \Omega$$

The current in the cord is thus 120 V/6.2 Ω = 19.35 A. This current results in a voltage drop of twice 19.35 A · 0.1 Ω, leaving 120 V − 3.87 V = 116.13 V for the appliances. The current in each appliance comes out to be 5.8 A, 1.9 A, and 11.6 A, respectively. Note that the three currents add up to the total current through the cord, as required by KCL.

The moral of this story is that one ought to be careful about using extension cords, especially when they are long and of narrow gauge (high resistance) and when using powerful appliances. Some appliances—not the resistive kind in this example, but those involving motors—will run less efficiently and may eventually even be damaged when supplied with a voltage much less than the nominal 120 V. Worse yet, the extension cord may become very hot and pose a fire hazard.

2.3.4 The Superposition Principle

In addition to applying Kirchhoff's laws and scaling circuits up or down, a third analysis tool is based on the *superposition principle*. This principle applies to circuits with more than one voltage or current source. It states that the combined effect—that is, the voltages and currents at various locations in the circuits—from the several sources is the same as the sum of individual effects. Knowing this allows one to consider complicated circuits in terms of simpler components and then combining the results.

In power systems, the superposition principle is used to conceptualize the interactions among various generators and loads. For example, we may think of the current or power flow along a transmission link due to a "shipment" of power from one generator to one consumer, and we add to that the current resulting from separate transactions in order to obtain the total flow on that link. With voltages held fixed, the currents become synonymous with power flows, and we can add and subtract megawatt flows superimposed along various transmission links. This procedure is illustrated in Section 6.1.5 on Loop Flow.

For a simple example where we can deal with currents and voltages explicitly, consider the circuit in Figure 2.6. This circuit has two power sources. The first source, labeled S_1, is a battery that functions as a *voltage source*, delivering 12 volts. The second source, S_2, is a *current source* that delivers 1.5 amperes. This type of source is less familiar in power systems than in electronics; it has the property of always delivering a specific current regardless of the resistance in the circuit connected to it, while allowing the voltage at its terminals to vary (as opposed to the more familiar voltage source that specifies terminal voltage while the current depends on resistance in the circuit). We introduce the current source here because it makes for a good illustration of the superposition principle.

Suppose we wish to predict the voltage level v and current i at the locations identified in the diagram. It is not immediately obvious what the voltages and currents at various points in this circuit should be as a result of the combination of the two sources. The superposition principle is an indispensable analytic tool here: it states that we can consider separately the voltage and current that would result from each individual source, and then simply add them together. This principle applies regardless of the circuit's complexity or the number of power sources; it also holds true at any given instant in a circuit with time-varying sources.

In our example, the voltage v that would result from only S_1—written as $v(S_1)$, indicating that v for now is only a function of S_1—is determined from the relative magnitudes of the two resistances in the circuit, R_1 and R_2, while ignoring the presence of the current source. Since R_2 at 2 Ω represents one-third of the total resistance in this simple series circuit, 2 Ω + 4 Ω, the voltage across R_2 that we want to find is simply one-third of the total:

$$v(S_1) = 12 \text{ V} \cdot 2 \text{ } \Omega/(4 \text{ } \Omega + 2 \text{ } \Omega) = 12 \text{ V} \cdot 1/3 = 4 \text{ V}$$

Ignoring the current source technically means setting the current through it to zero, or replacing it with an open circuit. Having gotten rid of the extra circuit branch, the current $i(S_1)$ through the resistor R_2 that would result from S_1 alone is easy to find with Ohm's law:

$$i(S_1) = 12 \text{ V}/(4 \text{ } \Omega + 2 \text{ } \Omega) = 2 \text{ A}$$

Next, we ignore the voltage source, meaning that we set the voltage difference across it to zero, or replace it with a short circuit. The current $i(S_2)$ through R_2 based on the current source alone is found from the relative magnitudes of the resistances in each branch: since R_2 has half the resistance of R_1, twice the current will

2.3 KIRCHHOFF'S LAWS

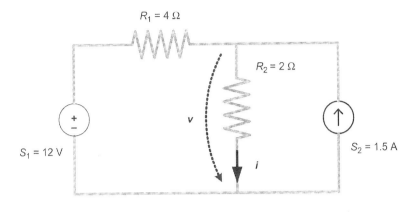

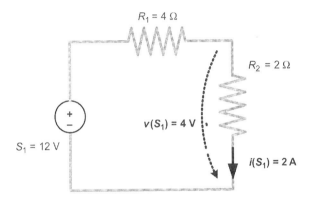

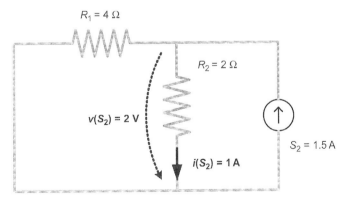

Figure 2.6 Superposition.

flow through it, or two thirds of the total:

$$i(S_2) = 1.5 \text{ A} \cdot 4\,\Omega/(4\,\Omega + 2\,\Omega) = 1.5 \text{ A} \cdot 2/3 = 1 \text{ A}$$

The voltage $v(S_2)$ due to the current source alone is again found via Ohm's law:

$$v(S_2) = 1 \text{ A} \cdot 2\,\Omega = 2 \text{ V}$$

We can now superimpose the contributions from the two sources and find

$$v = v(S_1) + v(S_2) = 4 \text{ V} + 2 \text{ V} = 6 \text{ V}$$

and

$$i = i(S_1) + i(S_2) = 2 \text{ A} + 1 \text{ A} = 3 \text{ A}$$

Although this simple example could have been solved without the use of superposition, the technique is vital for analyzing larger and more complex circuits, including those with time-varying sources.

2.4 MAGNETIC CIRCUITS

In Section 1.5 we introduced the notion of a magnetic field as a pattern of directional forces resulting from the movement of electric charge. We first described fields as analytic artifacts, or maps indicating what would happen to a test object situated in a particular space. We also argued that the field can be appropriately regarded as a physical entity in and of itself, despite the fact that it is devoid of material substance. In the context of magnetic circuits, the latter way of conceptualizing the field is more apt. Here we definitely think of the magnetic field as a "thing" in its own right: something that is present or absent, or present in a particular amount. And, because the presence of a magnetic field indicates that an object can be moved by it—that is, physical work can be done on the object—we consider the field (magnetic or electric) as containing stored or potential energy.

We also stated in Section 1.5 that the magnetic field in strict, formal terms represents the density of another quantity, the *magnetic flux*, ϕ. This flux is a measure of something imagined to flow, for example, flowing in circles around a current-carrying wire. A magnet can be represented in terms of a continuous, more or less circular (depending on the magnet's shape) flow of magnetic flux along the familiar "lines" of the magnetic field. This flux is denser (the lines closer together) inside the magnet and very close to it, and it becomes less dense with increasing distance. In this representation, a solid bar magnet is basically indistinguishable from an electromagnet created by a coil of wire (except for some subtleties around the edges).

The flux representation establishes a crucial property of magnets: namely, that they always appear as having two poles, rather than occurring as a single north or south "monopole." This property can be elegantly expressed in the mathematical statement that the magnetic flux is always continuous, just as if it were a material flowing. Flux lines neither begin nor end anywhere, but perpetually travel around in closed loops. If one enclosed a space with a hypothetical boundary (such as a balloon), no more and no less flux could enter than leave this boundary, since otherwise the flux would have to be created or destroyed within the enclosed space.[10] If a single magnetic north (or south) pole existed, it would violate this condition, since flux would emanate from it (or be absorbed by it) in all directions.

The flux, then, is the "stuff" that must always come back around. In this way, it is analogous to an electric current traveling through a closed circuit, which, as we have seen specifically in Sections 1.3 and 2.3, also has the property that "what goes in must come out." This analogy extends in such a way that we can speak of magnetic circuits that obey similar rules as electric circuits. In an electric circuit, the flow of charge can be associated with identifiable material particles; what is flowing in the magnetic circuit is just the abstract "flux." Thus, the concept of a magnetic circuit functions similarly as an analytic device that keeps track of the quantity of flux and also relates it to the properties of the materials that provide the circuit's physical path. Figure 2.7 illustrates a magnetic circuit.

Analogous to the electrical resistance, there is a property called *reluctance* that indicates the relative difficulty or ease with which the magnetic flux may traverse an element within a magnetic circuit. Reluctance is denoted by the symbol $\mathscr{R}$. The basic material property in this case is the magnetic *permeability*, which is analogous to conductivity, and is denoted by μ (Greek lowercase mu). Also in keeping with the analogy, the reluctance of a magnetic circuit element is given by

$$\mathscr{R} = \frac{l}{\mu A}$$

where l is the length and A the cross-sectional area of the element.

The permeability μ can be regarded as a material's propensity to carry magnetic flux in response to an externally applied magnetic field. This is like a conductor's propensity to allow a current to flow through it in response to an applied electric potential drop (electric field; see Section 1.5). We can visualize permeability in a vague sense as the ability of the material's particles to align with an externally applied magnetic field. However, there will be some flux even in the absence of a substantive medium. Thus, the permeability of a vacuum is not zero; rather, it is assigned a value (so as to keep all other units and measurements consistent) of $\mu_0 = 4\pi \times 10^{-7}$ tesla $\cdot$ meters/ampere, and μ_0 is called the *permeability constant*.

The fact that magnetic flux can exist anywhere gives rise to an important practical difference between electric and magnetic circuits: for an electric circuit, because the conductivity of metal is so many orders of magnitude greater than that of the surrounding air, it is a very good approximation to say that all the charge remains confined to the conducting material—such a good approximation, in fact, that we rarely stop to consider it. Magnetic flux, on the other hand, is "messy" to contain

---

[10]In mathematical notation, this statement constitutes one of the famous Maxwell's equations.

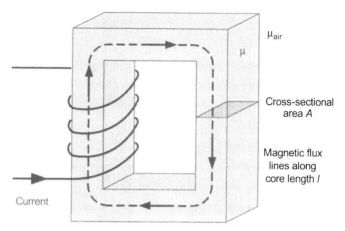

Figure 2.7 A magnetic circuit.

because the permeability of air is still some appreciable fraction of that of the best magnetic materials. Thus, the amount of *leakage flux*, or flux "spilling out" over the edges, is often significant and must be explicitly considered in engineering analyses.

For an actual material under the influence of a magnetic field, the permeability is not constant, but rather changes with the field strength or the degree of magnetization. Furthermore, the permeability depends on the material's recent history, that is, whether it is in the process of being increasingly magnetized or demagnetized.[11] Thus, the value of μ over a range of conditions during the actual operation of magnetic elements must be obtained empirically. Such data or *magnetization curves* are published for the various materials in common use for these purposes, primarily different types of iron and steel.

Quantitatively, we can write the relationship

$$\mathbf{B} = \mu \mathbf{H}$$

where **B** is the flux density through the medium (flux per area), which is the same as the familiar magnetic field, and **H** is called the *magnetic field intensity* or *magnetic field strength*.[12] **B** and **H** are written in boldface notation to indicate that they are vector quantities, that is, they have an associated direction, which in this case is the same for both.

As we know, an electric current "chooses" to flow along the path of least resistance. More rigorously speaking, the amount of current flowing through any one

---

[11]The property of a material to follow a different "path" during magnetization and demagnetization is known as *hysteresis*.

[12]These terms are given here for the sake of completeness: in the power engineering context, phenomena are less often described in terms of fields or field intensity, while the terms flux and permeability are frequently encountered.

2.4 MAGNETIC CIRCUITS

circuit branch is inversely proportional to its resistance, that is, directly proportional to its conductance. Similarly, magnetic flux tends toward regions of high magnetic permeability. Thus, while the total amount of flux is constrained by boundary conditions (just like the total amount of current through all circuit branches might be fixed), its density can be distributed throughout space, depending on local variations in magnetic permeability. In this way, the magnetic *field* or flux *density* can be concentrated in a certain region, as if the flux lines were "gathered up." This effect is used in almost all electric power devices that rely on magnetic fields, by "guiding" the magnetic field through appropriately shaped pieces of iron or steel.

However, unlike an electric current that is confined exclusively to the conducting material of a circuit, the confinement of magnetic flux inside the high-permeability material is always less than perfect. Consequently, an important issue in the design and operation of devices using magnetic flux is the so-called *leakage flux*. This leakage flux is simply the difference between the total amount of flux produced by an electric current and the amount that is successfully confined within the desired region.

The generation of magnetic flux by an electric current can be described in terms of a *magnetomotive force* (mmf for short), which is analogous to the voltage or electromotive force in an electrical circuit. The mmf depends on the amount of current and the configuration of the wire carrying it: specifically, it is not the shape of the wire that matters, but the number of times that the area in question is encircled by this wire in loops or "turns." Thus,

$$\text{mmf} = Ni$$

where i is the current and N the number of turns.

It is also possible to write an equation for flux analogous to the relationship between voltage and current for a magnetic circuit, where the flux is given by the ratio of mmf to the magnetic reluctance of the region within the turns of wire:

$$\phi = \frac{\text{mmf}}{\mathcal{R}}$$

Unlike electric resistance, which for many materials remains constant to a good approximation over a range of voltages and currents (Ohm's law), the permeability, and thus the reluctance of magnetic materials, varies as the magnetic flux increases and the material becomes "saturated." In general, the relationship between mmf and flux over great ranges of values is therefore not straightforward and must be determined experimentally. However, the reluctance $\mathcal{R}$ shares with the electric resistance the property that it is additive for elements connected in series, and thus the behavior of an entire magnetic circuit can be derived from its components.

Magnetic flux lines that pass through the enclosed area of a turn of wire are said to *link* this turn. In general, the *flux linkage* of an element is a measure of the extent to which this element is interacting with magnetic flux in its vicinity; it is denoted by λ (lambda). This interaction refers to two symmetrical processes: (1) the production of magnetic flux from electric current in the element, and (2) the reverse process, where a current through the element is induced by magnetic flux linking it. The single measure of λ applies to both phenomena simultaneously. For a coil of wire, the flux linkage is given approximately by the product of the number of turns and the flux through them:

$$\lambda = N\phi$$

This formula is approximate because it assumes that all the flux lines intersect every turn, while in reality, there will be some leakage flux that only interacts with some parts of the coil.

Finally, in anticipation of Section 3.2, where we discuss the *inductance* (L) as a crucial property of electric circuit elements, we can state that the flux linkage is also

$$\lambda = Li$$

where i is the current through the element. Thus, we can think of inductance as a measure of how much magnetic flux linkage is associated with a given amount of current for a particular device.

Magnetic circuits play a role in electric power systems primarily in the context of generators and transformers, where all the transmitted energy temporarily resides in the form of magnetic fields. Flux lines are used to describe how these devices work, and the analysis of the magnetic circuits they contain, including leakage flux, is crucial in their design. Magnetic flux, and electrical interactions that result from it (specifically, *mutual inductance*), is also important for transmission lines.

CHAPTER 3
AC Power

3.1 ALTERNATING CURRENT AND VOLTAGE

Many of the important technical characteristics of power systems have to do with their use of alternating current (a.c.) instead of direct current (d.c.). In a d.c. circuit, the polarity always remains the same: the potential always stays positive on one side and negative on the other, and the current always flows in the same direction. In an a.c. circuit, this polarity reverses and oscillates very rapidly. For power systems in the United States, the *a.c. frequency* is 60 hertz (Hz) or 60 cycles per second, meaning that the direction of voltage and current are reversed, and reversed back again, 60 times every second.

3.1.1 Historical Notes

The main reason for using a.c. in power systems is that it allows raising and lowering the voltage by means of transformers. As we will see in Section 6.2, transformers cannot be operated with d.c. It is still possible to change the voltage in a d.c. circuit, but it requires far more sophisticated and expensive equipment that had not been invented in the early days of electric power. The first power systems, which operated on d.c., were therefore limited to rather low transmission voltages: although the generators could have been designed to produce power at a higher voltage, safety considerations at the customer end dictated that the voltage be kept low. Consequently, line losses were a major problem and in effect limited the geographic expansion of power systems. After the transformer was introduced in the 1880s, d.c. and a.c. systems spent some years in fierce competition (the "Battle of the Currents"), with Edison and Westinghouse as prominent advocates on either side. The major obstacles in the way of the alternating current approach—namely, concerns about the safety of high-voltage transmission, as well as the challenge of designing an a.c. motor—were largely resolved by the mid-1890s.[1]

---

[1] For a thorough and fascinating historical discussion of the development of electric power systems, see Thomas P. Hughes, *Networks of Power: Electrification in Western Society, 1880–1930* (Baltimore: Johns Hopkins University Press, 1983).

Electric Power Systems: A Conceptual Introduction, by Alexandra von Meier
Copyright © 2006 John Wiley & Sons, Inc.

The choice of frequency for a.c. power represented a compromise among the needs of different types of equipment. During the early years of a.c. systems, numerous different frequencies ranging from 25 to $133\frac{1}{3}$ cycles were used.[2] For generators, lower frequencies tend to be preferable because this requires fewer magnetic poles inside the rotor (see Section 4.1), though this constraint became less significant as high-speed steam turbines supplemented and replaced slow-moving hydroturbines and reciprocating steam engines. For transmission, lower frequencies are especially desirable because a line's *reactance* increases with frequency and constrains the amount of power that can be transmitted on a given line (see Sections 3.2 and 6.5). For loads, on the other hand, higher frequencies are often preferable. This is particularly true for incandescent lamps, whose flickering becomes more and more noticeable to the human eye at lower frequencies. After due consideration of the different types of equipment already in use and the prospects for adapting new designs, efforts to standardize power frequency finally resulted in convergence to a 60 cycle standard in the United States and 50 cycles in Europe.

3.1.2 Mathematical Description

A sine wave represents the cyclical increase and decrease of a quantity over time. The oscillation of voltage and current in an a.c. system is modeled by a sinusoidal curve, meaning that it is mathematically described by the trigonometric functions of sine or cosine. In these functions, time appears not in the accustomed units of seconds or minutes, but in terms of an angle.

A sinusoidal function is specified by three parameters: *amplitude, frequency*, and *phase*. The amplitude gives the maximum value or height of the curve, as measured from the neutral position. (The total distance from crest to trough is thus twice the amplitude.) The frequency gives the number of complete oscillations per unit time. Alternatively, one can specify the rate of oscillation in terms of the inverse of frequency, the *period*. The period is simply the duration of one complete cycle. The phase indicates the starting point of the sinusoid. In other words, the phase angle specifies an angle by which the curve is ahead or behind of where it would be, had it started at time zero. Graphically, we see the phase simply as a shift of the entire curve to the left or right. The phase angle is usually denoted by ϕ, the Greek lowercase phi.

Expressing time as an angle allows us to take a sine or cosine of that number. For example, the sine of 30 degrees is $\frac{1}{2}$, but there is no such thing as the sine of 30 seconds. The argument of a sinusoidal function (the variable or object of which we find the sine) must be dimensionless, that is, without physical dimension like time, distance, mass, or charge. An angle, though measured in units of degrees or radians, has no physical dimension; it really represents a ratio or fraction of a whole. This is consistent with the fact that a sine function represents a relationship between two quantities: in a right triangle, the sine of one angle is the ratio of lengths

---

[2] See Hughes, *Networks of Power*, pp. 127*ff.*, and Benjamin G. Lamme, "The Technical Story of the Frequencies," *IEEE Transactions* **37**, 1918.

of two of the sides (the reader may recall "opposite over hypotenuse" for sine, and "adjacent over hypotenuse" for cosine). Time as an angle means time as a certain fraction of a whole.

Turning time into an angle or fraction is logical only because in an oscillation, time is cyclical; the process repeats itself. We do not care whether we are on our first or four-hundredth swing, but rather about *where* in the oscillation we find ourselves at a given instant. One complete oscillation, the duration or period of which would be 1/60th of a second for 60 Hz, is taken to correspond to a full circle of 360 degrees. Any angle can be understood, then, as specifying a fraction or multiple of that complete oscillation.

Plotted against angle on the horizontal axis, the height of the sine curve is simply the value of the sine for each angle, scaled up by a factor corresponding to the amplitude. As the angle is increased, it eventually describes a complete circle, and the function repeats itself.

In the context of sinusoidal functions, angles are often specified in units of *radians* (rad) rather than degrees. Radians refer to the arc described by an angle. The conversion is simple. Since the circumference of an entire circle is given by $2\pi r$, where r is the radius and π (pi) = 3.1415, 2π radians correspond to 360°. (The radius is left out since the size of the circle is arbitrary; in this way, we are only referring to the angle itself.) Any fraction of a radian, then, represents a fraction of a circle, or number of degrees: π rad = 180° or one-half cycle; $\pi/2$ rad = 90° or one-quarter cycle, and so on. Figure 3.1 illustrates a sine wave with both units of angle.

The frequency of a sinusoidal function is often given in terms of radians per second, in which case it is called an *angular frequency*. Angular frequencies are usually denoted by ω, the Greek lowercase omega, as opposed to f or ν (Greek lowercase nu) for frequency. The angular frequency corresponding to 60 cycles/s is

$$\omega = 60 \text{ cycles/s} \cdot 2\pi \text{ radians/cycle} = 377 \text{ rad/s}$$

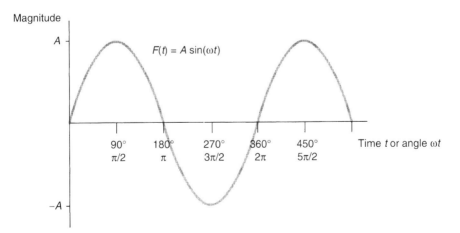

Figure 3.1 A sine function plotted against angle $f(t) = A \sin(\omega t)$.

52 AC POWER

For 50 cycles, $\omega = 314 \text{ rad/s}$. An alternating current as a function of time can be written as the following sinusoidal function:

$$I(t) = I_{max}\sin(\omega t + \phi_I)$$

The quantity I_{max} is the maximum value or amplitude of the current. Since the value of a sine or cosine varies between $+1$ and -1, the actual current oscillates between $+I_{max}$ and $-I_{max}$. The time t is measured in seconds, and, when multiplied by the angular frequency ω, gives a number of radians. In the simplest case, shown in the upper portion of Figure 3.2, there is no phase shift, meaning that ϕ is zero and the curve starts at zero. To illustrate the more general case, Figure 3.2 also

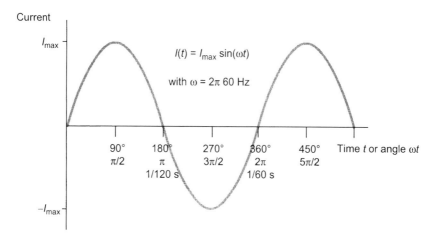

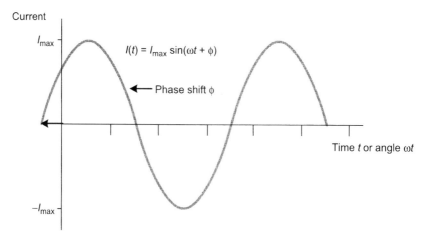

Figure 3.2 Sinusoidal alternating current.

shows the curve shifted over to the left by an angle ϕ. Note that the phase shift affects neither frequency nor amplitude of the curve; it simply amounts to a difference in what time is considered "zero" in comparison to some other curve, which will become important later when we consider the relative timing of current and voltage. The same discussion as for current holds for the voltage, which is written

$$V(t) = V_{\max}\sin(\omega t + \phi_V)$$

The subscripts on the phase angles are there to indicate that current and voltage do not necessarily have the same phase; that is, their maximum values do not necessarily coincide in time (Section 3.3 addresses the phase shift between current and voltage).

In practice, current and voltage waveforms do not conform to this precise mathematical ideal; in some cases, interpreting them as sinusoidal requires a vivid imagination. While they will always be periodic (i.e., repeating themselves), they may not be round and smooth like the proper sinusoidal curve. The discrepancy results in part from the internal geometry of the generators and in part from *harmonic distortion* caused by loads and other utility equipment. The degree of conformance to the shape of a sine curve, or "good waveform," is one aspect of power quality. Poor waveform may not be a problem, though, unless there are sensitive loads.

3.1.3 The rms Value

For most applications, we are only interested in the overall magnitude of these functions. Conceivably, we could just indicate the amplitude of the sine wave, but this would not represent the quantity very well: most of the time, the actual value of the function is much less than the maximum. Alternatively, we could take a simple arithmetic average or mean. However, since a sine wave is positive half the time and negative the other half, we would just get zero, regardless of the amplitude; this average would contain no useful information. What we would like is some way of averaging the curve that offers a good representation of how much current or voltage is actually being supplied: a meaningful physical measure; something of an equivalent to a d.c. value. Specifically, we would like average values of current and voltage that yield the correct amount of power when multiplied (see Section 3.4). Fortunately, such an average is readily computed: it is called the *root mean square* (*rms*) value.

The rms value is derived by first squaring the entire function, then taking the average (mean), and finally taking the square root of this mean. Squaring the curve eliminates the negative values, since the square of a negative number becomes positive. Figure 3.3 illustrates this process with the curves labeled $V(t)$ and $V^2(t)$—though we could just as well have chosen the current $I(t)$—where the squared sinusoid, while retaining the same basic shape, is now compressed in half. If we arbitrarily label the vertical axis in units such that the amplitude

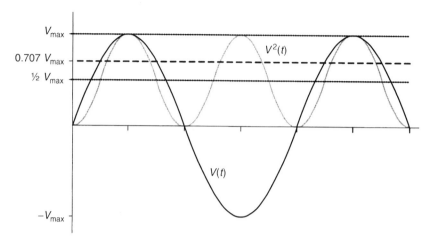

Figure 3.3 Derivation of the rms value.

$V_{max} = 1$, it is obvious that the squared wave has the same amplitude ($1^2 = 1$). Because the squared curve resides entirely in the positive region, it is now possible to take a meaningful average. Indeed, because the curve is still perfectly symmetric, its average is simply one half the amplitude. The only counterintuitive step now consists of renormalizing this average value to the original curve before squaring, which is accomplished by taking the square root: basically, we are just going backwards and undoing the step that made the curve manageable for averaging purposes. Since $\frac{1}{2}$ is less than 1, its square root is greater than itself; it comes to $\sqrt{\frac{1}{2}} = 1/\sqrt{2} = 0.707$. Thus, the rms value of a sine curve is 0.707 of the original amplitude.

Utility voltages and currents are almost always given as rms values. For example, 120 V is the rms voltage for a residential outlet. Note that when the rms voltage and current are multiplied together, the product gives the correct amount of power transmitted (see Section 3.3).

Example

If 120 V is the rms value of the household voltage, what is the amplitude?
Since $V_{rms} = 1/\sqrt{2} \cdot V_{max}$, $V_{max} = \sqrt{2} \cdot 120\,V = 169.7\,V$.

The maximum instantaneous value of the voltage is also of interest because it determines the requirements for electrical insulation on the wires and other energized parts. In fact, one argument against a.c. in the early days was that it would be less economical due to its insulation requirements, which are greater by a factor of $\sqrt{2}$ than those for d.c. equipment transmitting the same amount of power. For current, the instantaneous maximum is relatively uninteresting because current limitations are related to resistive heating, which happens cumulatively over time.

3.2 REACTANCE

In Chapter 1, we discussed electrical resistance as the property of a material or electric device to resist the flow of direct current through it. *Reactance* is the property of a device to influence the relative timing of an alternating voltage and current. By doing so, it presents a sort of impediment of its own to the flow of alternating current, depending on the frequency. Reactance is related to the internal geometry of a device and is physically unrelated to the resistance. There are two types of reactance: *inductive reactance*, which is based on *inductance*, and *capacitive reactance*, based on *capacitance*. Finally, *impedance* is a descriptor that takes into account both resistance and reactance. Resistance, reactance, and impedance are all measured in ohms (Ω).

3.2.1 Inductance

The basic inductive device is a coil of wire, called an *inductor* or a *solenoid*. Its functioning is based on the physical fact that an electric current produces a magnetic field around it. This magnetic field describes a circular pattern around a current-carrying wire; the direction of the field can be specified with a "right-hand rule."[3] When a wire is coiled up as shown in Figure 3.4, it effectively amplifies this magnetic field, because the contributions from the individual loops add together. The sum of these contributions is especially great in the center, pointing along the central axis of the coil. The resulting field can be further amplified by inserting a material of high magnetic permeability (such as iron) into the coil; this is how an electromagnet is made.

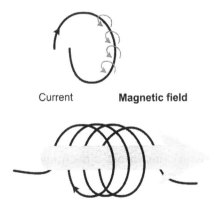

Figure 3.4 A basic inductor, or solenoid.

---

[3] If the direction of one's right thumb corresponds to the current flow (where current is taken by convention to flow from positive to negative potential), the curled fingers of that hand indicate the direction of the magnetic field.

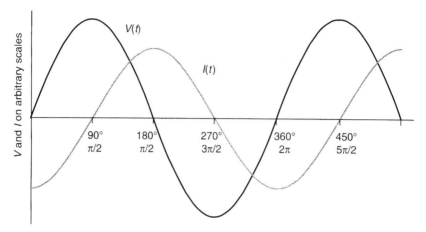

Figure 3.5 Current lagging voltage by 90°. These functions could be written as $V(t) = V_{max} \sin(\omega t)$ and $I(t) = I_{max}\sin(\omega t - \pi/2)$. Note that the relative amplitudes are unimportant, because they are measured on different scales.

When such a coil is placed in an a.c. circuit, a second physical fact comes into play, namely, that a *changing* magnetic field in the vicinity of a conducting wire *induces* an electric current to flow through this wire. If the current through the coil oscillates back and forth, then so does the magnetic field in its center. Because this magnetic field is continuously changing, it induces another current in the coil. This induced current is proportional to the *rate of change* of the magnetic field. The direction of the induced current will be such as to *oppose* the change in the current responsible for producing the magnetic field. In other words, the inductor exerts an inhibitive effect on a change in current flow.[4]

This inhibitive effect results in a delay or *phase shift* of the alternating current with respect to the alternating voltage.[5] Specifically, an ideal inductor (with no resistance at all) will cause the current to *lag* behind the voltage by a quarter cycle, or 90°, as shown in Figure 3.5.

This result is difficult to explain intuitively. We will not attempt to detail the specific changes in the current and magnetic field over the course of a cycle. One thing that can readily be seen from the graph, though, is that the current has its maximum at the instant that the magnetic field changes most rapidly.

As the magnetic field increases and decreases during different parts of the cycle, it stores and releases energy. This energy is not being dissipated, only repeatedly exchanged between the magnetic field and the rest of the circuit. This exchange process becomes very important in the context of power transfer. Because the induced current in an inductor is related to the change in the field per unit time,

---

[4]This result can be derived through right-hand rules, or simply by considering the law of energy conservation: if the induced current were in the same direction as the original (increasing) current, it would amplify the magnetic field that produced it, which would in turn increase the induced current, and so on, indefinitely—clearly an impossible scenario.

[5]The shift between voltage and current is a shift in time, but is expressed in terms of an angle.

the frequency of the applied alternating current is important. The higher the frequency, the more rapidly the magnetic field is changing and reversing, and thus the greater the induced current with its impeding effect is. The lower the frequency, the easier it is for the current to pass through the inductor.[6]

A direct current corresponds to the extreme case of zero frequency. When a steady d.c. voltage is applied to an inductor, it essentially behaves like an ordinary piece of wire. After a brief initial period, during which the field is established, the magnetic field remains constant along with the current. An unchanging magnetic field exerts no further influence on an electric current, so the flow of a steady direct current through a coil of wire is unaffected by the inductive property.

Overall, the effect of an inductor on an a.c. circuit is expressed by its *reactance*, denoted by X (to specify inductive reactance, the subscript L is sometimes added). The inductive reactance is the product of the angular a.c. frequency[7] and the *inductance*, denoted by L, which depends on the physical shape of the inductor and is measured in units of henrys (H). In equation form,

$$X_L = \omega L$$

Thus, unlike resistance, the reactance is not solely determined by the intrinsic characteristics of a device. In the context of power systems, however, because the frequency is always the same, reactance is treated as if it were a constant property.

When describing the behavior of electrical devices in the context of circuit analysis, we are generally interested in writing down a mathematical relationship between the current passing through and the voltage drop across the device. For a resistor, this is simply Ohm's law, $V = IR$, where the resistance R is the proportionality constant between voltage and current. It turns out that the inductance L also works as a proportionality constant between current and voltage across an inductor, but in this case the equation involves the *rate of change* of current, rather than simply the value of current at any given time. Readers familiar with calculus will recognize the notation dI/dt, which represents the *time derivative* or rate of change of current with respect to time. Thus, we write

$$V = L\frac{dI}{dt}$$

meaning that the voltage drop V across an inductor is the product of its inductance L and the rate of change of the current I through it. This equation is used in circuit analysis in a manner analogous to Ohm's law to establish relationships between current and voltage at different points in the circuit, except that it is more cumbersome to manipulate owing to the time derivative.

---

[6]The discriminating response of inductors to different frequencies is put to use in electronics. Electronic signals generally contain a multitude of frequencies. When such a signal is applied to an inductor, the lower frequencies are conducted preferentially. For this reason, inductors are also referred to as "low-pass filters."

[7]To obtain correct numerical results, it is important to remember that the angular frequency is measured in radians/second (not cycles/second). The angular frequency with its implicit factor of 2π appears in this formula because of the way the units of inductance are defined.

Casually speaking, we might say that this equation characterizes an inductor as the kind of thing that does not like to see current change—indeed, it resists (or "reacts to") any change in current. This is consistent with the idea that any change in current will cause a changing magnetic field that acts to oppose the new current. The greater the inductance, the more voltage is required to effect a change in current. Conversely, if the current is forced to change dramatically—say, by being interrupted elsewhere in the circuit—we will observe a voltage spike caused by the inductor's reaction to this change. We can also identify the relationship between voltage and the rate of change of current in Figure 3.5, where the voltage curve has a maximum at the instant that the current curve shows the steepest increase.

Although we used a coil-shaped object to introduce the concept of inductance, the same principles also apply to devices other than coils. The coil makes for the strongest inductance and is easiest to understand because the contributions to the magnetic field from different sections of wire add together in a fairly obvious direction. We do not attempt to extend this type of analysis to objects of different shapes, except to say that inductance exists there, too, and that it depends on an object's geometry. In power systems, the most important example of inductors that are *not* coil-shaped are transmission and distribution lines (see Section 6.2).

3.2.2 Capacitance

The other type of reactance is *capacitive reactance*, whose effect is opposite that of inductive reactance. The basic capacitive device is a *capacitor*.

A capacitor consists of two conducting surfaces or plates that face each other and are separated by a small gap (Figure 3.6). These plates can carry an electric charge; specifically, their charges will be opposite. By having an opposite charge on the opposing plate, very nearby but not touching, it is possible to collect a large amount of charge on each plate. We might say that the charge "sees" the opposite

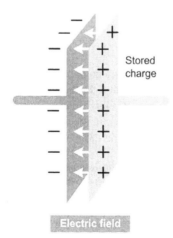

Figure 3.6 A basic capacitor.

charge across the gap and is attracted to it, rather than only being repelled by its like charge on the same plate. In physical terms, there is an electric field across the gap that serves to hold the accumulation of charge on the plates. The gap could simply be air, but is often filled with a better insulating (*dielectric*) material to prevent sparks from bridging the gap (though even such an insulating material can fail or leak charge; this is known as *dielectric breakdown*). For compactness, the plates are often made out of pliable sheets that are folded or rolled up inside a container. Like inductance, capacitance also occurs in devices that do not conform to an ideal shape; coaxial cables are an important example. Our discussion extends to all devices with capacitance.

When presented with a d.c. voltage, a capacitor essentially behaves like a gap in the circuit. Initially, there is a brief period of building up the charge on both plates. But once the charge is built up and cannot go across, the capacitor acts as an open circuit, and no current will flow.

An alternating current, however, *can* get across the capacitor. Recall from Section 1.1 that although a current represents a flow of charge, individual electrons do not actually travel a significant distance through a conductor; rather, each electron transmits an impulse or "push" to its neighbor. Because this impulse can be transmitted across the gap by means of the electric field, it is not necessary for electrons to physically travel across. This transmission only remains effective as long as the impulse (in other words, the voltage) keeps changing, because once the charge has accumulated on the capacitor plate and a steady electric field is established, there is nothing more to transmit. Indeed, the current flow across a capacitor is proportional to the *rate of change* of the electric field, which corresponds to the rate of change of the voltage across the capacitor. As the voltage oscillates, the electric field continually waxes and wanes in alternating directions.

The greater the frequency, the more readily the current is transmitted, since the rate of change of the voltage will be greater.[8] The capacitive reactance, denoted by X or X_C, is given by the inverse of the product of the angular frequency and the capacitance, which is denoted by C and measured in farads (F). The capacitance depends on the physical shape of the capacitor; it increases with the area of the plates and with decreasing separation between them (since greater proximity of the charges will cause a stronger electric field), as long as there is no contact. In equation form,

$$X_C = -\frac{1}{\omega C}$$

The equation shows that the magnitude of the capacitive reactance (neglecting the negative sign) increases with decreasing ω, as it should. It also increases with decreasing capacitance. This is intuitive because a decrease in capacitance means that the plates are becoming less effective at supporting an electric field to transmit anything. The negative sign in the equation has to do with the effect of capacitance

---

[8]In electronics, capacitors are therefore used as "high-pass filters."

60 AC POWER

on a circuit being opposite that of an inductor, as we see shortly. Thus, when inductive and capacitive reactance are added together, they tend to cancel each other (see the example in Section 3.2.3).

Like an inductor, a capacitor causes a phase shift between current and voltage in an a.c. circuit, but the crucial point is that the shift is in the *opposite direction*. A pure capacitance causes the current to *lead* the voltage by 90°, as shown in Figure 3.7. We can see that the moment of greatest current flow coincides with the most rapid change in the voltage. A capacitor also stores and releases energy during different parts of the cycle. This energy resides in the electric field between the plates. The storage and release of energy by a capacitor occurs at time intervals opposite to those of an inductor in the same circuit (this is consistent with the phase shift being in opposite directions). A capacitor and an inductor can therefore exchange energy between them in an alternating fashion. Like an ideal inductor, an ideal capacitor (without resistance) only exchanges but never dissipates energy.

Also analogous to an inductor, there is an equation relating voltage and current for a capacitor. Here it is the voltage V that appears in terms of a rate of change with respect to time, and the capacitance C is the proportionality constant linking it to the current I:

$$I = C\frac{dV}{dt}$$

This equation characterizes a capacitor as a thing that resists changes in voltage, which makes sense if we consider the capacitor plates as a large reservoir of charge: it takes a lot of current to effect a change in the potential. Again, the equation is consistent with the graph (Figure 3.7) that shows the alternating current reaching its maximum value at the instant that the voltage changes most rapidly.

Like inductance, capacitance is geometry dependent and occurs to some extent with electrical devices of any shape. For example, there is a certain amount of

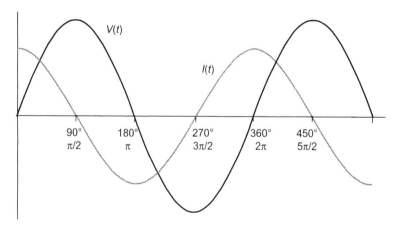

Figure 3.7 Current leading voltage by 90°.

capacitance between a transmission line and the ground. The parallel-plate capacitor is the strongest and simplest case, and it becomes much more difficult to derive a capacitance value for objects with different shapes. Such calculations are treated in standard electrical engineering texts.

> **COMPLEX REPRESENTATION**
>
> *Complex numbers* are a concise way to mathematically represent two aspects of a physical system at the same time. This will be necessary for describing *impedance* as a combination of resistance and reactance in the following section (readers familiar with complex notation may skip ahead).
>
> A complex number contains a *real part*, which is an ordinary number and directly corresponds to a measurable physical quantity, and an *imaginary part*, which is a sort of intangible quantity that, when projected onto physical reality, represents oscillatory behavior.
>
> An imaginary number is a multiple of the imaginary unit quantity $\sqrt{-1}$. This quantity is denoted by i for imaginary in mathematics, and j in electrical engineering so as to avoid confusion with the label for current. The definition of $j = \sqrt{-1}$ is another way of saying that $j^2 = -1$. It also implies that $1/j = -j$.
>
> This entity j embodies the notion of oscillation, or time-varying behavior, in its very nature. Consider the equation $x^2 = -1$. There is no real number that can work in this equation if substituted for x. Pick any positive number for x, and x^2 is positive. Pick any negative number for x, and when you square it, the result is also positive. Pick zero, and x^2 is zero. So we devise an abstract object we call j—not a real number, as we know it, but a thing which, it turns out, can also be manipulated just like a regular number. We define j as that thing that makes the equation $x^2 = -1$ true.
>
> You can think of j as the number that cannot decide whether it wants to be positive or negative. In fact, the equation $x^2 = -1$ can be translated into the logical statement, "This statement is false." The statement cannot decide whether it is true or false; it flip-flops back and forth. The imaginary j is therefore in essence "flippety."[9]
>
> The number j makes the most sense when we represent it graphically, as is usually done in electrical engineering. Consider the number line with positive and negative real numbers, the positive numbers extending to the right and the negative numbers to the left of zero. Now think of multiplying a positive number by -1. What does this operation do? It takes the number from the right-hand side of the number line and drops it over to the left. For example,

---

[9] I owe this term to Heinz von Foerster, as rendered in the transcript of the 1973 AUM conference at Esalen Institute with G. Spencer Brown (www.lawsofform.org/aum). For an unconventional and profound theoretical treatment of $\sqrt{-1}$, see G. Spencer Brown, *Laws of Form* (New York: Crown Publishing, 1972).

$3 \cdot -1 = -3$. In other words, we can think of the multiplication by -1 as a *rotation* by 180 degrees about the origin (the zero point).

Now we can take the result (say, -3) and multiply it again by -1. We get $-3 \cdot -1 = 3$. In other words, the number was again rotated by 180°. Successive applications of the "multiplication by -1" operation result in successive rotations.

So far, we have restricted our imagination to numbers that lie on the real number line. Now suppose we "think outside the box" and ask the following question: What if there were an operation that rotates a number not by 180°, but by only 90°? Rotating by 90° does not immediately seem meaningful, because it takes us off the number line into uncharted territory. But we do know this: If performed twice in succession, a rotation by 90° corresponds to a complete 180° rotation. In other words, a rotation by 90°, performed twice, gives the same result as multiplying by -1.

But this leads us directly to the equation $x^2 = -1$, which asks for the number that, when multiplied by itself, becomes negative. If we define "multiplying by j" as "rotating by 90°" and "multiplying by j^2" is the same as "multiplying by j and then multiplying by j again," then to multiply by j^2 is to rotate by 180°, which is exactly the same thing as multiplying by -1. In this sense we can say that j^2 and -1 are the same.

Based on this rotation metaphor, j is conceptualized as the number that is measured in the new "imaginary" direction 90° off the real number line. We can now extend this concept of being "off the real number line" to an entire new, imaginary number line at right angles to the real one, intersecting at the origin or zero point. Along this new axis we can measure multiples of j upward and $-j$ downward.

With these two directions, we have in effect converted the one-dimensional number line into a two-dimensional plane of numbers, called the *complex plane*. This plane is defined by a real axis labeled Re (the old number line) and an imaginary axis labeled Im (the new imaginary number line at right angles to the real one). We can now conceive of numbers that lie anywhere on this complex plane and represent a combination of real and imaginary numbers. These are the *complex numbers*. The complex number $C = a + jb$ refers to the point a units to the right of the origin and b units above. We say that a is the *real part* and b is the *imaginary part* of C. In Figure 3.8, $a = 3$ and $b = 4$.

Rather than specifying the real and imaginary components explicitly, another way of describing the same complex number is in reference to an arrow (vector) drawn from the origin to the point corresponding to the number. The length of the arrow is the *magnitude* of the complex number. This magnitude is a real number and is usually denoted by vertical lines, as in $|C|$. In addition, the *angle* between the arrow and the real axis is specified, which we denote by ϕ. The representation in terms of magnitude and angle is called the *polar* form.

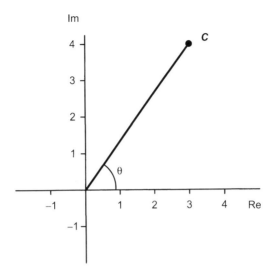

Figure 3.8 The number $C = 3 + j4$ in the complex plane.

These alternative representations can be converted into one another by using any two of the relationships

$$|C|^2 = a^2 + b^2 \qquad \sin\phi = b/C \qquad \cos\phi = a/C \quad \text{or} \quad \tan\phi = b/a$$

Adding or subtracting complex numbers is easily done in the rectangular component format: one simply combines the real parts and the imaginary parts, respectively. To multiply two complex numbers, both components of one are multiplied by both components of the other, and the results are added (like in a common binomial expression). However, multiplying, and especially dividing, complex numbers is much easier in the polar format: the magnitudes are multiplied together (or divided) and the angles are *added* (for division, the divisor angle is *subtracted*).

The imaginary j and all the complex numbers that spring from it do not have the same utilitarian properties that real numbers do. You cannot eat j eggs for breakfast, and you cannot be the jth person in line at the post office. Nevertheless, complex numbers as operational devices do obey rules of manipulation that qualify them as "numbers" in the mathematical sense, and their special properties in these manipulations make them useful tools for representing certain real phenomena, especially phenomena that involve flippety, such as alternating current.[10]

---

[10] More about the conceptualization of $\sqrt{-1}$ can be found in George Lakoff and Rafael Núñez, *Where Mathematics Comes From: How the Embodied Mind Brings Mathematics into Being* (New York: Basic Books, 2000), who introduce the role of metaphor in mathematics from a linguistic perspective.

3.2.3 Impedance

The combination of reactance and resistance that describes the overall behavior of a device in a circuit is called the *impedance*, denoted by Z. However, Z is not a straightforward arithmetic sum of R and X. Mathematically speaking, Z is the *vector sum* of R and X in the *complex plane*. A boldface **Z** may be used to indicate a vector or complex number with a real and an imaginary component. As shown in Figure 3.9, the impedance Z is a complex number whose real part is the resistance and whose imaginary part is the reactance:

$$\mathbf{Z} = R + jX$$

Any device found in an electric power system has an impedance. For different devices and different circumstances, the resistive or reactive component may be negligible, but it is always correct to use Z.

Impedances can be combined according to the same rules for series and parallel combination that we showed in Section 2.1 for pure resistances (although the arithmetic grows tedious rather quickly). Qualitatively, we can note that inductive and capacitive reactance tend to cancel each other, whether they are combined in series or parallel.

When written in the polar format, the angle ϕ of the impedance has an important physical significance: it corresponds to the phase shift between current and voltage produced by this device. By convention, when the reactance is inductive and the current is lagging, ϕ is positive. When the reactance is capacitive and the current is leading, ϕ is negative. Thus, what appears as an angle in space in the triangle of Figure 3.9 can also be interpreted as an angle in time.

Example

An electrical device contains a resistance, an inductance, and a capacitance, all connected in series. Their values are $R = 1\,\Omega$, $L = 0.01H$ and $C = 0.001F$, respectively. At an a.c. frequency of 60 cycles, what is the impedance of the device?

The inductive reactance is $X_L = \omega L = 377$ rad/s $\cdot 0.01H = 3.77\,\Omega$. The capacitive reactance is $X_C = -1/\omega C = -1/377$ rad/s $\cdot 0.001F = -1/0.377 = -2.65\,\Omega$. Because the reactances are in series, they are added to find the combined reactance: $X = X_L + X_C = 3.77 - 2.65 = 1.12\,\Omega$. The impedance is the complex sum of the resistance and the combined reactance: $Z = 1\,\Omega + j1.12\,\Omega$. In the polar form, this corresponds to $|Z| = 1.5\,\Omega$, $\phi = 48.2°$. Casually speaking, this impedance would simply be referred to as 1.5 Ω.

3.2.4 Admittance

The inverse of the complex impedance is called *admittance*, denoted by Y. The complex Y is decomposed into its real and imaginary parts, the *conductance* G

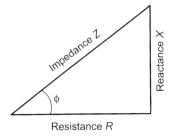

Figure 3.9 The complex impedance Z, with resistance R in the real direction and reactance X in the imaginary direction.

and the *susceptance B*:

$$Y = G + jB$$

We have already encountered the conductance in Section 1.2.2 as the inverse of resistance, for the case of a pure resistor without inductance. In the complex case, however, it is not true that G is the inverse of R and B the inverse of X; rather, we require that

$$\mathbf{Y} = 1/\mathbf{Z}$$

Because **Y** and **Z** are complex (vector) quantities, this entails two things: first, that the magnitudes of Y and Z are reciprocal of each other,

$$Y = 1/Z$$

and second, that the orientation (angle) in the complex plane remain the same. By performing some algebra,[11] we can derive the magnitudes of G and B, respectively:

$$Y = \frac{1}{Z} = \frac{1}{R+jX} = \frac{(R-jX)}{(R+jX)(R-jX)} = \frac{(R-jX)}{R^2 + jRX - jRX + X^2}$$
$$= \frac{(R-jX)}{R^2 + X^2} = \frac{(R-jX)}{Z^2} = \frac{R}{Z^2} - j\frac{X}{Z^2} = G + jB$$

where

$$G = \frac{R}{Z^2} \quad \text{and} \quad B = -\frac{X}{Z^2}$$

This means that while a greater impedance is associated with a smaller admittance and vice versa, the relationship between G and B (considering only their magnitudes,

---

[11] It is easiest to divide complex numbers in polar notation. Working explicitly with the real and imaginary components, however, we use the technique of multiplying both numerator and denominator by the complex conjugate of the denominator, which conveniently eliminates the denominator's imaginary part.

not the negative sign) is directly proportional to the relationship between R and X. Thus, a device whose reactance outweighs its resistance also has a susceptance that outweighs its conductance.

Example

Consider a transmission line that has a resistance $R = 1\ \Omega$ that is small compared to its reactance $X = 10\ \Omega$. What, approximately, are the conductance and susceptance?

The magnitude of Z is very close to 10 (by the Pythagorean theorem, we find $Z = \sqrt{R^2 + X^2} = \sqrt{101} = 10.05$). From $G = R/Z^2$ we obtain $G \approx 1/100$ or 0.01 mho, and $B = X/Z^2 \approx 0.1$ mho. Thus, the susceptance is ten times greater than the conductance, just as the reactance is ten times the resistance.

3.3 POWER

3.3.1 Definition of Electric Power

Power is a measure of energy per unit time. Power therefore gives the *rate* of energy consumption or production. The units for power are generally watts (W). For example, the watt rating of an appliance gives the rate at which it uses energy. The total amount of energy consumed by this appliance is the wattage multiplied by the amount of time during which it was used; this energy can be expressed in units of watt-hours (or, more commonly, kilowatt-hours).

As we saw in Section 1.3, the power dissipated by a circuit element—whether an appliance or simply a wire—is given by the product of its resistance and the square of the current through it: $P = I^2 R$. The term "dissipated" indicates that the electric energy is being converted to heat. This heat may be part of the appliance's intended function (as in any electric heating device), or it may be considered a loss (as in the resistive heating of transmission lines); the physical process is the same.

Another, more general way of calculating power is as the product of current and voltage: $P = IV$. For a resistive element,[12] we can apply Ohm's law ($V = IR$) to see that the formulas $P = I^2 R$ and $P = IV$ amount to the same thing:

$$P = IV = I(IR) = I^2 R$$

Example

Consider an incandescent light bulb, rated at 60 W. This means that the filament dissipates energy at the rate of 60 W when presented with a given voltage, which we assume to be the normal household voltage (120 V). The power equals the

---

[12] In the general case (where there may also be reactance), Ohm's law becomes $V = IZ$, so the substitution is not as straightforward. It will still be true, though, that $P = I^2 R$ gives the power dissipated by resistive heating.

voltage applied to the light bulb times the current through it. The current is 0.5 A: 60 W = 0.5 A · 120 V. While the power rating is specific to a given voltage, the one property of the light bulb that is always the same is its resistance (in this case, 240 Ω). We could determine the resistance from Ohm's law, 120 V = 0.5 A · 240 Ω, and verify that the power also corresponds to I^2R: 60 W = $(0.5 \text{ A})^2 \cdot 240 \; \Omega$.

Now consider a transmission line. We must distinguish between the power *dissipated* and the power *transmitted* by the line. The dissipated power is simply given by $P = I^2R$. We could also write this as $P = IV$, but that would be less convenient for two reasons. First, although it is tempting to think of a power line as just a resistive wire, it actually has a significant reactance. Because of the phase shift involved, the presence of reactance means that multiplying I and V together will be more complicated. On the other hand, taking the square of the current magnitude is always easy, regardless of phase shifts.

Second, if we tried to calculate dissipated power by using $P = IV$, we would have to be very careful about which V to use: remembering that Ohm's law refers to the voltage drop *across* a resistor, or between either end, we recognize that the V must be the voltage difference between the two ends of the line, otherwise known as the "line drop" (see Section 1.3.3). This line drop is distinct from the line voltage, which specifies the line voltage with respect to ground, or between one conductor (phase) and another. Typically, the line drop would be a few percent of the line voltage, but it is usually not known precisely.[13] For these reasons, thermal losses are better calculated using $P = I^2R$, and are often referred to as "I^2R losses."

When we ask about the power transmitted by the line, we can think of the line as extended terminals (like battery terminals). The power that is available to a load connected to this line can be calculated with the formula $P = IV$, but now V refers to the line voltage, which is that seen by the load between the two terminals. We say that the power has been transmitted by the line at the voltage V.

We can enhance our intuitive understanding of $P = IV$ by remembering that voltage is a measure of energy per unit charge, while the current is the flow rate of charge. The product of voltage and current therefore tells us how many electrons are passing through, multiplied by the amount of energy each electron "carries." Energy is carried in the sense that an electron that has been propelled to a higher voltage level has the potential to do more work as it returns to ground. In terms of the units, we can see that the charge cancels, and we are left with the proper units of power:

$$\text{Charge/Time} \cdot \text{Energy/Charge} = \text{Energy/Time}$$

---

[13]Ideally, if transmission lines had no resistance at all, there would be zero line drop.

3.3.2 Complex Power

Applying the simple formula $P = IV$ becomes more problematic when voltage and current are changing over time, as they do in a.c. systems. In the most concise but abstract notation, power, current, and voltage are all complex quantities, and the equation for power becomes

$$\mathbf{S} = \mathbf{I}^*\mathbf{V}$$

where S is the *apparent power* and the asterisk denotes the *complex conjugate* of the current I, meaning that for purposes of calculation, the sign (positive or negative) of its imaginary component is to be reversed. All this ought to make very little sense without a more detailed discussion of complex quantities and their representation by *phasors*. In the interest of developing a conceptual understanding of a.c. power, let us postpone the elegant mathematics and begin by considering power, voltage, and current straightforwardly as real quantities that vary in time.

The fundamental and correct way to interpret the statement $P = IV$ when I and V vary in time is as a statement of *instantaneous* conditions. Regardless of all the complexities to be encountered, it is *always* true that the *instantaneous* power is equal to the instantaneous product of current and voltage. In other words, at any instant, the power equals the voltage times the current *at that instant*. This is expressed by writing each variable as a function of time,

$$P(t) = I(t) \cdot V(t)$$

where the t is the same throughout the equation (i.e., the same instant).

However, instantaneous power as such is usually not very interesting to us. In power systems, we generally need to know about power transmitted or consumed on a time scale much greater than $1/60$ of a second. Therefore, we need an expression for power as averaged over entire cycles of alternating current and voltage.

Consider first the case of a purely resistive load. Voltage and current are in phase; they are oscillating simultaneously. The average power (the average product of voltage and current) can be obtained by taking the averages (rms values) of each and then multiplying them together. Thus,

$$P_{ave} = I_{rms} V_{rms} \quad \text{(resistive case)}$$

Power for the resistive case is illustrated in Figure 3.10.

But now consider a load with reactance. The relative timing of voltage and current has been shifted; their maxima no longer coincide. In fact, one quantity is sometimes negative when the other is positive. As a result, the instantaneous power transmitted or consumed (the product of voltage and current) is sometimes negative. This is shown in Figure 3.11. We can interpret the negative instantaneous power as saying that power flows "backwards" along the transmission line, or out of

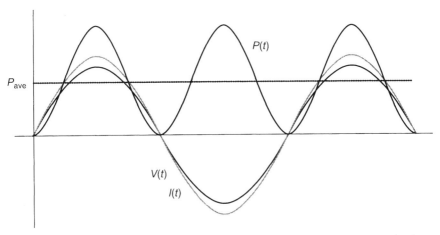

Figure 3.10 Power as the product of voltage and current, with voltage and current in phase.

the load and back into the generator. The energy that is being transferred back and forth belongs to the electric or magnetic fields within these loads and generators. Since instantaneous power is sometimes negative, the average power is clearly less than it was in the resistive case. But just how much less? Fortunately, this is very easy to determine: the average power is directly related to the amount of phase shift between voltage and current. Here we skip the mathematical derivation and simply state that the reduction in average power due to the phase shift is given by the cosine of the angle of the shift:

$$P_{ave} = I_{rms} V_{rms} \cos \phi$$

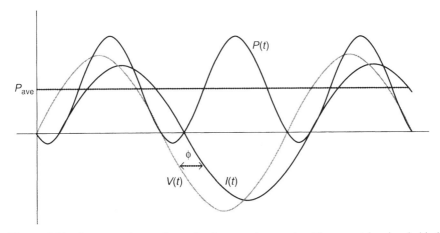

Figure 3.11 Power as the product of voltage and current, with current lagging behind voltage by a phase angle ϕ.

The factor of cos ɸ is called the *power factor*, often abbreviated p.f. This same equation can also be written as

$$P_{ave} = \frac{1}{2} I_{max} V_{max} \cos \phi$$

which is identical because each rms value is related to the maximum value (amplitude) by a factor of $1/\sqrt{2}$.

This equation is true for any kind of load. In the special case where there is only resistance and no phase shift, we have ɸ = 0 and cos ɸ = 1, so there is no need to write down the cos ɸ, and we get the formula from the previous page. In another special case where the load is purely reactive (having no resistance at all), the phase shift would be ɸ = 90° and cos ɸ = 0, meaning that power only oscillates back and forth, but is not dissipated (the average power is zero).

The average power corresponds to the power actually transmitted or consumed by the load. It is also called *real power, active power* or *true power*, and is measured in watts.

There are other aspects of the transmitted power that we wish to specify. The product of current and voltage, regardless of their phase shift, is called the *apparent power*, denoted by the symbol S. Its magnitude is given by

$$S = I_{rms} V_{rms}$$

Although apparent and real power have the same units physically, they are expressed differently to maintain an obvious distinction. Thus, the units of apparent power are called volt-amperes (VA).

Apparent power is important in the context of equipment capacity. Actually, the crucial quantity with respect to thermal capacity limits is only the current. In practice, though, the current is often inconvenient to specify. Since the operating voltage of a given piece of equipment is usually quite constant, apparent power is a fair way of indicating the current. The point is that apparent power is a much better measure of the current than real power, because it does not depend on the power factor. Thus, utility equipment ratings are typically given in kVA or MVA.

Finally, we also specify what we might intuitively think of as the difference between apparent and real power, namely, *reactive power*. Reactive power is the component of power that oscillates back and forth through the lines, being exchanged between electric and magnetic fields and not getting dissipated. It is denoted by the symbol Q, and its magnitude is given by

$$Q = I_{rms} V_{rms} \sin \phi$$

Again, note how the equation converges for the resistive case where ɸ = 0 and sin ɸ = 0, as there will be no reactive power at all. Reactive power is measured in VAR (also written Var or VAr), for volt-ampere reactive. We can represent power as a vector in the complex plane: namely, an arrow of length S (apparent power) that makes an angle ɸ with the real axis. This is shown in Figure 3.12. The angle ɸ is the same as the phase difference between voltage and current.

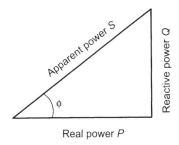

Figure 3.12 The complex power S, with real power P in the real and reactive power Q in the imaginary direction.

The projection of the apparent power vector onto the real axis has length P and corresponds to the real power; the projection of apparent power onto the imaginary axis has length Q and corresponds to reactive power. This agrees with the factors of $\cos \phi$ and $\sin \phi$ in the formulas for P and Q, respectively. In mathematical terms, S is the vector sum of P and Q. In this sense, it is completely analogous to the complex impedance Z, which is composed of the resistance R in the real and the reactance X in the imaginary direction.

Note that when Q and S are pointing upward as in Figure 3.12, ϕ is positive and the power factor is said to be lagging (in agreement with the current lagging behind the voltage). For a leading power factor, ϕ would be negative and Q and S would point downward.

Example

Consider a vacuum cleaner that draws 750 W of real power, at a voltage of 120 V a.c. and a power factor of 0.75 lagging. How much current does it draw?

Since the real power is given by the apparent power times the power factor, the apparent power equals $750 \div 0.75 = 1000$ VA $= 1$ kVA. The rms current is the apparent power divided by the rms voltage: 1000 VA $\div 120$ V $= 8.33$ A.

When we say that a load "draws power," we mean that as a result of its internal characteristics (the impedance), when presented with a given voltage, a certain amount of current will flow through this device, and accordingly a certain amount of power will be dissipated or exchanged. Just as a load draws real power in relation to its resistance, it draws reactive power in relation to its reactance. In fact, the ratio of resistance to reactance *determines* the ratio of real to reactive power drawn by a load. In other words, the angle ϕ in Z is the same as the angle ϕ in S.

Specifically, inductive loads are said to "consume" reactive power, whereas capacitive loads are said to "supply" reactive power. This is merely a terminological convention, and a rather misleading one. Recall that inductors and capacitors produce opposite phase shifts. Either type of shift causes reactive power to oscillate through the circuit. But because of the difference in timing, the contributions of inductance and capacitance to reactive power are opposite: at the instant that the inductor magnetic field absorbs energy, the capacitor electric field in the same circuit releases energy. Conversely, at the instant that the magnetic field releases energy, the electric field absorbs it. Although on average neither inductor nor capacitor gains or loses energy, their effects are complementary.

Following the law of energy conservation, the amount of energy going *into* the circuit must equal the energy coming *out* of the circuit at every instant. In principle, therefore, inductance and capacitance in a circuit must always be matched. Of course, this is not always given by design. But nonetheless, a circuit will behave in such a way as to provide equal absorption and release of reactive power at any instant. The preferable way to satisfy the reactive power balance is by adjusting the a.c. power source to compensate for the load's circulation of reactive power.[14] Thus, in operational terms, the problem of managing reactive power is analogous to that of managing real power: just like the utility must supply the precise amount of real power that is demanded at any instant, the utility must compensate for the precise amount of reactive power that is being circulated at any instant. In practice, electric loads are dominated by inductance, not capacitance, and utilities therefore associate supplying real power with compensating for inductive reactance (a lagging current). This operational perspective explains the use of the physically improper terminology of "consuming" and "supplying" reactive power.

Example

For the preceding example, how much reactive power does the vacuum cleaner draw?

With p.f. = 0.75 lagging = $\cos \phi$, the phase shift is $\phi = 41.4°$, and $\sin \phi = 0.661$. Thus, the reactive power is $1000 \cdot 0.661 = 661$ VAR. Because the power factor is lagging, the vacuum cleaner is said to "consume" reactive power.

What is the impedance of the vacuum cleaner?

The magnitude of the impedance is given by the voltage divided by the current (rms values) according to Ohm's law: 120 V ÷ 8.33 A = 14.4 Ω. The power factor, being the ratio of real to apparent power, corresponds to the ratio of resistance to impedance. Therefore, the magnitude of the resistive component is $R = 0.75 \cdot 14.4$ Ω = 10.8 Ω. The reactive component is proportional to the reactive power, or $\sin \phi$, and is given by $X = 0.661 \cdot 14.4$ Ω = 9.51 Ω. We can express the impedance as the complex sum of its components: $Z = 10.8 + j9.51$ (Ω). Alternatively, we can express it in terms of its magnitude and angle: $Z = 14.4$ Ω, 41.4°. Because ϕ and X are positive, the reactance is inductive (rather than capacitive).

The vast majority of loads are inductive rather than capacitive. Motors of all kinds (pumps, refrigerators, air conditioners, power tools) are the most common inductive load; ballasts for fluorescent lighting are another example.

---

[14]The alternative is to allow voltage and frequency changes in the circuit, which will occur so as to enforce compliance with the law of energy conservation. The same is true for the conservation of real power. Trouble in maintaining the reactive power balance will manifest primarily as variations in the voltage, whereas over- or undersupply of real power primarily results in frequency fluctuations.

The different power factors of individual loads combine into an aggregate power factor for the set of loads. The calculation is based directly on the combination of impedances in series and parallel, as introduced in Section 2.1. Although straightforward in principle, because it involves complex numbers, this calculation becomes quite tedious in practice, and we do not attempt to reproduce it here. Qualitatively, we can note that if two loads of different power factors are connected in parallel, their combined power factor will be somewhere in the middle, and it will be closer to that of the bigger load. Just as an equivalent resistance can be defined for any combination of individual resistors, one can specify a power factor for a set of loads at any level of aggregation, from a single customer to a local area or the entire utility grid. Overall utility power factors tend to be between unity and 0.8 lagging; a typical value would be 0.9 lagging.

While all the reactive power "consumed" in the grid has to be "supplied" by generators or capacitive devices, there is some operational discretion in how these contributions to reactive power are allocated. This allocation is addressed in Section 4.3 on "Operational Control of Synchronous Generators," and again in Section 6.6 on "Voltage Control."

3.3.3 The Significance of Reactive Power

A low power factor is undesirable for utilities in terms of operating efficiency and economics. Most customers, especially small customers, are only charged for the real power they consume.[15] At the same time, the presence of reactive power oscillating through the lines and equipment is associated with additional current. While reactive power as such is not consumed, it nonetheless causes the utility to incur costs, both in the form of additional losses and in the form of greater capacity requirements. Owing to its property of occupying lines and equipment while doing no useful work, reactive power has been referred to jokingly as "the cholesterol of power lines."

> **Example**
>
> To illustrate the effect of the power factor on line losses, consider a load of 100 kW at the end of a several-mile-long 12 kV distribution line. Suppose the line's resistance is 10 Ω. If the power factor is 0.8 lagging, the apparent power drawn by the load is 125 kVA, and the reactive power is 75 kVAR. The current to this load is 125 kVA $\div$ 12 kV = 10.4 A. The distribution line losses due to this load are given by $I^2R = (10.4 \text{ A})^2 \cdot 10 \, \Omega = 1.08$ kW.[16] This is significantly more than we might have estimated just on the basis of real power demand: using only 100 kW as the power, we would have obtained a current of 8.33 A and losses of only 0.69 kW.

---

[15] When large customers are charged for reactive power, it is generally not on a per-kVAR basis, but by means of a rate schedule for real power that is scaled up according to the customer's power factor.

[16] There may be other loads on the same line, but we take advantage of the superposition principle to consider this load and its associated current and losses individually. Losses due to other loads would simply be added.

In order to minimize losses and at the same time maximize available equipment capacity, utilities take steps to improve the power factor with capacitors or other means of "VAR compensation." Preferably, this compensation is placed near the load, so as to minimize the distance that reactive power must travel through the lines. VAR compensation is discussed further in Section 6.6 on voltage control, particularly with regard to the relationship between reactive power and voltage.

Example

For the preceding example, how much of a reduction in line losses could be achieved by improving the power factor to 0.9, assuming that real power remains unchanged?

The apparent power is now 100 kW ÷ 0.9 = 111 kVA; reactive power is 54.5 kVAR. The current is 111 kVA ÷ 12 kV = 9.25 A, and line losses are $(9.25 \text{ A})^2 \cdot 10 \, \Omega = 0.856$ kW. Losses on this distribution line have been reduced by 226 W as a result of increasing the power factor from 0.8 to 0.9. Over the course of a year, 226 W · 8766 h = 1981 kWh would be saved.[17]

How much capacity is freed up on the substation transformer that supplies this line?

The transformer capacity is limited by current, which is related to apparent power. The difference in apparent power before and after the power factor correction is 125 kVA − 111 kVA = 14 kVA. This result is a first approximation. In order to refine it, we would also consider the effect of the loss reduction, since the transformer must supply the energy that is lost "downstream" from it. The contribution from losses on this line is small (on the order of 1%), increasing the saved capacity to about 14.2 kVA. To calculate the answer precisely, we would also have to take into account the line's reactance.

Line losses are *real* power losses and result in physical heating of the lines, even when they are attributable to reactive power consumption. They are not to be confused with the quantity termed *reactive losses* in power flow analysis (see Chapter 7). Reactive losses are the difference between the reactive power supplied by generators and that consumed by loads, and are measured in VAR. The extra "consumption" of reactive power is due to the inductive reactance of the lines and transformers in the system. From the standpoint of energy conservation, reactive losses are irrelevant. They are important, however, in the context of planning and scheduling reactive power generation.

Reactive power can arise as an issue in the context of renewable energy and energy conservation. Specifically, wind turbines may be considered problematic

---

[17] It is a common habit, especially among eager students of science and engineering, to write down numerical answers to a greater degree of precision than they are actually known, given the degree of uncertainty of the input variables. In this example, the answer should be more realistically stated as "2 MWh." One reason for carrying extra significant figures is to avoid rounding errors if numbers are to be used in further calculations. The reason they are shown here, though, is to keep the numbers recognizable so that readers can easily compare their results.

in some particular instances because their induction generators consume, but cannot supply, reactive power (see Section 4.5).

Among energy conservation technologies, fluorescent lights of older vintages are often culprits of low power factor. The inductance in fluorescent lamps is due to the ballast, a device inside the fixture that limits current flow and may also step up the voltage. These ballasts can be made with an inductor coil (the traditional magnetic ballast), or, as is today's standard, with solid-state technology (electronic ballast). Without correction, their power factors can be as low as 0.6 for electronic and 0.5 for magnetic ballasts. With corrective devices (e.g., a small capacitor) built in at some extra cost, the power factor is improved to 0.8–0.9 for magnetic and about 0.9 for electronic ballasts. By comparison, an incandescent lamp has a power factor of 1.0 (since it is a pure resistor), but of course it consumes much more real power to produce the same amount of visible light (for example, 60 W versus 15 W for an equivalent fluorescent).

Thus, the conversion from incandescent to fluorescent is always a good idea in overall energy terms. For a customer who is not charged for reactive power, the power factor is irrelevant in any case. From the utility's standpoint, however, a marked reduction of power factor would entail some transmission and distribution cost (line losses as well as capacity).

3.4 PHASOR NOTATION

3.4.1 Phasors as Graphics

In most practical situations when dealing with a.c. circuits, we are interested in average as opposed to instantaneous values of current, voltage, and power. In other words, we want to know what happens over the course of many cycles, not within a single cycle. Thus, we describe current and voltage in terms of rms values. Rms values provide a measure of each sine wave's amplitude (scaled by a factor of $\sqrt{2}$), but no information about the timing—the frequency and phase—of the wave. Taking the frequency as a given, the main aspect of timing that has practical significance is the phase difference between voltage and current.

This information is captured in the magnitudes of real, reactive, and apparent power, which also represent averages over time: real power is, in fact, the mathematical average instantaneous power delivered by a circuit, and reactive power is a measure of the amount of power oscillating throughout the circuit, which is also constant over time. Apparent power is their geometric sum. These averages appear as simple numbers measured in units of volts, amps, watts, VARs, and VA that represent practically useful and comparatively tangible information. Drawing watts, VARs, and VA as sides of a triangle recovers the crucial information about timing, the angle ϕ, which in itself is more abstract and to which there is rarely a need to refer explicitly.

This need arises, however, in the context of analysis where the timing of events within an a.c. cycle is of the essence. Here, for purposes of efficient calculation, it is useful to keep track at all times of both magnitude and phase of every variable.

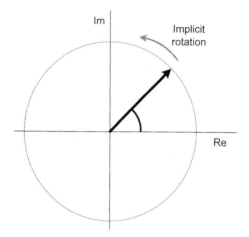

Figure 3.13 A phasor.

This is where *phasor notation* comes in. Illustrated in Figure 3.13, a phasor is a shorthand way to characterize a sine wave, specifying its magnitude (we will use the convention that states the rms value as opposed to the full wave amplitude) and angle (in relation to a reference, usually voltage).

To create a phasor, we map the sine wave onto a circle—that is, instead of visualizing something that goes up and down over time (e.g., the voltage increases and decreases), we visualize it as going around in a circle. So, instead of tracing out a wave that keeps going and going along the time axis, we imagine the situation as circling around and around, returning to the same point on the circle over and over again.

This circle, representing the evolution of, for example, voltage over time, is traced in the complex plane. Here's how it makes sense: consider the sine wave that describes voltage (or current or power), where the vertical axis on the graph represents the real, physical quantity. The horizontal axis represents time. Now, for purposes of visualization, imagine adding to this graph a third dimension: a third axis that extends out of the paper. This third dimension is the imaginary axis. When looking straight at the sine wave—or making a physical measurement—we cannot see an imaginary component; we see only its projection onto the real plane, the plane of the paper. But treating the wave as a complex entity, we can say that it also extends in the imaginary, front–back direction. It does so by describing a *helix*—the shape of an extended spring, for example, or a DNA molecule.

We can now imagine that the physical quantity (say, voltage or current) evolves in time by spiraling around this helix. From the side, a helix looks precisely like a sine wave. This is like compressing the imaginary front–back dimension into the plane of the paper and looking at only the *real* projection of the helix: we see a real, physical quantity changing, up–down, over the course of time. But looked at from the end, the helix appears as a circle, with the quantity going around and around. This viewing angle amounts to compressing the time dimension into the plane of the paper. The complex notation thus encapsulates the time-varying character of the physical quantity without having to make any explicit reference to time. This is how circular motion in the complex plane comes to represent sinusoidal oscillation in the physical world.

The complex mapping affords some great conveniences. Whereas an object tracing a sine wave accelerates up and down, the circular motion has a constant rotational speed. Thus, a sinusoidally varying voltage is represented as an arrow of constant length, spinning around at the constant frequency ω, where ω is the *angular frequency* that assigns the unit circle's circumference, 2π, to a complete revolution of 360 degrees.

Because in a.c. power systems we can generally assume voltages and currents to be oscillating at the same constant frequency, we can ignore this circular spinning to the extent that it will be the same for all quantities, and they are not spinning *in relation to each other*. Thus, we can capture any moment in time and simply indicate a quantity as a fixed point on the circle, or an arrow from the origin to that point. This arrow is called the *phasor*. We know that it is supposed to be spinning, but we do not have to draw that. What we really care about are two aspects of the oscillating quantity we are representing: its *magnitude*, and its relative timing or *phase* in relation to another sinusoidal function, as, for example, in the phase difference between voltage and current. The magnitude is represented in the phasor diagram simply as the length of the arrow. But the phase shift is now also straightforward to represent, namely, as an angle within the circle in the complex plane.

Consider the voltage and current waves in Figure 3.14 below. The current lags the voltage by 30 degrees, meaning that the peak current always occurs with a delay of one-twelfth of a cycle after the voltage peak. The notion that the voltage is zero at time zero and peaks at exactly 90 degrees was an arbitrary choice, amounting to when we decide to start counting time, and choosing that starting point so as to be convenient for computation. But once we decided on how to label the timing for voltage, we are committed to describing the timing of current properly *in relation to* that voltage timing.

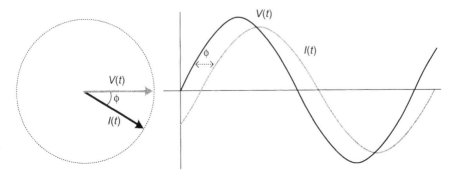

Figure 3.14 Voltage and lagging current phasor.

On the phasor diagram, we draw the voltage as an arrow—the voltage phasor. We imagine that this arrow spins around the circle counterclockwise, but we just capture for our diagram any snapshot in time. For convenience, let us draw the voltage with a zero angle at this moment, so that it points right along the horizontal axis. In addition to the timing or phase, we also care about the amplitude of the wave, or the magnitude of the voltage. We indicate this magnitude simply by the length of the arrow, which corresponds to either the maximum or rms value.

Now we draw the current phasor in relation to the voltage phasor. Knowing it will reach every point in the oscillation 30 degrees or 1/12 cycle *after* the voltage in time, we draw the arrow so that it appears 30 degrees *behind* the voltage (assuming the counterclockwise direction as forward). The result is as shown. Note that in this case, although the length of the arrow represents magnitude, the relative length of the two phasors on the diagram does not convey information because voltage and current are measured in different units, meaning that the two arrows can be drawn to different scales (just like the relative amplitudes of the two sine waves).

In this way, we can combine any number of sinusoidally varying quantities (of the same frequency!) as phasors on the same diagram in the complex plane. We can also include in the same diagram static quantities like impedance that map onto the complex plane. The beauty of this method is that we can actually perform mathematical operations among these quantities by graphically combining them. How and why this works will make more sense once we consider the phasors as complex exponential functions, which are mathematical shorthand for sine waves.

3.4.2 Phasors as Exponentials

A phasor written in exponential notation conveys all the relevant information about the desired quantity—for example, voltage—without need for a diagram. The voltage phasor becomes $Ve^{j\phi}$, where V is the magnitude of the voltage and ϕ is the voltage phase angle. The current phasor would be $Ie^{j\phi}$, with I the current magnitude and ϕ now the current phase angle. (Sometimes different letters or subscripts like ϕ_V and ϕ_I are used to avoid confusing the phase angles with one another.) The number e is the natural exponential base.[18] Engineers also use an even more abbreviated notation that simply drops the e and the j (assuming everyone knows they are implicit) so that $Ve^{j\phi}$ becomes $V\angle\phi$, where the $\angle$ symbol means "angle." Translated into a phasor diagram, V corresponds to the length of the voltage phasor and ϕ to the angle it makes with the x axis.

---

[18]The number e is known as the "natural base" for logarithms or exponential functions, because it allows describing natural growth or decay phenomena with unique simplicity. In short, the function e^x is that which, for any value of x, equals its own slope or rate of change. This property can also be expressed in terms of sines and cosines (see Footnote 20). Its numerical value, $e = 2.718281828459\ldots$, is rarely seen in print (as with $\pi = 3.1415926\ldots$, the sequence of digits never repeats). A very readable explanation of e and its interesting history can be found in Eli Maor, *e: The Story of a Number* (Princeton, NJ: Princeton University Press, 1994).

What is the significance of the exponential here? It serves as a shorthand for a sinusoidal wave, using the relationship known as Euler's equation:

$$e^{j\phi} = \cos\phi + j\sin\phi$$

It is not immediately obvious that the expressions on either side of this equation should represent something moving in a circle, but they do. The combination $\cos\phi + j\sin\phi$ indicates a superposition of two sinusoidal waves, one a quarter cycle ahead of the other, with one oscillating in the real and the other in the imaginary direction. Together, the two waves make up the helix we visualized earlier. This helix looks like a circle head-on and like a sinusoidal up-and-down motion from any side.

That the expression $e^{j\phi}$ should be equivalent to this combination of waves is a hard concept to swallow. A mathematician might prove $e^{j\phi} = \cos\phi + j\sin\phi$ by rewriting both sides of the equation in terms of a *series expansion*, where the exponential and the sinusoids are expressed as sums over an infinite number of terms that keep getting smaller (thus converging to a finite value), and show that the two series are the same. While rigorous, this approach does little for our intuitive understanding or visualization of the problem. For this, we must familiarize ourselves a bit with the function's behavior.

To try to make sense of the behavior of $e^{j\phi}$, we first note that if it were not for the imaginary j, then we would simply have exponential growth: e^{ϕ} would just get bigger and bigger as ϕ increases. But the j in the exponent does a very strange thing. As in simple multiplication, j has the effect of "rotating" the value of the expression in the complex plane.

Depending on their degree of familiarity with exponentials and complex numbers, readers may wish to experiment by plugging in some sample values for ϕ (which can be any number), raise e to the power of imaginary ϕ, and plot the results. We immediately encounter a practical problem: How does one plug an imaginary exponent into a calculator? And what does it even *mean* to raise a number to an imaginary exponent?[19] You do need a scientific calculator for this, and enable the complex mode in which the machine expects every number to consist of two separate coordinates, a real and an imaginary part (consult your owner's manual or a mathematically minded friend). However counterintuitive, exponentiation with complex numbers is in fact a perfectly rigorous procedure in that the rules of operation are well-defined and the results unique. When the complex results are displayed in their real and imaginary components on a calculator, they can be plotted as such on the complex plane. Some sample points are shown in Figure 3.15.

The key observation is this: As ϕ increases, the value of $e^{j\phi}$ shows alternatingly a greater real or a greater imaginary component, with both real and imaginary

---

[19]Clearly, you cannot multiply a number by itself an imaginary number of times. This is why we must abandon the accustomed metaphor of exponentiation as multiplication of objects in favor of the rotation-in-a-plane metaphor. Lakoff and Núñez discuss this problem in satisfying depth.

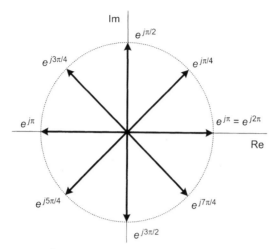

Figure 3.15 $e^{j\phi}$ for some values of ϕ, plotted in the complex plane.

components oscillating between positive and negative values. However, the combination of real plus imaginary stays confined in a certain way, such that the distance from the origin of all the plotted $e^{j\phi}$ points (the *magnitude* of $e^{j\phi}$) remains the same (1) for all ϕ. Thus, $e^{j\phi}$ is like an arrow of constant length, rotating counterclockwise about the origin as we let ϕ increase. For some value of ϕ—and this value, not coincidentally, is 2π, the number of radians in a full circle—$e^{j\phi}$ comes back around to where it started, at 1 in the real direction.

This image gives us a visual way to understand what is so special about the number e. Try the same process using some number other than e as the base, say, $2^{j\phi}$ or $3^{j\phi}$. In each case, the plot will start with 1 for $\phi = 0$, since any number raised to the 0th power is 1. Also, in each case the results will wander around the complex plane counterclockwise. But with any number smaller than e, the results will diminish in magnitude, and the plot will spiral inward toward the origin as ϕ increases. With any number greater than e, the plot will spiral outward to infinity. The number e is that special number which, when raised to any imaginary exponent, circles around forever.[20] And so the phasor $e^{j\phi}$ comes to denote a quantity like alternating current that keeps going around and around at a steady magnitude, meaning that it physically oscillates with a constant amplitude.

3.4.3 Operations with Phasors

Phasor notation makes no explicit mention of the frequency of the implicit waves, because we assume that we are operating in a situation where this frequency is

---

[20]This property is related to the fact that at any point along the circular rotation, one coordinate always indicates the other's rate of change (save for an occasional minus sign). Readers familiar with calculus will recognize this phenomenon in terms of the sine and cosine functions being derivatives of each other.

known and that it is the same for all the sinusoidally varying quantities. Instead, phasors emphasize only the *phase angle* of each quantity, or the precise timing of the zero (or the maximum) as compared to a reference. To establish that reference, we conventionally define the timing of one particular voltage wave as having a zero phase angle. In a simple a.c. circuit with one voltage and one current, we would arbitrarily set the voltage phase angle to zero; the current phase angle, if current is lagging behind voltage by θ degrees, will be negative θ (Greek lowercase theta). We write the phase angle with an "angle" symbol as $\angle -\theta$. (Note that θ plays exactly the same role as ϕ above; both letters are commonly used to denote phase angle, and it helps to be accustomed to seeing both.)

The phasor also specifies each quantity's magnitude, and here there are two conventions: either the wave's maximum value (amplitude), or its rms value (amplitude divided by $\sqrt{2}$) can be stated. While theoretical texts tend to favor the former convention, power systems engineers prefer the latter (which will be used here). Magnitude and phase angle are then written side-by-side as two coordinates to give a complete phasor, indicated in boldface: $\mathbf{V} = V_{rms}\angle 0°$ and $\mathbf{I} = I_{rms}\angle -\theta$. We omit the subscript from now on, with the understanding that all magnitudes represent rms values.

When we represent a phasor graphically as a vector or an arrow, the magnitude represents the length of the arrow and the phase angle is the angle between the arrow and the reference axis (upward or counterclockwise from the x axis corresponds to a positive angle, downward to a negative angle). The fact that we are drawing a phasor quantity in two dimensions means we have exactly two pieces of information about the quantity. Choosing phasor notation to specify a point on a plane simply means choosing an angular coordinate system as opposed to Cartesian (x and y) coordinates.

Although it is constant over time, the impedance vector can be treated along with voltage and current. But now instead of decomposing the impedance Z into resistance R and reactance X (its Cartesian components), we write it as a magnitude (the length of Z) and a phase angle. For the impedance itself, this angle means a ratio of components (as seen in Figure 3.9), not a time, but as we know this same ratio physically determines the time lag of current with respect to voltage, and is therefore consistent with the use of angles as representing time. We can write $\mathbf{Z} = Z\angle \theta$.

Now we can begin to appreciate the conveniences of phasor notation. Ohm's law becomes

$$\mathbf{V} = \mathbf{IZ}$$

or

$$V\angle 0° = I\angle -\theta \cdot Z\angle \theta$$

which in one succinct statement captures the relationship among the respective magnitudes of voltage, current, and impedance as well as the impact on the time lag between voltage and current due to the reactive component of impedance.

82 AC POWER

The statement as written here illustrates a fundamental rule of doing arithmetic with phasors: When multiplying two or more phasors together, first multiply their magnitudes to find the magnitude of the result. (Thus, $V = IZ$.) To find the angle of the product, *add* the angles of the factors. Thus, $-\theta + \theta = 0$. The reader may deduce correctly the rule for dividing phasors: divide their magnitudes and *subtract* the angle in the denominator from that in the numerator.

The process of adding and subtracting phase angles works because they actually represent exponents of the same base, e. Thus, if we were to rewrite the previous equation in longhand, it would be

$$Ve^{j\theta V} = Ie^{j\theta I}Ze^{j\theta Z} = IZe^{j(\theta I + \theta Z)}$$

Therefore, $\theta V = \theta I + \theta Z$.

Adding angles is a straightforward procedure to perform graphically. Thus, in order to multiply the I and Z phasors together, we draw a new vector at an angle with the x axis corresponding to the sum of the two angles, and with a length corresponding to the product of lengths.

Note that because this product has different physical dimensions or units (voltage) than its factors (current and impedance), the new arrow can be drawn on the same diagram, but its length cannot be compared to phasors representing other units. If two or more phasors are to be added together—which is done graphically with amazing ease by placing them end to end—they must represent quantities of the same dimension drawn to the same scale.[21]

Consider, for example, the a.c. circuit in Figure 3.16. The phasor diagram is drawn in terms of voltage. It simultaneously shows several things about the circuit: First, it illustrates that the total voltage drop around the circuit adds up to zero, as required by Kirchhoff's voltage law (KVL) (see Section 2.3.1). It does this very simply by placing voltage phasors head to tail, adding them just like voltage drop is added in series along the circuit path, and requiring that the destination or sum correspond to the voltage drop measured around the other side of the circuit. (Note that a phasor's position on the graph is arbitrary, so we can shift it around as long as we preserve its length and angle.)

Second, the phasor diagram disaggregates the resistive and reactive components of the impedance in the circuit and their respective impacts on voltage drop. Note that the two arrows IR and IX each have units of volts, indicating the difference in voltage across the (idealized) resistive and reactive circuit elements, respectively. Owing to the two-dimensional representation, we see not only the effect of each circuit element on voltage magnitude, but phase as well. Across the ideal resistor,

---

[21] A physical *dimension* means the kind of thing being measured, such as length, time, or energy. The *units* of that measurement are specific to the dimensionality, but there may be many different units or scales for the same thing. For example, a length can be expressed in units of inches, meters, or miles, but all of these have the same dimension. While measurements in different units but the same dimension can be added or subtracted if proper care is taken with unit conversions—for example, one inch plus one meter equals 1.0254 meters—there is no mathematical meaning to the question: What is one inch plus two minutes? Likewise, we cannot add a voltage to a current, numerically or graphically, because they have different dimensions.

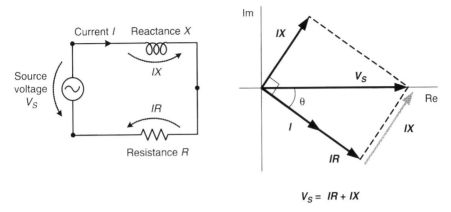

Figure 3.16 Phasor addition.

the voltage and current phasors point in the same direction: the length of the current phasor in the diagram is arbitrary, but its angle is zero relative to the voltage IR because a pure resistor does not sustain a phase shift. Yet we know that, because of the inductor's presence in the same circuit, the current (which must be the same everywhere along the series path) will lag the source voltage V_S. Reconciling these facts, we graphically decompose the source voltage (which is the same as the voltage drop across the resistor–inductor combination) into a component collinear with current (IR across the resistor) and one perpendicular to current. The perpendicular component IX describes the voltage drop across the inductor, obeying the requirement that current through an ideal inductor must lag the voltage across it by 90° (note the right angle between IX and I). The angle θ is determined by the relative magnitudes of R and X (compare Figure 3.9) so that $IR + IX = V_S$.

If we allow all the phasors to rotate and plot their projection onto the horizontal (real) axis, they generate sine waves for the respective voltages. These sinusoidally varying voltages IR and IX with their distinct magnitude and phase could be physically measured across each circuit element if each element modeled here in fact corresponded to a distinct physical object, rather than one abstracted property (resistance or reactance) of a real object (like a wire coil) that actually manifests both properties in the same space.

The phasor diagram in Figure 3.16 offers a visual explanation of how the phase angle or power factor of a circuit is determined by the combination of its elements. Note that each individual element could be drawn in any arbitrary direction—for example, were we dealing with a resistance alone, we probably would have chosen to make it horizontal—but that the combination of elements in the circuit dictates the phasors' *relationship to each other*. Thus, the relative magnitude of resistance and reactance in the same circuit determines the proportions of the triangle and the angle between voltage and current. Because the essential features of the physical relationship in time are strictly preserved, the phasor diagram is a powerful device for combining variables from different portions of an a.c. circuit and for analyzing their interactions. The representation shown in the figure for a simple series circuit works in an analogous way for parallel circuit branches, except that we would choose to represent all goings on in terms of current rather than voltage, using Kirchhoff's current law (KCL) for the addition of currents at each node.

Finally, let us turn to the phasor representation of power. Complex power S can be written in vector form as the product of two complex numbers or phasors, voltage and current:

$$S = I^*V$$

where the asterisk denotes the complex conjugate of the current I, meaning that for purposes of calculation, the sign (positive or negative) of its imaginary component is to be reversed. This reversal has to do with the definition of the positive direction of angles in circuit analysis. By convention, a lagging current has a negative phase angle (as compared to the voltage, which is taken to have a zero phase angle), and the resulting power has a positive phase angle:

$$\begin{aligned} S &= I_{rms} \angle -\theta^* \cdot V_{rms} \angle 0° \\ &= I_{rms} \angle \theta \cdot V_{rms} \angle 0° \\ &= I_{rms} V_{rms} \angle \theta \end{aligned}$$

Thus, the phase angle θ of complex power S is the same as the angle by which current is lagging voltage. This is the same result we claimed without further justification in Section 3.3.2, where the angle ϕ in the triangle in Figure 3.12 is the same as the phase angle ϕ between the sine waves in Figure 3.11.

Because complex, real, and reactive power all have the same physical dimension (energy per time), despite their different labels (VA, watts, and VARs) for identification purposes, they can usefully be combined in a phasor diagram. Indeed, Figure 3.12 is precisely such a diagram, except that, when introducing it, we did not mention any implicit rotation of the arrows, nor is it necessary to retain the visual image of rotation in order to apply the diagram for calculation purposes in practice. Considering the importance, simplicity, and explanatory richness of that very basic drawing, we can appreciate why phasors are an electrical engineer's most essential tool for visualizing a.c. power.

CHAPTER 4

Generators

An electric generator is a device designed to take advantage of *electromagnetic induction* in order to convert movement into electricity. The phenomenon of induction (introduced in Section 1.5.4) can be summarized as follows: an electric charge, in the presence of a magnetic field in relative motion to it—either by displacement or changing intensity—experiences a force in a direction perpendicular to both the direction of relative motion and of the magnetic field lines. Acting on the many charges contained in a conducting material—usually, electrons in a wire—this force becomes an *electromotive force* (*emf*) that produces a voltage or potential drop along the wire and thus causes an electric current (the *induced current*) to flow.

A generator is designed to obtain an induced current in a conductor (or set of conductors) as a result of mechanical movement, which is utilized to continually change a magnetic field near the conductor. The generator thus achieves a conversion of one physical form of energy into another—energy of motion into electrical energy—mediated by the magnetic field that exerts forces on the electric charges. In this sense, a generator is the opposite of an electric motor, which accomplishes just the reverse: the motor converts electrical energy into mechanical energy of motion, likewise mediated by the magnetic field. As far as the physical principles are concerned, electric generators and motors are very similar devices; in fact, an actual generator can be operated as a motor and vice versa. To achieve the best possible performance, however, there are many subtleties of design that specialize a given machine for one or the other task.[1] These subtleties have to do almost exclusively with geometry (though the choice of materials may also be important). Indeed, geometry is what distinguishes the many different types of specialized generators: the particular way in which the conducting wires are arranged within the generator determines the spatial configuration of the magnetic field, which in turn affects the precise nature of the current produced and the behavior of the machine under various circumstances.

---

[1] One example of a situation where it is practical to install a single machine for both generating and motoring is in the case of pumped hydroelectric storage or tidal power plants, where large water pumps are operated reversibly as turbine generators.

Electric Power Systems: A Conceptual Introduction, by Alexandra von Meier
Copyright © 2006 John Wiley & Sons, Inc.

86 GENERATORS

A comprehensive discussion of the many specific types of generators in common use would far exceed the scope of this text. This chapter instead concentrates on three cases: first, we discuss a greatly simplified device that can work as a motor or generator (a small version of which the reader may actually build) in order to illustrate the basic principles of generator function. Second, we describe at some length the standard type of generator used in utility power systems, the *synchronous generator*. Rather than going into the details of its construction and the many subtle variations among specific designs, we focus on the operation of generators in the system context, emphasizing the means by which synchronous generators control such variables as voltage, frequency, real and reactive power, and stressing in particular the interaction among generators, which is fundamental to the overall performance and stability of an alternating current (a.c.) system. Third, we briefly describe the *induction generator*, which is used in some specific applications such as wind turbines and which has some distinct and important properties.

Unlike most engineering texts on this subject, we do not discuss in any detail the ramifications of various design choices for generators, nor do we use mathematical derivations beyond the initial description of the basic induction phenomenon. Rather, the goal is to develop a conceptual understanding of the workings of a generator, including its operating constraints and limitations, so as to appreciate its function from the perspective of the power system as a whole.

This chapter also omits discussion of the *prime mover*, or whatever energy source pushes the turbine, because those aspects of power plants are well covered in standard texts on energy. The most common assumption for utility-scale generators is that in some sort of boiler water is turned to high-pressure steam, and that while being guided to expand through a set of fanlike turbine blades, this steam forces the turbine to rotate. For our purposes of understanding the electromagnetic phenomena inside a generator, it is irrelevant what boils the water; it could be burning hydrocarbons, fissioning atoms, or concentrated sunshine, or the turbine might simply be pushed by cold water running downhill. All we assume in this chapter is that any generator's *rotor*, or the part that rotates, is mechanically connected to something—usually by being mounted on a single rotating steel shaft—that continually exerts a force and expends energy to make it turn.

4.1 THE SIMPLE GENERATOR

For the purpose of developing a conceptual understanding of the generator, let us begin by considering a greatly simplified setup that includes only a single wire and a bar magnet. The objective is to induce a current in the wire; in a real generator, this wire corresponds to the *armature*, or the conductors that are electrically connected to the load. This can be accomplished by moving the wire relative to the magnet, or the magnet relative to the wire, so that the magnetic field at the location

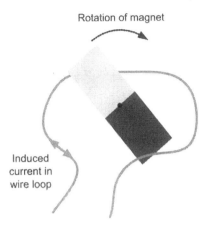

Figure 4.1 The simple generator.

of the wire increases, decreases, or changes direction. In order to maximize the wire's exposure to the magnet, since the field decreases rapidly with distance, we form a *loop* of wire that surrounds the bar (with just enough space in between). Now we can produce an ongoing relative motion by rotation: either we rotate the magnet inside the loop, or we hold the magnet fixed and rotate the loop around it. The analysis of the magnetic field is essentially the same in either case, but since the first type of arrangement turns out to be generally more practical, we shall choose it for our example (see Figure 4.1). We now have a bar magnet spinning around inside a loop of wire. How do we analyze the effect of this rotating magnetic field in terms of induced current? It might seem as though this would require some tedious analysis of magnetic field lines, including their directions with respect to the various sections of wire, and so forth. Fortunately, it is possible to use a comparatively simple approach. The key is to apply the notion of *magnetic flux* (see Sections 1.5.3 and 2.4), which equals the magnetic field multiplied by the area it crosses. Stated differently, the magnetic field represents the density of the flux in space. The flux is a convenient quantity in this context because the total flux emanating from the bar magnet does not change. The flux is also more easily related to the induced current, especially in a loop of wire. It turns out that the induced current in a closed loop is directly proportional to the flux *linking* (i.e., going through) this loop, irrespective of the spatial distribution of the flux inside the enclosed area. This means that there is no need to consider individually every point of the wire and its particular relationship to a magnetic field somewhere; rather, the entire loop and all the flux through it can be treated as compact quantities, joined in a straightforward relationship: the amount of current induced in the loop of wire will be directly proportional to the rate of change of the magnetic flux linking it.

How, then, can we specify the rate of change of the flux over time as our bar magnet spins? In actuality, the shape of the magnet would come into play, accounting for whatever particular bending of the magnetic field lines occurs, especially at

88 GENERATORS

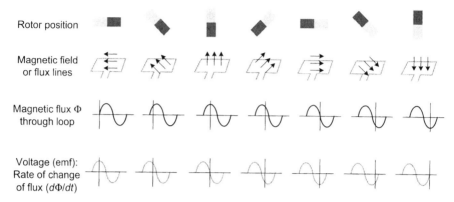

Figure 4.2 Changing flux and emf versus time.

the edges of the poles. For the sake of this example, however, we imagine some sort of ideal magnet that produces a uniform flux across the entire area enclosed by the loop. The *flux linkage* of this loop then varies according to the intersected area, which is in turn described by a *sinusoidal* mathematical function (see discussion in Section 3.1). When the magnet points in a direction parallel to the plane of the loop, none of its flux intersects the loop; we would say that the flux linkage of the loop at that time is zero. As the magnet rotates, more and more of its flux goes through the loop, reaching a maximum when the magnet is perpendicular to the plane of the loop. As it continues to rotate, the flux linkage decreases and becomes zero again. Then the magnet points across the loop in the opposite direction; we would now say that the flux linkage is negative. With continued rotation, the flux over time could be plotted out as an oscillating function, resembling more or less closely the mathematical sine or cosine wave. This progression is illustrated in Figure 4.2.

The induced voltage or emf and current in the wire are in turn given by the rate of change of the flux.[2] Technically, we should say the *negative* rate of change, though this sign convention is only to remind us that, because of energy conservation, any induced current will have a direction so as to oppose, not enhance, the changing magnetic field that created it. Regardless of direction, the emf is greatest while the flux is *changing* the most, which actually occurs at the moment when the value of the flux is zero. As the flux reaches its maximum (positive or negative), its value momentarily does not change (in other words, the very top of the sine curve is flat), and the emf at that instant is zero. Like the magnetic flux, the induced voltage and current will also change direction during this cycle: for one part of the cycle, with the positive flux decreasing and the negative flux increasing,

---

[2]Assuming that the load connected to the generator is purely resistive (see Section 3.2.3), the voltage and current in the wire will be exactly proportional to each other at all times. They can thus be used interchangeably in the following discussion, and, for simplicity, we will only refer to the current.

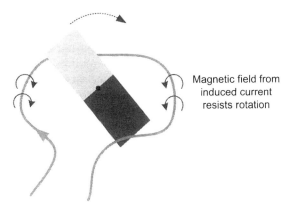

Figure 4.3 Simple generator with armature reaction.

the emf will be negative (with respect to the conventional reference direction), and for the other part, with the negative flux decreasing and the positive flux increasing, it will be positive. Mathematically, this emf and its resulting current could be plotted as another (more or less) sinusoidal wave, but offset from the first.[3] We have generated an alternating current! The frequency of this a.c. is the same as the rotational frequency of the spinning magnet; that is, one complete oscillation of voltage or current corresponds to one complete revolution of the magnet.

Since the wire or armature is carrying a voltage and current, it is in fact carrying electric power: a load could be connected to it, and the moving charges in the wire would do physical work while traversing that load, driven by the induction phenomenon in the generator. Thus, in abstract terms, energy is being transferred from the generator to the load. Where did this energy come from, and how did it get into the wire? From common sense, we know that energy was "put into" the generator by an external source of mechanical energy: something spinning the magnet. For a real generator, this energy source is the *prime mover*, which might be a steam turbine with a shaft connected to the generator's "magnet." But spinning a magnet in itself does not imply expending any energy. Rather, the magnet must offer some resistance to being spun; there must be something to push against.[4] Where does this resistance come from?

The answer is that the magnetic field of the rotating magnet is pushing against a second magnetic field that is the result of the induced current in the wire. This is illustrated in Figure 4.3. As described in Section 1.5.3, a current-carrying wire is surrounded by a magnetic field proportional to this current. For a straight wire, the field lines describe a circular pattern around it; when the wire is bent into a loop, these lines add together to form a magnetic field pointing straight down the middle,

---

[3]Readers familiar with calculus will recall that the derivative of the sine is a cosine, or a curve of the same shape, only shifted by 90°.
[4]In terms of physics, the mechanical work done equals the force applied multiplied by the distance an object is being pushed. If the magnet offered no resistance to being pushed, it would be impossible to exert a consistent force on it, and therefore impossible for it to do any work.

perpendicular to the plane of the loop. But, as we observed, the current in our wire is alternating. This results in an alternating magnetic field, changing intensity and direction in direct proportion to the current flowing; in a real generator this field is called the *armature reaction*.

The two magnetic fields—the spinning magnet's own field and the armature reaction created by the induced current—interact with each other. This interaction is like the repulsive force that we feel when we try to push the like poles of two magnets together. It is not immediately obvious how to visualize this force changing over time as the bar magnet rotates and the induced field waxes and wanes, but we can imagine two magnets that resist being superimposed with their poles in the same direction. Thus, as the bar magnet rotates toward a field in the same direction, it will be repelled or slowed down in its rotation.

The force in this case will actually be a pulsating one, reaching a maximum at the instant of maximum current as well as a momentary zero when the current is zero. In any case, as we would predict based on considerations of energy conservation (namely, that energy cannot be created or destroyed), the force on the magnet will always act in such a direction as to retard, not accelerate, its motion.

To summarize: an external force spins the bar magnet, resulting in a magnetic field or flux that changes over time. This changing magnetic field induces an alternating voltage and current in a loop of wire surrounding the space. The a.c. in turn produces its own magnetic field, which acts to retard the motion of the spinning magnet. In this way, the interaction of the magnetic fields mediates the transfer of energy from mechanical movement into electricity. When the generator is operated in reverse as a motor, the armature field is produced by an externally supplied current, but the interaction between the two magnetic fields is completely analogous. Now, the force between the two fields acts on the magnet in the center and forces it to spin, converting electrical into mechanical energy.

In the simplest case, an electric generator produces a.c. However, if direct current (d.c.) is desired as the output, there is a relatively straightforward way of *rectifying* or converting the a.c. output into d.c. In this case, it is preferable to reverse the arrangement and hold the magnet stationary while the armature rotates. The rotating ends of the wire in which the current is generated are then connected to the load by sliding contacts, in the form of brushes or slip rings, that reverse the connection with each half turn. Thus, while the magnitude of the current still oscillates, its directionality in the terminals going out to the load always remains the same. An identical setup with sliding contacts is used to operate d.c. motors.

THE PAPER CLIP MOTOR

Building something with our own hands often provides a new quality of insight, not to mention fun. With a few inexpensive materials, you can build your own d.c. electric motor. The process of fiddling with your motor to get it to work well illustrates the principles of physics as no textbook description can, and watching it actually spin gives tremendous satisfaction. Putting your paper clip motor together takes only a few minutes, and it is worth it!

Materials Needed

2 paper clips
1 small, strong magnet (from Radio Shack; most refrigerator magnets are too weak)
1 C or D battery
1 yard of 20-gauge (AWG 20) *coated* copper wire
2 rubber bands or tape
1 small piece of sandpaper

Make a tidy wire coil by wrapping it 10 times or so around the battery. Leave a few inches of wire at both ends. Tighten the ends around the coil on opposite sides with an inch or more of wire sticking out straight from the coil to form an axle on which the coil will spin (see Figure 4.4).

Attach the magnet to the side of the battery. It may stick by itself, or you may want to secure it with a rubber band or tape. The magnet's north or south pole should point directly away from the battery (this is the way the magnet naturally wants to go).

Bend the two paper clips and attach them with a rubber band or tape to both ends of the battery so as to form bearings on which the axle rests. The clips need to be shaped so that they make good electrical contact with the battery terminals, allow enough room for the coil to spin in front of the magnet, and keep the axle in place with a minimum of friction.

After checking the fit of the axle on the bearings, use the sandpaper to remove the red insulation coating on ONE end of the wire so that it can make electrical contact with the paper clip. At the other end, remove the coating on only HALF the wire by laying the wire coil flat down on a table and sanding only the top side. This will interrupt the electrical contact during half the coil's rotation, which is a crude way to reproduce the effect of commutator brushes.

(Ideally, the direction of current flow through the coil should be reversed with every rotation, which would then deliver a steady torque on the coil in one direction; this is what commutator brushes do. If direct current were allowed to flow continuously, the direction of the torque on the coil from the changing magnetic flux would reverse with every half-turn of the coil. Simply interrupting the current for half a turn interrupts the torque during just that period when it would be pulling the wrong way. Once the spinning coil has enough momentum, it will just coast through the half-turn without power until it meets the correct torque again on the other side.)

When you place the coil on the bearings with the contact side down and current flowing, you feel it being pulled in one direction by the interaction of the fields of the permanent magnet and the coil (the "armature reaction"). Now give the coil a little shove with your finger and watch it spin!

Figure 4.4 The paperclip motor. (From http://www.motors.ceresoft.org.)

4.2 THE SYNCHRONOUS GENERATOR

4.2.1 Basic Components and Functioning

Having characterized the essence of the generator's functioning in the simplified version just discussed, we need only make some "cosmetic" changes in order to turn our description into that of a real generator. Let us begin by clarifying the nomenclature. The rotating assembly in the center of the generator is called the *rotor*, and the stationary part on the outside the *stator*. In the majority of designs, and the only ones considered here, the rotor contains the magnet and the stator the armature that is electrically connected to the load. One simple rationale for this choice is that the armature typically carries much higher voltages, where fixed and readily insulated connections are preferable to the sliding contacts required for the rotating part of the machine. Furthermore, making the armature the stationary outside part of the machine provides more space for those windings that carry the most current.

Our first modification of the hypothetical simple generator concerns the rotor. With the exception of some very small-scale applications, using a permanent bar magnet to spin around is impractical because such magnets' fields are relatively weak compared to their size and weight. Instead, we mimic the permanent magnet by creating a magnetic field through a coil of wire (a *solenoid*, as described in Section 3.2) wound around an iron or steel core of high *permeability* (see Section 2.4) that enhances the magnetic field (Figure 4.5). This conducting coil is called the *rotor winding*, and its magnetic field the *rotor field* or *excitation field*. As will become relevant later, such an electromagnet has additional advantages

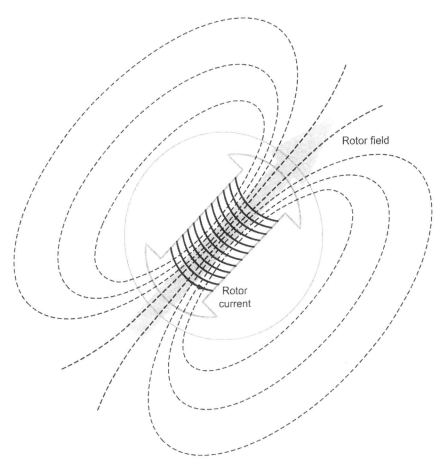

Figure 4.5 Cylindrical rotor and its magnetic field.

over a permanent magnet in the rotor, namely, that it is adjustable in strength (one need only vary the amount of current flowing through the coil in order to vary the field strength proportionately), and that it allows geometric configurations that create the effect of not just two, but many more magnetic poles. During basic operation, the rotor field is held constant and thus requires a constant direct current, the *rotor current* or *excitation current*, to support it. This rotor current is supplied by an external d.c. source called the *exciter* (see Section 4.2.2).

Next, consider the conductor in the stator. Instead of using just a single loop, we increase the emf or voltage generated in the conductor by winding it many times around successively, creating what is called the *armature* or *stator winding*. Each turn of the conductor adds another emf in series along the length of wire, and thus the voltages are additive. In theory, the magnitude of the generated voltage is quite arbitrary; in practice, utility generators operate in the neighborhood of 10–20 kV (see Section 4.2.2).

The many turns of the individual conductor are arranged in a staggered fashion, which makes the curve or *waveform* of the outgoing voltage and current resemble a mathematical sinusoid with reasonable accuracy. The details are beyond the scope of this text, so let us simply note that the specific arrangement of the windings through precisely machined slots in the armature is one important design aspect of generators.

Another revision to our initial model is that instead of a single conductor, a set of *three* conductors makes up the armature winding in standard utility equipment. These three conductors are not electrically connected to each other, but together they constitute three *phases* of a power circuit that correspond to the three wires we are accustomed to seeing on transmission lines. The phases are conventionally labeled A, B, and C. This three-phase design and its rationale are discussed in some detail in Section 6.3. Each phase carries an alternating voltage and current offset, or shifted in time from the others by one-third of a cycle (120°, where 360° corresponds to a complete oscillation). The phases are wound such that they are also 120° apart spatially on the stator, as shown in Figure 4.6. This spatial configuration is responsible for the time delay of one-third cycle, since it takes the rotor that much time to pass by the given points on the armature. Aside from the engineering advantages the three-phase system offers for power transmission, it provides for a much smoother conversion of energy in the generator. While in our initial, single-phase version the force between the magnetic fields pulsated during the rotation, the three-phase winding provides for a uniform force or *torque* on the generator rotor.

The uniformity of the torque arises because the magnetic fields resulting from the induced currents in the three conductors of the armature winding combine spatially in such a way as to resemble the magnetic field of a single, rotating magnet. This phenomenon can be illustrated by vector addition, or the geometric combination of

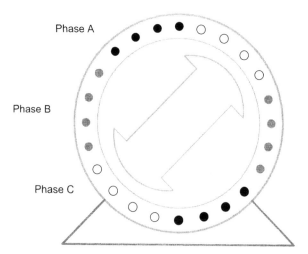

Figure 4.6 Schematic arrangement of three-phase stator winding.

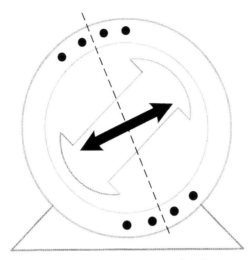

Figure 4.7 Contribution to armature reaction from one phase.

directional quantities, where these quantities (the three magnetic fields) vary over time. Their relative timing, one-third of a cycle apart, is crucial for the result. The alternating magnetic field due to just one phase is shown in Figure 4.7. The addition of all three is illustrated in Figure 4.8, which shows the three magnetic fields of the armature windings as vectors (arrows whose length corresponds to the intensity of the field) at some distinct moments during the cycle.[5] Owing to geometry, the sum of the three oscillating fields remains constant in magnitude, but changes direction in a cyclical fashion dictated by the relative timing of the three constituent fields;[6] in this case, the apparent motion is counterclockwise.[7]

The armature reaction of a three-phase generator thus appears as a steady, rotating field called the *stator field*. The stator field spins at the same frequency as the rotor, meaning that the two fields move in synchronicity and maintain a fixed position relative to each other as they spin; this is why this type of generator is called *synchronous*. As we see in Section 4.3.2, the exact relative position of the rotor and stator fields can be adjusted operationally and relates to the generation of reactive power or the generator's power factor (see Section 3.3).

---

[5]One cycle here is synonymous with one complete revolution of the rotor (not shown), or one complete oscillation of the alternating voltage or current, or one complete revolution of the resulting combined magnetic field that now constitutes the armature reaction.

[6]For readers not familiar with vector addition, we note that the sum of two or more vectors indicates the combined effect of the fields or forces represented, simultaneously showing direction and magnitude. This sum can be obtained geometrically by placing the arrows to be added tip-to-end and connecting the origin with the final destination.

[7]A clockwise motion would be obtained by reversing the relative timing of the three phases; for example, Phase B would reach its maximum one-third of a cycle before instead of after Phase A. This is referred to as *positive sequence* versus *negative sequence*.

96   GENERATORS

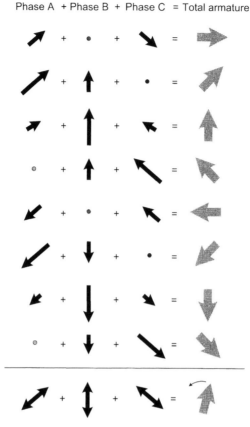

Figure 4.8 Three-phase armature reaction.

While the rate of rotation of the rotor and the frequency of alternating current in the synchronous generator are the same, they are conventionally specified in different units: a.c. frequency is measured in hertz (Hz), equivalent to cycles per second, and rotor rotation is usually given in revolutions per minute (rpm). Since a minute equals 60 seconds, a rotational frequency of 60 Hz corresponds to $60 \times 60 = 3600$ rpm.

A complication arises in situations where such a high rotational frequency is impractical or impossible to obtain from a given prime mover or turbine. This applies especially to hydroelectric turbines, because water simply does not flow downhill that fast.[8] The solution is to alter the design of the rotor so as to increase the rate of change of its magnetic field compared to the rotation of the entire assembly. This is accomplished by arranging the rotor windings in such a way as to

---

[8] Another intriguing case is that of very large steam turbines where, at 3600 rpm, the tangential velocity on the outside edge of the turbine blades would exceed the speed of sound. Unlike helicopter blades, it is not feasible to engineer an efficient steam turbine to withstand the associated mechanical stresses.

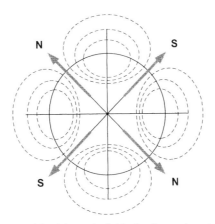

Figure 4.9 Magnetic field of a four-pole rotor.

produce the effect of more than two magnetic poles. Then, each physical revolution of the rotor results in more magnetic poles moving past the armature windings, where each passing of a north and south pole corresponds to a complete "cycle" of a magnetic field oscillation. For example, if the rotor has eight magnetic poles instead of two, each armature winding is exposed to four complete cycles of magnetic field reversal during one rotor revolution. Therefore, the rotor need only spin at one-quarter the revolutions (900 rpm) in order to still produce an a.c. frequency of 60 cycles per second. This relationship is usually specified in the form of the following equation,

$$f = \frac{np}{120}$$

where f is the a.c. frequency in Hz, n is the rotational rate of the rotor in rpm, and p is the number of magnetic poles. The factor of 120 comes from 60 seconds per minute and two poles in a single magnet.

In theory, the number of poles on the rotor can be any even number, but powers of two (4, 8, 16, 32) are especially common. Figure 4.9 illustrates a four-pole rotor and its magnetic field lines. To produce these poles, the rotor windings are configured in coils pointing outward from the center of the rotor, wound on more or less protruding cores; this type of arrangement is known as a *salient pole* rotor. Two-pole generators use a *cylindrical* or *round rotor* with rounded edges instead, whose field more closely resembles the bar magnet of our initial example (Figure 4.5). This round shape is important for minimizing drag (air resistance) at high rotational speeds.

4.2.2 Other Design Aspects

As mentioned earlier, the terminal voltage of most utility generators is on the order of 10 kV. In general, a reasonable choice of generator voltage is bounded by two considerations. High voltage requires more insulation between the terminals and the generator casing, and can represent a hazard of arcing or flashover. Ultimately, there is also a space limitation, as the voltage is produced by many armature conductor turns in series. Low voltage, on the other hand, becomes increasingly inefficient for large amounts of power. The reasoning here is the same as for transmission lines: since power is the product of current and voltage, transmitting the same power at a lower voltage requires a proportionately greater current to flow. Current, however, is associated with resistive heating of the conductors, and thus with thermal losses (recall that resistive heating is proportional to the square of the current). Aside from wasting energy, the heating of the conductors inside a generator limits its output capacity. It is therefore desirable to use the highest practicable voltage level.

Even so, the conductors on both the rotor and stator still need to carry very large currents and must be designed to do so safely. For example, consider a utility-sized generator producing 100 MW at 10 kV on three phases. Because $P = IV$, the current in each phase exceeds 3000 amperes. Since the limiting factor is the dissipation of heat (I^2R), both conductor diameter for lowering the resistance and the cooling of the conductor are important. The cooling system is indeed a very important component, especially of large generators. A quick calculation shows that if, say, a 100-megawatt (MW) generator has an efficiency of 99%, the 1% lost represents 1 MW of heat—equivalent to several hundred residential space heaters in the space of a single room!—continuously released into the generator's environment.

While smaller generators can be passively or actively air-cooled (the latter using fans to force the air), the cooling medium of choice for larger generators is hydrogen gas. Hydrogen conducts heat well and creates relatively little drag at high speeds. The hydrogen coolant is kept at an overpressure inside the generator so as to prevent infiltration of air, since a mixture of hydrogen and oxygen may ignite. It was the development of hydrogen cooling systems that provided for the increase in the size and efficiency of state-of-the-art generators in the 1940s and 1950s.

All the conductors inside a generator are coated with a thin layer of insulation to force the current to flow around the loops rather than short-circuiting across windings or through the rotor or stator iron. This insulating layer can be thin because, although the currents carried by these conductors are very large, the potential differences between adjacent windings are relatively small. The temperature tolerance of the insulation material sets an important limit on the currents to which the generator windings may be subjected, because overheating can cause embrittlement and failure.

The geometry of generator design is primarily concerned with guiding and utilizing the magnetic fields in the most efficient way possible. This involves detailed analysis of the air gap between rotor and stator, which is the actual locus of the interaction between the rotor and stator fields, where the physical force is transmitted. It becomes especially important here to consider fringe effects such as leakage flux and distortions around the edges of magnetic materials, since magnetic flux does not remain perfectly confined to the high-permeability regions. Furthermore, one must account for *eddy currents* that develop in localized areas within a conducting or magnetic material.[9]

Finally, there are several options for providing the d.c. field current to the generator rotor. For most large, synchronous generators, this is done with an auxiliary d.c. generator called the *exciter*. This exciter in turn requires its own field current, which it may draw from its own output in a process called *self-excitation*,[10] or it may contain a permanent magnet. While the exciter need only provide sufficient energy to overcome heating losses in the generator rotor, it is convenient to mount it on the same shaft along with the a.c. generator and the turbine and simply draw some of the mechanical turbine power for its purpose. Alternatively, a synchronous generator could draw its field current from a battery, but this is not generally practical. The third option is to draw directly upon the a.c. grid for the excitation current by *rectifying* it, that is, converting it from a.c. to d.c. Note that this latter approach only allows for the generator to be started up with the help of the external a.c. system providing the proper voltage; such a generator will not have the capability to start up in the event of a system blackout known as *black-start capability*.

4.3 OPERATIONAL CONTROL OF SYNCHRONOUS GENERATORS

Ultimately, the operational control of interconnected synchronous generators in a power system must be understood in terms of the interactive behavior among the generators, as discussed in Sections 4.3.3 and 4.3.4. Let us begin, however, by introducing the basic concepts for controlling an individual generator. While we distinguish many specific quantities within a generator—fields, currents, and voltages—these variables are so highly interdependent that there are only two real control variables in actual operation, or two ways to adjust a generator's behavior. These variables are the rotational frequency of the generator, which is related to the real power it supplies, and the voltage at its terminals (referred to as the generator *bus voltage*), which is related to the *reactive power* it supplies.

The term *bus* is very important in the analysis of power systems. Derived from the Latin *omnibus* (for all) the *busbar* is literally a bar of metal to which all the appropriate incoming and outgoing conductors are connected. To be more precise, the busbar consists of three separate bars, one for each phase. Called bus for short, it provides a reference point for measurements of voltage, current, and power flows.

---

[9]Electrical eddy currents are those that flow locally within a conductor rather than traveling its full length, much like pools of whirling water in a stream.

[10]It is not obvious that a generator should be able to supply its own excitation, since it cannot start to produce an excitation current without a magnetic field. The answer is that after an initial external magnetization of the "virgin" generator, a small amount of residual magnetization remains as a memory in the core material even after the generator is shut off. This small residual field is sufficient to interact in a mutually amplifying way with its own emf once a torque is applied, so as to reconstitute the full magnetic field.

In power flow analyses that encompass larger parts of the grid, buses constitute the critical points that must be characterized, while the detailed happenings "behind" the bus can be ignored from the system point of view. For a generator, voltage and current measurements at its bus are the definitive measure of how the generator is interacting with the grid.

4.3.1 Single Generator: Real Power

Real power output is controlled through the force or torque exerted by the prime mover, for example, the steam turbine driving the generator rotor. Intuitively, this is straightforward: if more electrical power is to be provided, then something must push harder. The rotor's rate of rotation has to be understood as an equilibrium between two opposing forces: the torque exerted by the turbine, which tends to speed up the rotor, and the torque exerted in the opposite direction by the magnetic field inside the generator, which tends to slow it down. The slowing down is directly related to the electric power being supplied by the generator to the grid. This is because the magnetic field that provides the retarding effect (the armature reaction) is directly proportional to the current in the armature windings, while the same current also determines the amount of power transmitted.

For example, if the load on the generator suddenly increases (someone is turning on another appliance), this means a reduction in the load's impedance, resulting in an increased current in the armature windings, and the magnetic field associated with this increased current would slow down the generator. In order to maintain a constant rotational frequency of the generator, the turbine must now supply an additional torque to match. Conversely, if the load is suddenly reduced, the armature current and thus its magnetic field decreases, and the generator would speed up. To maintain equilibrium, the turbine must now push less hard so that the torques are equal and the rotational frequency stabilizes.

The torque supplied by the prime mover is adjusted by a governor valve (Figure 4.10). In the case of a steam turbine, this increases or decreases the steam flow; for a hydro turbine, it adjusts the water flow.[11] This main valve can be operated manually (i.e., by deliberate operator action) or, as is general practice, by an automated control system. In any situation where a generator must respond to load fluctuations, either because it is the only one in a small system or because it is designated as a *load-following* generator in a large power system, automatic governor control will be used; in this case, the generator is said to operate "on the governor." The automatic governor system includes some device that continually monitors the generator frequency. Any departure from the set point (e.g., 3600 rpm) is translated into a signal to the main valve to open or close by an appropriate amount. Alternatively, a generator may be operated at a fixed level of power output (i.e., a fixed amount of steam flow), which would typically correspond to its maximum load (as for a *base-load* plant); in this case, the generator is said to operate "on the load limit."

---

[11] It is usually a good approximation to say that torque depends only on flow rate, which assumes a constant pressure of the steam or falling water. In reality the pressure would tend to vary inversely with flow rate, but not much.

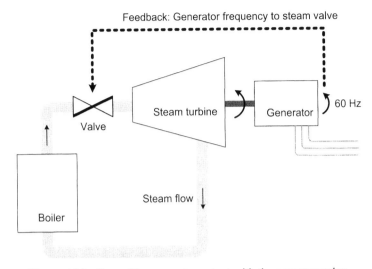

Figure 4.10 Controlling generator output with the governor valve.

Various designs for governor systems are in use. Older ones may rely on a simple mechanical feedback mechanism such as a flywheel that expands with increasing rotational speed due to centrifugal force, which is then mechanically connected to the valve operating components. Newer designs are based on solid-state technology and digitally programmed, providing the ability to govern based on not just the frequency measured in real time but its time rate of change (i.e., the slope). This allows anticipation of changes and more rapid adjustment, so that the actual generator frequency ultimately undergoes much smaller excursions. In any case, such a governor system allows the generator to follow loads within the range of the prime mover's capability, and without direct need for operator intervention.

4.3.2 Single Generator: Reactive Power

The other dimension of generator control has to do with voltage and reactive power, which are controlled by the field current provided to the rotor windings by the exciter. This effect is rather less obvious than the relationship between rate of rotation and real power, and to understand it we must further discuss the magnetic fields within the generator.

To begin, we can readily appreciate that increasing the d.c. current in the rotor will result in a corresponding increase in the rotor magnetic field, which in turn increases the electromotive force in the armature's windings.[12] The important

---

[12] Of course, the emf is a function of the *rate of change* of the magnetic flux, which will increase with greater magnitude of the rotor field. This rate of change could also be increased by a higher rotational frequency (as it is in the induction generator; see Section 4.5), but for the synchronous generator we assume frequency to remain constant during normal operation.

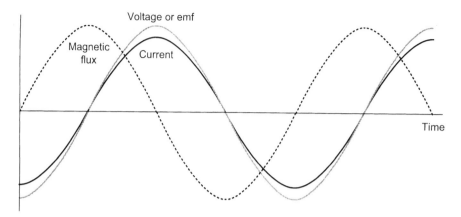

Figure 4.11 Flux, armature voltage, and current versus time, unity power factor.

point to realize here is that a generator's emf first manifests as a voltage, or potential difference between the generator's terminals, and that this emf is what causes current to flow through the windings and to the load. But the magnitude of this current is determined by the load impedance. Therefore, we must think of the armature current as reflecting what happens in the load, not the generator itself. The terminal voltage, on the other hand, is determined by the generator and is almost completely independent of the load. (Although a very large load may cause a voltage reduction if it exceeds the generator's capability to sustain the desired voltage, an absence of load does not act to increase the voltage because it does not, in itself, create a potential difference between the generator's terminals.) Thus, the generator terminal voltage is primarily a function of the emf in the armature, controlled by the d.c. rotor field current: increasing the field current will increase the generator bus voltage, and decreasing the field current will decrease generator bus voltage.

What does this have to do with reactive power? Here it becomes necessary to discuss in more detail the geometry of the magnetic field and the relative timing or phase difference of the alternating voltage and current.[13] Let us construct a graph that shows the variation of the stator voltage and current over time in relation to the rotation of the rotor (refer to Figures 4.11 and 4.12). We can represent these phenomena for the entire stator by referring only to a single phase, even just a single winding, in the armature. The horizontal axis indicates time, marked in terms of the rotor angle (i.e., the position of the rotor at any given instant). The interval depicted, ranging from 0° to 360°, thus corresponds to one complete revolution of the rotor. To represent other windings, we would only need to shift the zero mark on the time axis. The vertical axis indicates voltage or current, whose scales in the illustration are arbitrary since they are in different units.

---

[13] As the reader will recall from Section 3.3, reactive power refers to the transfer of power back and forth between circuit components that is associated with a shift in the relative timing of voltage and current.

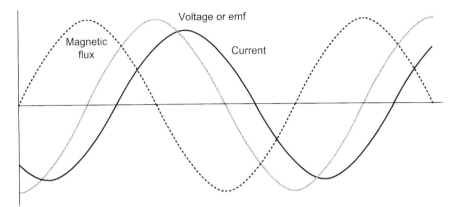

Figure 4.12 Flux, armature voltage, and current versus time, lagging power factor.

The first curve shows the magnetic flux (the rotor field) as seen by the armature winding. This flux varies sinusoidally, with its maximum occurring at the time that the rotor field is pointing in a direction perpendicular to the plane of the winding. The second curve shows the (negative) rate of change of this flux, which can also be read as the emf or voltage in the armature winding. The third curve shows the current in the armature winding, which also can be read as the magnitude of the (stator) magnetic field due to this one winding. The armature current, including its temporal relationship to the voltage, is determined by the load connected to the generator. In Figure 4.11, the load is purely resistive, or the power factor unity. Consequently, the timing of current and voltage coincides. Figure 4.12 shows a more typical operating condition with a lagging power factor, where the load includes some inductive reactance (such as electric motors). The lagging power factor means that the current lags or is delayed with respect to the voltage. Accordingly, the third curve is shifted here from the second curve, so that its maximum occurs some fraction of a cycle later. In this situation, the generator is said to supply VARs or reactive power to the load (see Section 3.3.2).

The opposite situation, shown in Figure 4.13, is less often encountered in practice. It corresponds to a leading power factor, where the generator is taking in VARs. Here the load has capacitive rather than inductive reactance, and, as a result, the third curve is shifted in the other direction so that its maximum occurs somewhat *ahead* of the voltage curve.

Let us now graph the same phenomena in a different way, one that accounts for the whole armature and takes advantage of the synchronicity, or the fact that events in the armature remain in step with the rotor's rotation. Rather than charting voltage and current over time, as observed at a particular location (one winding), we chart the position of the rotor and stator magnetic fields in space, as observed at one instant in time. Recall that the alternating magnetic fields of the stator (armature) windings combine geometrically so as to resemble a single, rotating magnetic field of constant magnitude. From the point of view of any given winding, the

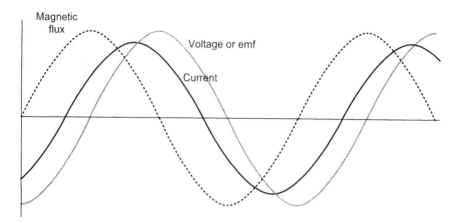

Figure 4.13 Flux, armature voltage, and current versus time, leading power factor.

maximum effect of the rotor magnetic field occurs at a specific time interval after the rotor field has "passed by," because that is when its maximum rate of change is observed. At unity power factor, this time interval corresponds to 90°, or one-quarter of a cycle. With a lagging power factor, the time interval will be somewhat longer than 90°, since the armature current experiences an additional delay due to the reactive load; with a leading power factor, it will be somewhat shorter than 90°, since the armature current is accelerated with respect to the induced emf by the capacitive load. However, no matter what the angle between rotor and stator field, this angle remains fixed as both fields rotate in synchronicity. Thus, we can adequately characterize the situation by drawing two arrows representing the two fields at any arbitrary moment during their rotation. This is done in Figures 4.14 to 4.16 for unity, lagging, and leading power factors, respectively.

The significance of the stator field's angle in relation to the rotor field is that it affects the amount of physical force or torque exerted on the rotor. This force is

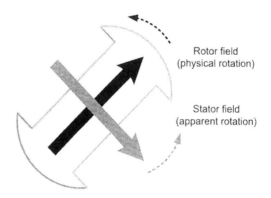

Figure 4.14 Rotor and stator field, unity power factor.

4.3 OPERATIONAL CONTROL OF SYNCHRONOUS GENERATORS   105

Figure 4.15 Rotor and stator field, lagging power factor.

greater the more perpendicular the fields are to each other; if they were pointing in the same direction, there would be no force between them at all. It therefore makes sense, for analytic purposes, to decompose the stator field into two components, that is, to treat the stator field as the sum of two separate phenomena that have distinct physical effects: one component that is perpendicular to the rotor field, and one that is parallel. This decomposition is easy to perform graphically. We need only draw two arrows, one perpendicular and one parallel to the rotor field, which, when placed tip-to-end, add up to the original stator field. The process is illustrated in Figure 4.17 for the case of lagging power factor.

The perpendicular component represents the extent to which the rotor field is physically acting to exert force on the stator field, thereby enabling the conversion of mechanical power. The parallel component, on the other hand, is the one that acts in conjunction with the rotor field. Thus, the parallel component of the rotor field effectively represents an *addition* to the stator field. In the case of a lagging power factor, this is a negative addition (subtraction) since the parallel component of the stator field points in the opposite direction. Thus, when the power factor lags

Figure 4.16 Rotor and stator field, leading power factor.

106 GENERATORS

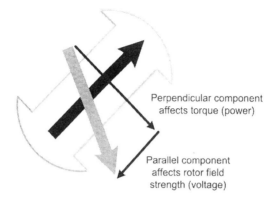

Figure 4.17 Decomposition of the stator field.

and the generator is supplying reactive power, the stator field acts in part to diminish or weaken the rotor field. For a leading power factor, the stator field acts in part to increase the rotor field.

As we said earlier, the magnitude of the rotor field determines the induced emf in the armature, and thus the generator bus voltage. This establishes the connection between reactive power and voltage. Suppose a single generator is supplying a resistive load, and suddenly some purely reactive, inductive load (an ideal inductor) is added. There is no change in the real power (watt) requirements, only an increase in reactive power (VAR) drawn. The armature current (determined by the load) now has added to it a component that lags 90° behind its voltage and previous current. This addition amounts to a slight increase in the overall magnitude as well as an overall phase shift of the armature current in the lagging direction. Consequently, the stator field takes a new position with respect to the rotor field, now lagging behind it by some angle greater than 90°. Since the stator field grows slightly in magnitude (owing to the increased current), its perpendicular component (related to real power) does not change, but now it also has a component parallel and opposite to the rotor field, diminishing it. This diminished rotor field now produces a lower generator voltage.

An analogous process occurs if we add a capacitive instead of an inductive load. If, before the addition, the load was purely resistive, then the chain of events is as follows: the armature current has added to it a current 90° ahead of its voltage and previous current, which constitutes an overall increase in magnitude and shift in the leading direction; the stator field is now less than 90° ahead of the rotor field and thus acts to strengthen the rotor field; the generator voltage increases; the power factor is now negative (leading), and the generator is said to consume reactive power (supplied by the load). It should be pointed out, though, that this hypothetical situation is rarely encountered in practice, since most actual loads are inductive in nature. Thus, the addition of a purely capacitive load in a real situation would probably just compensate for larger, inductive loads in the system and therefore act to move the power factor closer to unity: armature current would be reduced, the stator field more perpendicular to the rotor field, and generator voltage increased.

Without any further action, a change in generator voltage will result in a change of real power supplied to the load. For example, if inductive load is added, the generator voltage drops, and with that real power would also decrease. But here, finally, is where operator action comes in. In response to the lower bus voltage, the operator or an automatic control system will increase the rotor field current (the exciter current) to compensate for its diminution by the stator field and return the voltage to its original value. Conversely, if generator bus voltage increases due to a reduction in reactive load, the field current is appropriately reduced by operational control.

Most generating units are operated on an automatic voltage-control mechanism that maintains the generator bus voltage constant at a particular set point and continually varies the field current, as required by the load. When it becomes desirable to make a deliberate change in the field current (for example, to reallocate various generators' reactive power contributions), the operator typically intervenes by changing the voltage set point and allowing the automatic mechanism to adjust the field current accordingly. It is important to recognize that in either case, reactive power and generator voltage cannot be controlled independently. Given a particular voltage that we choose for a generator to supply, the VARs produced by this generator (and thus the generator power factor) are dictated by the load.

To a first approximation, the rotor field current consumes no energy, and changing it therefore has no direct bearing on the amount of real power generated. In reality, of course, there are thermal losses associated with rotor and stator currents as well as the current on transmission lines, and this lost energy has to come from somewhere. In this way, an increased reactive power requirement implies a small, indirect increase in real power generation, though not necessarily by the same generator supplying the reactive power. Another complication may arise if a generator is operating at its thermal limit, and its real power output needs to be curtailed in order to accommodate the additional armature current due to reactive power (see Section 4.4 on operating limits). Reactive power therefore does ultimately bear on real power generation, along with an associated monetary cost. It is important, though, to distinguish these second-order effects (which, in a sense, result from imperfections in the system) from the fundamental, first-order relationships between real power and frequency, on the one hand, and reactive power and voltage, on the other hand, which arise from the essential properties of a.c. generation. The first-order approximation is good enough that it is used not only for pedagogical purposes, but relied upon in practice for predicting power system behavior.

4.3.3 Multiple Generators—Real Power

In the preceding sections, we considered an individual generator supplying a load and discussed the operational changes necessary to accommodate varying real and reactive power demands. In utility power systems, however, the interconnection of many generators substantively affects their operation. Any change made to any

108 GENERATORS

individual generator, whether torque or rotor current, has repercussions for all other generators operated synchronous with the grid. We will first examine these generators' relationship in terms of real power, which has to do with rotational frequency and their synchronicity, and then, in the following section, turn to the problem of how reactive power is allocated among generators, which relates to their respective bus voltages.

All interconnected synchronous generators in an a.c. system not only rotate at exactly the same frequency, but are also in step with each other, meaning that the timing of the alternating voltage produced by each generator coincides very closely. This is a physical necessity if all generators are simultaneously to supply power to the system. As we will show, if any one generator speeds up to pull ahead of the others, this generator immediately is forced to produce additional power while relieving the load of the others. This additional power contribution results in a stronger armature reaction and greater restraining torque on the turbine, which tends to slow down the generator until an equilibrium is reached. Conversely, if one generator slows down to fall behind the others, this change will physically reduce this generator's load while increasing that of the others, relieving the torque on its turbine and allowing it to speed up until equilibrium is again reestablished.

Equilibrium here means that a generator's rotational frequency is constant over time, in contrast to the transient period during which the generator gains or loses speed. While all generators will settle into such an equilibrium at the system frequency, usually within seconds following a disturbance, this equilibrium may reside slightly ahead or behind in terms of the phase, or the exact instant at which the maximum of the generated voltage occurs. This variation of the precise timing among voltages as supplied by different generators (or measured at different locations in the grid) is referred to as the *power angle* (Figure 4.18) or *voltage angle*,

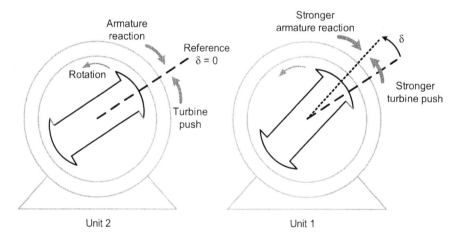

Figure 4.18 The power angle.

4.3 OPERATIONAL CONTROL OF SYNCHRONOUS GENERATORS

often denoted by the symbol δ (lowercase delta). Each generator's power angle is directly related to its share of real power supplied: the more ahead the power angle (expressed as a greater positive angle), the more power the generator is producing compared to the others.[14]

The power angles may only vary by a small fraction of a cycle, or else synchronicity among the generators is lost. This problem is referred to as *stability* (see Section 8.3). For interconnected generators, loss of synchronicity means that the forces resulting from their electrical interaction no longer act to return them to a stable equilibrium, as we assume in the present discussion, making their coordinated operation impossible. In terms of power system management, such a condition clearly spells disaster; it is therefore fortunate that, for most systems under normal circumstances, synchronicity naturally follows from the inherent physics of generators.

We now examine why an increase in one generator's phase angle corresponds to an increase in real power delivered to the system by that generator, and what the forces are that tend to keep the generators in sync. For simplicity, we shall only consider the interaction with one nearby generator, not the system as a whole. This approximation is especially good for two generating units at the same power plant. It is justified physically in that the impedance between two very nearby generating units is small compared to the impedance of the remaining system and the load (i.e., the path between them is a preferred path for any current), meaning that these units will interact more strongly with each other than with the rest of the system. Although more complicated, the same analysis can in principle be applied to the entire system. We refer here to a hypothetical case of two generators, Units 1 and 2, at Plant A and observe their interaction as the power generation level on Unit 1 is increased.[15]

We begin by increasing steam flow at Unit 1. The additional forward torque on the turbine causes it to speed up. This acceleration of the rotor causes it to move slightly ahead compared to the rotor in Unit 2, meaning that the maxima of the emf or voltage produced in each phase of the armature windings at Unit 1 occur slightly ahead of those in Unit 2: the voltage or power angle δ is increased slightly. Note that the two units are still considered *synchronous* in that they remain in step with each other and their movement remains interdependent; only one marches a small fraction of a step in front of the other. Indeed, the power angles would be identical only if both units supplied the same amount of power to a load exactly in the middle.

This change in timing of the voltage results in a net difference between the voltage of Unit 1 and the system, represented by Unit 2, as measured at any given instant. Let us refer to this difference between the slightly ahead (V_1) and normal (V_2) voltages as the *difference voltage*. If we graph the difference voltage over time, we obtain a curve that is approximately 90° out of phase with both (see

---

[14]This relationship is observed quantitatively in power flow analysis.
[15]The terms *unit* and *generator* are synonymous here, since a generating unit refers to a single turbine–generator assembly.

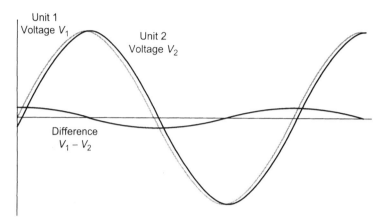

Figure 4.19 Power angle and circulating current.

Figure 4.19).[16] From the perspective of Unit 1, this difference voltage is positive at those times when V_1 (its own voltage) is greater than V_2, and negative when V_1 is less than V_2; this is shown in the figure. From the perspective of Unit 2, of course, we would draw the difference voltage just the opposite way and call it negative when V_1 is greater than V_2, positive when V_1 is less than V_2.

This difference voltage is associated with a current that flows between the two generators. This current is called a *circulating current* because it essentially circulates between the two units. (There is also a circulating current between Unit 1 and the other generators in the system, but it is smaller because of the higher impedance between them; this is why representing the whole system only by Unit 2 is a reasonable approximation.) The timing of this circulating current is crucial. We must recognize two things: First, because Units 1 and 2 have opposite perspectives (they see a voltage of opposite sign), the circulating current measured in the armature windings of each generator will be opposite. Casually speaking, we could say that the current is leaving one side as it enters the other. Second, we recognize that the impedance of the circuit comprising two nearby generators is essentially all inductive reactance and very little resistance, since the wires are thick and coiled. Therefore, the circulating current will lag by almost 90° the difference voltage that induced it to flow.

The result, as shown in Figure 4.20, is a circulating current that is approximately in phase with the regular initial current generated by Unit 1, but approximately 180° *out of phase* (just opposite) with the same initial current generated at Unit 2. Thus, the circulating current *adds* to the armature current at Unit 1, but *subtracts* from the same current at Unit 2. Accordingly, one would observe an increase in the magnitude of the armature current at Unit 1 and a decrease in armature current at Unit 2. Since the magnitude of the generator voltage has not been changed in either

---

[16]The smaller the power angle or relative phase shift, the closer to 90° will the difference curve be out of phase with the two originals. In practice, the power angle is a small fraction of a cycle.

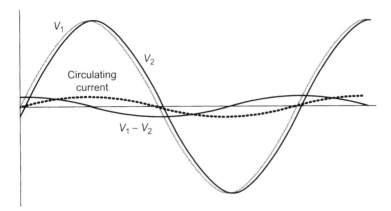

Figure 4.20 Effect of circulating current on power generation.

case, and since power is the product of voltage and current, this means that the (real) power output of Unit 1 increases, while the power output of Unit 2 decreases. Effectively, Unit 1 pulling ahead in phase means that it will "push harder" and literally take some of the load off of Unit 2. In fact, the load reduction will be shared among all the generators in the system, with the most significant change in the closest units.

It is important, then, to recognize that all interconnected synchronous generators will be affected in principle by an increase or decrease in power output of any one generator. Suppose that, in our example, Unit 1 pulls ahead of the others by increasing its turbine power output, while no other changes are made in the system. As we have shown, circulating current between generators will flow so as to take load off the other generators. In response to the reduced mechanical resistance, these generators will tend to speed up. It is reasonable to assume that their governor systems sense the increasing speed and accordingly reduce the turbine power output. However, if no such corrections were made, the frequency of the entire system would increase as a result of total generation exceeding the total load. Conversely, if the total power generated were less than the load, all the generators would slow down.[17]

When a synchronous generator is connected to an energized system that is already operating at the specified frequency, a process of *synchronization* is required, also referred to as *paralleling* the generator to the grid. In this process, the generator is first brought up to its synchronous speed while still electrically disconnected. With instrumentation on both circuits, the frequency as well as the relative phase of the generator and the rest of the system are carefully compared, and small adjustments made on the generator speed to match the phase precisely. Once the match is achieved, the electrical connection is established by closing

---

[17] As stated earlier, this coupling among interconnected synchronous generators can be broken (losing synchronicity) if an excursion from the equilibrium is too large. The precise value of the power angle at which synchronicity is lost depends on a combination of generator and system characteristics; this problem is discussed in Section 8.3 on power system stability.

a circuit breaker between the generator and its bus. Finally, with the generator "floating" at zero load, turbine steam flow and field current are increased until the generator is delivering its specified power output.

4.3.4 Multiple Generators—Reactive Power

From the principle of energy conservation, it is clear that the total amount of real power supplied by a set of interconnected generators is dictated by the load: since energy is neither created, destroyed, nor stored (in appreciable quantity) within the transmission system, the instantaneous supply must equal the instantaneous demand. If necessary, this principle will enforce itself: if operators tried to generate a different amount of power than is being consumed, the system's operating state—first frequency, and ultimately voltage, too—would change so as to make energy conservation hold true.

The energy associated with reactive power is similarly conserved. Although reactive power involves no *net* transfer of energy over time from generators to load, the instantaneous flow must still be accounted for. Specifically, the energy going into the electric or magnetic field of one device in a circuit during some part of the cycle must be coming out of some other device, which stores that energy during the complementary part of the cycle. Therefore, while the nomenclature of "generating" and "consuming" reactive power is an arbitrary convention, it remains true that the amount of reactive power generated at any instant must equal that consumed at the same instant. Thus, the total reactive power or VAR output of a set of interconnected generators is dictated by the load, just as real power is. The conservation of reactive power will also enforce itself, in this case primarily through changes in voltage.

Just as the allocation of real power generation among interconnected generators can be varied by means of the power angle or relative timing of the voltage, so can the allocation of reactive power. Here, the means of control is the magnitude of the supplied voltage in relation to other generators, which is in turn controlled by the rotor field current. If all generators in a system were generating the same bus voltage, their power factors would be the same, and thus their reactive power output would be in the same proportion to the real power they generate. It may be desirable to change this balance for any of a number of reasons: to maintain a certain voltage profile throughout the system; to minimize cost (because one unit might generate additional VARs at lower cost than another); or because some generators are operating at their capability limit and can only produce more VARs at the expense of real power. In general, though, it is not desirable to maintain a gross imbalance between VAR generation at different units, because, as we will show, a circulating current is associated with differences in voltage levels among generators, and this current entails losses in the generator armature windings and transmission lines.

To understand the interdependence of generator voltage levels and reactive power contributions, consider again two nearby generators, Units 1 and 2. Suppose also that we wish to increase Unit 1's contribution to the system's reactive

4.3 OPERATIONAL CONTROL OF SYNCHRONOUS GENERATORS

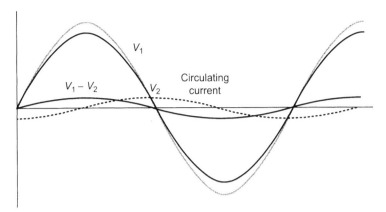

Figure 4.21 Circulating current with Unit 1 providing more reactive power.

power needs, while reducing that of Unit 2. This change is effected by increasing the rotor field current to Unit 1, which in practice would be accomplished by raising the voltage set point. The situation is illustrated in Figure 4.21 with V_1 slightly greater than V_2. Again, we can graph the change in voltage as a difference voltage, but now this difference is perfectly in phase with the "normal" voltage generated, and the change is therefore observed simply as an increase in the magnitude of that voltage. The increased voltage results in an increased current in the armature windings of Unit 1. To a first approximation, this current circulates in the local circuit composed of the two neighboring generator stators and the bus connecting them.

As before, we note that since the impedance of this circuit is almost all inductive reactance and no resistance, the circulating current is 90° out of phase with the voltage. Thus, it has no effect on the real power generated by either generator. Also, since Unit 2 has an "opposite" perspective, the circulating current in its armature is negative at the same time as it is positive in Unit 1. In Unit 1, the circulating current is lagging and thus coincides with the lagging component of the armature current that is associated with reactive power supplied to the load. In Unit 2, on the other hand, the same circulating current is observed as a leading current (since being just opposite or 180° apart brings it from 90° lagging to 90° leading).

Recall now that the magnetic field associated with a lagging or leading armature current acts to weaken or strengthen the rotor field, respectively. Thus, the circulating current weakens the rotor field in Unit 1 and strengthens that in Unit 2. Assuming that all units were operating at a somewhat lagging power factor to begin with, the magnitude of the armature current in Unit 2 is now less, since it is relieved of some of its lagging component by the leading circulating current. As a result, the reactive load on Unit 1 is now increased, while Unit 2 experiences a decrease in reactive load.

Let us suppose that the automatic voltage regulators in both units are out of service, and all changes to field current are made manually. In Unit 1, operators increase the field current to initiate the chain of events. The generator bus voltage increases, but not as much as one would have expected before taking the circulating current into account. Indeed, the higher they attempt to raise Unit 1's voltage compared to other units, the more field current will be required in order to achieve a further increase in rotor field strength and thus voltage. This condition is synonymous with supplying more reactive power. In Unit 2, the voltage also increases because the rotor field is strengthened by the circulating current and less reactive power is supplied. In response, the Unit 2 operators may lower the field current so that the voltage returns to its previous value.

In practice, with voltage regulators in service, Unit 1 operators would increase the voltage set point and allow the regulator to increase the field current automatically. The voltage regulator at Unit 2 would recognize the elevated voltage due to the circulating current and reduce the field current appropriately.

Overall, we might say that the circulating current has the effect of equalizing the voltages between generators, analogous to the way that a circulating current tends to equalize the rotational frequencies and real power output between generators. Again, this discussion extends qualitatively to other generators in the system, with the closest ones being affected most. The electrical interaction between generators thus results in a stabilizing force that tends to equalize voltages, analogous to the force that tends to equalize frequency.

Let us emphasize again that the total amount of reactive power supplied by all generating units is determined by the load, and that this reactive load can be shared among generators in whatever way is most economical, independent (within reasonable bounds) of real power allocation. Typically, because of the losses associated with circulating current, an economic allocation of reactive power will have generators operating at fairly similar power factors (p.f.). A simple example of how such an allocation might look is shown in Figure 4.22,

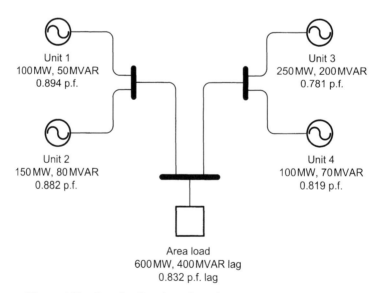

Figure 4.22 Sample allocation of reactive power among generators.

where four units, located at two power plants, are supplying an area load. (For clarity, the example ignores real and reactive line losses.) This example highlights the fact that the *system power factor*, which is determined exclusively by the load, is different from the individual *generator power factors*. Indeed, the plant operators have no way of knowing what the system power factor is, based on observations only at their own unit.

4.4 OPERATING LIMITS

The generating capacity of electric power plants is generally referred to in units of real power output, such as kilowatts or megawatts. This is appropriate in that the decisive constraint on how much power can be generated is the ability of the prime mover to deliver mechanical power to the turbine. In a typical plant design, the electric generator will be sized so as to be able to handle any amount of real power the turbine can provide.[18] However, generator performance limits are still important in the operational context, because they also apply to reactive power, which is independent of the prime mover. Indeed, the appropriate measure of capacity for the electric generator as such is not in terms of real power, but rather in units of *apparent power*, kilovolt-ampere (kVA) or megavolt-ampere (MVA). It is thus a combination of real and reactive power that must be considered in order to determine whether the generator is operating within its range or is in danger of becoming overloaded.

"Overloading" for a generator primarily means overheating due to high current, though some mechanical factors may also be relevant. Excessive temperature will cause the insulating material on the generator windings to deteriorate and thus lead to an internal fault or short circuit. Different rates of thermal expansion between the winding conductors and the core at excessive temperatures can also cause insulation damage through movement and abrasion. Depending on the particular operating condition, "hot spots" may develop on different components, which is problematic because the temperature cannot readily be measured everywhere inside the generator. Possible sources of mechanical damage under excessive loading include rotor vibration due to imperfect balance, vibration due to fluctuating electromagnetic forces on the components, and loss of alignment between turbine and generator shafts due to thermal expansion or distortion of the generator frame.

Any of these types of damage are irreversible in that the generator will not recover after the load is reduced. Therefore, rather than waiting for signs of distress

---

[18]Wind turbines are an interesting exception. Because of the wide variability of wind speed, it is unrealistic to install a wind generator capable of handling the maximum expected wind power—say, during a storm. Instead, proper design calls for an optimization considering the expected range and frequency of wind speeds, the cost of the generator in relation to its size, and the reduction in operating efficiency that occurs when the generator is producing much less than its rated output. It is also necessary to include some type of mechanical restraint that will shut down the wind turbine beyond a certain wind speed (the *cutout speed*) and prevent it from overloading the generator.

under high loads, generators are operated within limits specified by the manufacturer that will allow for some margin of safety to assure the integrity of the equipment. To a first approximation, these limits are indicated by the generator *rating* in kVA or MVA. Since apparent power is directly proportional to current, regardless of the relative proportions of real and reactive power, this is synonymous with a limit on the current in the armature or stator windings.

Interestingly, there has been some historical change in the implicit conventions for such ratings, not only for electric generators. In the engineering tradition, it has long been customary to provide substantial safety margins in the design of machinery, to the extent that an experienced operator can at times exceed the nominal ratings with confidence. Yet over the past few decades, the philosophy of building some slack into technical systems by generously oversizing components has increasingly given way to a more refined approach where, aided by more sophisticated instrumentation and computing, components can be matched more precisely to needs and specifications. While such refinement in design has obvious economic justifications, it does in some sense increase the vulnerability of the system, as in the case of a generator being operated at 100% of its nominal rating, which will now tend to have less tolerance for excursions from normal operating conditions.

In most situations, the current in the armature windings can be used as a criterion for the generator operating limit, indicated simply by a red line on the "generator current" display. Under certain operating conditions, however, it becomes necessary to observe more stringent criteria on generator loading, especially if there is little slack in the rating. These criteria have to do with heating of components other than the armature windings, which tends to become more prohibitive when operating at a power factor very different from unity. This comprehensive information about a particular generator is captured in a diagram called the *reactive capability curve*, which indicates a boundary on permissible combinations of real and reactive power output: all points inside the area bounded by the curve are achievable without risk of damage, and all points outside this area are prohibited.

The curve consists in part of the circle that describes a constant amount of apparent power, as in the kVA or MVA rating.[19] Within the normal operating range, where the power factor is relatively close to unity, this circle does indeed prescribe the operating limit, imposed by the resistive (I^2R) heating of the armature (stator) conductors as a function of apparent power. For the typical case shown in Figure 4.23, this operating range is defined between power factors of 0.85 lagging

---

[19] Recall from Section 3.3.2 that apparent power can be obtained by vector addition from real and reactive power, where real power is represented by a horizontal arrow and reactive power by a vertical one, which are placed tip to end. The length of the resulting vector indicates the magnitude of apparent power, and all the possible vectors of the same length (starting at the origin) describe a circle. Here, we are interested only in the right half of this circle that corresponds to positive real power generation; the omitted left half would correspond to operation as a motor.

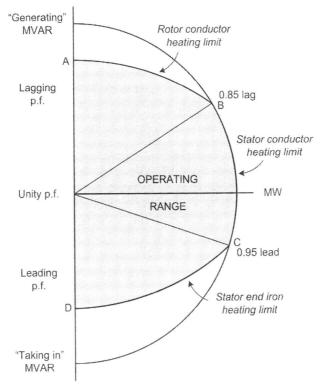

Figure 4.23 Reactive capability curve.

and 0.95 leading.[20] The applicable segment of the reactive capability curve is that between the points labeled B and C in the figure.

For more lagging power factors, it turns out that the operating limit is more stringent than what we would expect from extrapolating the circle. Instead, the limit is described by a line closer to the origin, connecting points A and B. Physically, it arises from the rotor field current or *overexcitation* required to supply large amounts of reactive power (i.e., maintaining the desired voltage despite a large reactive load). Beyond a certain power factor, heating of the rotor conductor due to this field current becomes more constraining than for the stator conductor.

In the opposite case of *underexcitation*, where the field current is reduced in response to the leading armature current beyond a certain power factor, a more stringent limit also applies. Here the concern is again about heating in the stator, but

---

[20]In the figure, this region is bounded by straight lines at angles corresponding to the specified power factors. Any point along a straight line from the origin indicates a combination of real and reactive power in the same ratio. The horizontal axis corresponds to a p.f. of unity (1.0); the vertical axis to a p.f. of zero (no real power generated at all).

this time because of eddy currents that tend to develop within the stator core iron. The part of the reactive capability curve that applies to this condition connects points C and D.

In general, it is the operator's responsibility to assure these limits are not exceeded. Outside the normal operating range (in this example, between power factors of 0.85 lagging and 0.95 leading), automatic controls are generally used to assist the operator. For instance, voltage regulators may provide limitations that prevent the field current from being increased or decreased beyond set limits, depending on real power output.

Generator operating limits also apply to voltage. In conventional U.S. utility practice, the tolerance for voltage is $\pm 5\%$. Thus, a generator nominally producing a bus voltage of 20 kV might in practice be operated between 19 and 21 kV. Staying within this range is more a matter of policy than of urgent technical necessity, since it generally takes a larger voltage excursion to actually damage equipment. However, accurate voltage is a key criterion of *power quality*, and is discussed further in Section 6.6.

4.5 THE INDUCTION GENERATOR

4.5.1 General Characteristics

An induction or *asynchronous* generator is one that operates without an independent source for its rotor field current, but in which the rotor field current appears by electromagnetic *induction* from the field of the armature current. The rotor field then interacts with the stator field to transmit mechanical torque just as it does in a synchronous generator, regardless of the fact that it was the stator field that created it (the rotor field) in the first place. This may seem reminiscent of pulling yourself up by your own bootstraps, but it does actually work. The catch is that some armature current must be provided externally; thus, an induction generator cannot be started up without being connected to a live a.c. system. Another practical concern is that, as we show later in this chapter, induction generators can only operate at leading power factors. For both reasons, their use is quite limited.

Their one important application in power systems is in association with wind turbines. In this case, induction generators offer an advantage because they can readily absorb the erratic fluctuations of mechanical power delivered by the wind resource. They also cost less than synchronous machines, especially in the size range up to one megawatt.

In terms of mechanical operation, the most important characteristic of the induction generator is that the rate of rotation is not fixed, as in the case of the synchronous generator, but varies depending on the torque or power delivered. The reference point is called the *synchronous speed*, which is the speed of rotation of the armature magnetic field (corresponding to the a.c. frequency) and also the speed at which a synchronous rotor would spin. The more power is being generated, the faster the induction rotor spins in relation to the synchronous speed; the difference is called the *slip speed* and typically amounts to several

4.5 THE INDUCTION GENERATOR

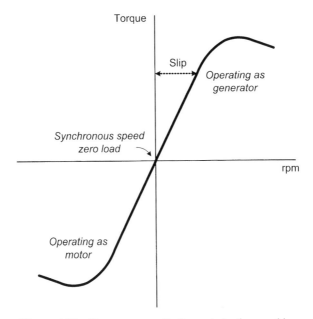

Figure 4.24 Torque versus slip for an induction machine.

percent.[21] The rotor may also spin more slowly than the armature speed, but in this case, the machine is generating negative power: it is operating as a motor! While induction machines are usually optimized and marketed for only one purpose, either generating or motoring, they are all in principle reversible. (The same is true for synchronous machines, though their design tends to be even more specialized.)

Figure 4.24 shows a curve of torque versus slip speed for a generic induction machine. Zero slip corresponds to synchronous speed, and at this point, the machine delivers no power at all: neglecting friction, it spins freely in equilibrium. This is called a *no-load condition*. If a forward torque is exerted on the rotor in this equilibrium state (say, by a connected turbine), it accelerates beyond synchronous speed and generates electric power by boosting the terminal voltage. If the rotor is instead restrained (by a mechanical load), it slows down below synchronous speed and the machine is operating as a motor. Now we call the torque on the rotor negative, and it acts to push whatever is restraining it with power derived from the armature current and voltage.

The synchronous speed of a given induction machine may be equal to the a.c. frequency (3600 rpm for 60 Hz; 3000 rpm for 50 Hz) or some even fraction thereof (such as 900 or 1800 rpm), depending on the number of magnetic poles, which in this case are created by the armature conductor windings instead of the rotor.

---

[21]The ratio of slip speed to synchronous speed, simply called the *slip*, is also sometimes expressed as a decimal between 0 and 1.

Note that unlike the synchronous generator, where the stator magnetic field has two poles but the rotor field may have any even number of poles, an induction generator must have the same number of poles in the rotor and stator field (because there is no independent excitation).

4.5.2 Electromagnetic Characteristics

The rotor of an induction machine consists of a set of conductors arranged in such a way that, when exposed to the armature magnetic field, a current will flow. This can be done with regular windings of insulated wire in what is called a *wound rotor*, or with a much simpler structure of conducting bars running parallel to the generator shaft that are connected in rings at the end, known as a *squirrel-cage rotor*. Squirrel-cage rotors are the most common since they are much less expensive to fabricate.

When the armature is connected to an a.c. source, it produces a rotating magnetic field just like the stator field of the synchronous generator. The rotor conductors form loops that are now intersected (*linked*) by a changing amount of magnetic flux. As the stator field rotates, the flux linking any given rotor loop is zero whenever the stator field points in a direction parallel to the plane of the loop, and reaches a maximum when the field is perpendicular to the loop. This changing flux induces a current to flow in each loop.

The timing of this current will vary from one loop to the next as the stator field makes a complete revolution. In a wound rotor, the emfs add together for every turn of a single conductor; rotors are typically wound for several phases. In a squirrel-cage rotor, each pair of opposite bars represents one phase. It may seem counter-intuitive that we can distinguish different currents flowing through conducting parts that are all electrically connected. In fact, they are eddy currents in a single conducting object. However, because of the object's shape, these eddy currents are very ordered, and their electromagnetic effect is similar to that of separate currents traveling along individual, insulated wires. Thus, the rotor current for each phase produces its own magnetic field, and the geometric combination of these fields into a single rotating field of constant strength works just like it did for the armature.

What is different for the rotor field, however, is the frequency. The frequency of alternating current within the rotor loops, regardless of how they are physically constructed, is given by the rate of change of the flux through the rotor loops. This rate of change depends not only on the a.c. frequency in the stator, but also the slip speed, or relative speed between the rotor's mechanical rotation and the apparent rotation of the stator field.

Suppose that the rotor is initially at rest, while the stator field rotates at 60 revolutions per second (rps) (assuming a two-pole stator). This start-up condition corresponds to a *slip*, or ratio of slip speed to synchronous speed, of unity (100%). In this case, the flux linkage through the rotor loops undergoes one complete reversal with each revolution of the stator field. Accordingly, an alternating current of 60 cycles (Hz) is induced in the rotor. The resulting rotor field also rotates at 60 cycles, which implies that it remains in a fixed position with respect to the stator field, as it does in a synchronous generator.

As we will show later, the position between rotor and stator field is such that a torque is exerted on the rotor. If the rotor is initially at rest, this torque will accelerate it in the same direction of rotation as the stator field. But as the rotor accelerates toward synchronous speed, the slip decreases.

Suppose the rotor has reached 57 rps, which corresponds to a slip speed of 3 rps (or a slip of 1/20). The spinning stator field now undergoes a complete reversal with respect to the rotor only three times per second, as the flux lines intersect the conducting loops from a different direction. Therefore, the current induced in the rotor now alternates at only three cycles.

From the perspective of the rotor, the resulting magnetic field—geometrically composed of the individual fields from each phase alternating at three cycles in different directions—appears to make a circular revolution three times per second. But how does the rotor field appear from the perspective of the stator? In this stationary reference frame, the apparent speed of the rotor field, three cycles, is added to the relative (mechanical) speed of the rotor itself, 57 cycles. The result is that the rotor field still revolves at synchronous speed, 60 cycles, just like the stator field!

The same reasoning applies for any arbitrary slip speed, positive or negative. Thus, unlike the *mechanical* rate of rotation of the rotor, which varies, the *magnetic* rotation of its field is always at synchronous speed and therefore remains in a steady relation to the rotating stator field.

As long as the rotor is mechanically spinning below synchronous speed, it experiences a changing flux in its loops, and thus has current induced in it. But as the mechanical rotor speed approaches that of the revolving stator field, the rate of change of this flux becomes less and less, as does the frequency of the induced current. Furthermore, the magnitude of the induced current diminishes, because it is proportional to the rate of change of magnetic flux. Thus, the rotor field gets weaker as the rotor approaches synchronous speed, and so does the torque between the rotor and stator magnetic fields. Finally, when the rotor reaches synchronous speed, the torque is zero.

If we now supply an external torque on the rotor, the machine speeds up beyond synchronous speed and operates as a generator. Now we have relative motion between the rotor loops and the revolving stator field, so that an alternating current is again induced. Because the relative motion is in the other direction (the rotor is revolving faster instead of slower than the stator field), the current is reversed, and the magnetic torque now acts to restrain rather than accelerate the rotor. This is the result we should expect based on energy conservation.

By forcing the rotor to maintain this faster speed, we are forcing a rotor field of a certain strength and direction to coexist with the stator field. This rotor field acts just like that of a synchronous generator: it induces an emf in the armature windings that will cause current to flow to the load and thus transmit electric power. Although there is already a current preestablished in the armature windings when the induction generator is first connected to the a.c. system, the induced current is additive, since the induced emf acts to strengthen the potential difference at the generator terminals. By contrast, when the induction machine is operating as a motor, the emf induced by the rotor field counteracts the existing potential difference, resulting in the motor "drawing" current from the a.c. grid.

The more the rotor speed deviates from the no-load equilibrium in either direction, the stronger the torque pushing it back toward synchronous speed. This relationship remains true up to a point, beyond which the torque diminishes (but still acts in the proper direction); this is shown by the decline in the slip–torque characteristic on the far right-hand side in Figure 4.24. An induction generator is operated in the region between zero and maximum torque, because, as the reader may convince herself, any operating point in this region is *stable* in the sense that an excursion will be associated with a change in torque that tends to restore the operating condition. Beyond the "knee" in the curve at maximum torque, where the torque decreases with increasing speed, the operating condition is *unstable* because an increase in rotor speed further reduces the restraining torque. This is called an *overspeed* condition, which will ultimately damage the generator and must be prevented by disconnecting the prime mover if the rotor speed exceeds a given value. Correspondingly, when the machine is operated as a motor, there is an unstable condition beyond the knee at which the motor simply stops working, for example, when stopping the motor of an electric toy with your hand.

The mechanical power transmitted is given by the product of torque and angular frequency of rotation (rpm). Machines can be designed to produce maximum torque and maximum power at different operating speeds, depending on the application.

Let us now return to the question of the relative orientation of the rotor and stator fields in an induction machine, which (at least in principle) explains the constraint on the power factor. Recall that, as seen from the stator's reference frame, the rotor magnetic field is spinning at synchronous speed, and the spatial relationship between the two fields therefore remains fixed. Because there is no independent excitation current to produce the rotor field, it can only come from two sources: the stator field, and the relative movement or slip between rotor and stator. As in Section 4.3.2, we can decompose the magnetic fields into two vector components. Again, we would decompose the rotor field into a component parallel to the stator field and one perpendicular to it. Recall also that when the rotor and stator fields are parallel, there is no torque on the rotor. This is the situation in the no-load condition with zero slip. We can then think of this parallel rotor field as the one created by the stator field.

When slip is introduced, the relative motion between the rotor and stator creates an additional component in the rotor field. This is the perpendicular component, which is associated with torque. When the machine is operated as a motor, with a mechanical force holding the rotor back, this rotor field component lags 90° behind the parallel or stator field.[22] When the machine is operated as a generator,

---

[22]This is because the rotor conductors have only inductive reactance and no appreciable resistance, and the rotor current therefore lags by 90° the emf induced in the rotor by the stator field as it slips.

with a mechanical force accelerating the rotor beyond synchronous speed, the rotor field component instead *leads* the stator field by 90°.

Because there is no way to adjust the rotor field by external means, and the only way to create the perpendicular component is through the relative motion or slip between rotor and stator field, the induction generator can only generate at leading power factor. This means that it "consumes" reactive power in the same way an inductive load does.[23] Consequently, in a power system, other generators or capacitors installed close to the induction machines must compensate for the difference and supply an appropriate amount of VARs to meet the requirements of both the induction generators and the load. By itself, an induction generator could only supply a load with capacitive reactance. This is not a relevant scenario, however, because without another a.c. source around, the induction generator could not get started in the first place.

4.6 INVERTERS

An *inverter* is a device that changes d.c. into a.c., meaning that it has to change the current's direction, or *invert* it, many times per second to produce the desired a.c. frequency. Inverters are used wherever there is a d.c. electric source, such as a battery, photovoltaic module or fuel cell, but a.c. power is needed to either serve a specific appliance or inject power into the grid. Because d.c. sources are often at a low voltage, inverters also include a transformer component to produce the desired a.c. output voltage (usually 120 V). Common applications include recreational vehicles, stand-alone systems for remote home power, and, increasingly, distributed generation within the electric grid (see Section 9.2.2). Inverters are also used as part of the process to convert a.c. frequencies, for example, the *wild a.c.* output from a variable-speed wind turbine to a steady 60 Hz.

Inverter technology has undergone significant advances over the past years, prompted by the increasing use of inverters in grid-connected applications where power quality is of significant technical and institutional concern. A wide range of models are now on the market that employ different methods of inversion and power conditioning. The key criterion for inverter performance is the *waveform*, or how closely the a.c. output resembles a mathematically ideal sinusoidal wave; this is quantified in terms of *harmonic content* or *total harmonic distortion* (THD) (see Section 8.4.3). Other differences among inverter designs involve size,

---

[23] The nomenclature can be confusing, because the terms "lagging" and "leading" are in a sense opposite for loads and generators. A generator said to operate at a lagging power factor supplies a load with lagging p.f., even though the generator is behaving like a leading load. Physically, it means that the relative timing of current and voltage are complementary in the generator and the load, so that the generator absorbs reactive power in the same instant as the load releases it. As for the labeling, the generator voltage is measured in the opposite direction of that in the load because power is injected rather than dissipated; if the instantaneous voltage and power at the load are "positive," those at the generator are "negative." But flipping a sine wave over by making it negative is equivalent to shifting it by 180°, which makes "lagging" as a generator equivalent to "leading" as a load.

124 GENERATORS

reliability, tolerance with respect to varying input, and efficiency, which ranges from below 60% to around 96% on the better units.

The oldest design is the *rotary inverter*, which essentially consists of a d.c. motor powering an a.c. generator. As the reader can imagine, this type of machine is large, heavy, and requires maintenance; it is also inefficient. Furthermore, it has poor frequency regulation because a d.c. motor's rotational speed is sensitive to the input voltage.

The advent of solid-state technology made it possible to invert current with rapidly switching electronic circuits instead of moving mechanical parts. The

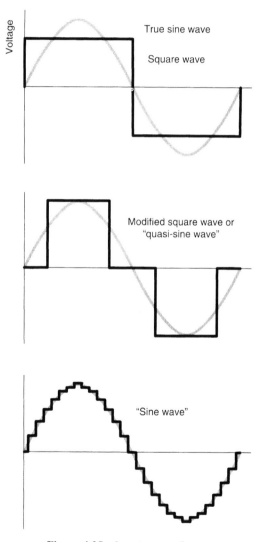

Figure 4.25 Inverter waveforms.

earliest such inverters, which appeared commercially in the 1970s, simply reversed the direction of voltage and current 120 times per second, creating a *square wave* of 60-Hz frequency whose magnitude could then be easily stepped up with a transformer. While this is a simple and efficient design, the square wave tends to create problems for motors because of the extra power contained in the "corners" of the wave, which is not physically usable for creating torque, and instead acts to overheat motor windings. In addition, the square waveform has a very high harmonic content that causes buzzing in electronic and especially audio equipment.

The *modified sine wave*, sometimes called a *quasi-sine wave*, represents an improvement over the square wave (see Figure 4.25). It can be produced in essentially the same way, using a switch and a transformer, except that the voltage and current are kept at zero for a brief moment between every reversal. Another technique is to first increase the d.c. voltage with a step-up converter (a device that creates a high voltage by assaulting an inductor coil with extremely fast-changing current obtained from rapid switching) and then chopping the output; this makes for reduced weight and increased efficiency. In either case, the width of the nonzero voltage pulse is made such that the area under the curve, which is proportional to the amount of power transmitted, is the same as it would be for a sinusoidal wave of equal amplitude, thus making the modified sine wave more compatible with motor loads. The harmonic content is still high, but not as high as for a square wave.

With more elaborate switching schemes, the modification can be carried further and the cycle divided into smaller segments, creating a stepped voltage function that now resembles a sine wave much more closely. One approach is called *pulse-width modulation* (PWM) and involves switching the voltage on and off very rapidly (at up to tens of kilohertz) but in intervals that get longer and shorter throughout the a.c. cycle, so as to create voltage and current pulses of varying width (Figure 4.26). From the standpoint of power transfer, the changing duration of the pulse (known as the *duty cycle*) has the same effect as a varying voltage magnitude, because we can think of the voltage and current as being averaged over a small portion of the a.c. cycle. Thus, during the portion of the cycle where the pulses are wide, the circuit effectively sees the full voltage amplitude; when the pulses are narrow,

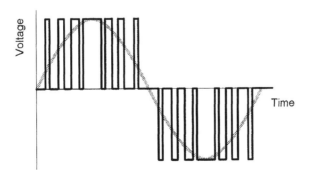

Figure 4.26 Pulse-width modulation

they appear to the circuit as a diminished voltage. In that sense, PWM effectively represents a signal of changing magnitude.

Another approach to creating a sine wave by rapid switching involves adding a set of transformers with different turns ratios whose outputs can be combined to yield different voltage magnitudes. For example, there are 27 possible combinations for the sum of three voltage terms that can each be either positive, negative, or zero. By comparison, modulating the pulse width can produce the equivalent of several hundred steps per cycle, which is therefore smoother and offers the lowest harmonic content. The transformer approach also makes for a heavier unit (due to the weight of the transformer cores) but yields the highest efficiency.

Either device is referred to as a "sine wave inverter" (without modifying adjectives) because the output is for most practical purposes indistinguishable from an ideal sine wave—indeed, it could well resemble a sine wave more closely than the existing utility waveform. Note that the illustrations in Figure 4.25 and 4.26 are simplified for clarity and show fewer steps than an actual inverter unit might produce. The output can also be further "cleaned up" after the switching process by filtering the harmonics, where a simple transformer winding can serve as a low-pass filter (see Section 3.2.1) that converts the high-frequency components of the wave into heat.

In addition, state-of-the-art inverters can perform a variety of functions to assure power quality, including voltage and power factor corrections, while permitting the voltage magnitude of the d.c. source to vary.[24] Finally, like any generation equipment, inverters must contain appropriate circuit breakers to protect both the equipment from overload and the grid from being accidentally energized during an outage.

---

[24]This is particularly useful in conjunction with photovoltaic cells, whose optimal operating voltage varies with sunshine and temperature conditions.

CHAPTER 5
Loads

In the context of electric circuits, the term *load* refers to any device in which power is being dissipated (i.e., consumed). From the circuit perspective, a load is defined by its *impedance*, which comprises a resistance and a reactance. The impedance of an individual device may be fixed, as in the case of a simple light bulb, or it may vary, for example, if an appliance has several operating settings.

In the larger context of power systems, loads are usually modeled in an aggregated way: rather than considering an individual appliance, "load" may refer to an entire household, a city block, or all the customers within a certain region. In the language of electric utilities, the term load therefore has attributes beyond impedance that relate to aggregate behavior, such as the timing of demand.

From a physical perspective, we would think of loads in terms of the electrical characteristics of individual devices. If we consider a load as being defined by its impedance, there are theoretically three types of loads: purely resistive loads, inductive loads, and capacitive loads. *Resistive loads* are those consisting basically of a heated conductor, whether a heating element in a toaster oven or a glowing filament in a light bulb. *Inductive loads* are the most common and include all types of motors, fluorescent lights, and transformers like those used in power supplies for lower-voltage appliances—basically, anything with a coil in it. *Capacitors*, by contrast, do not lend themselves for doing mechanical or other practical work outside electrical circuitry. We know of "capacitive loads" as standard components of electronic circuits, but not on the macroscale among utility customers. While capacitance occurs in small amounts deep within many appliances, it does not dominate the overall electrical appearance of these appliances to the power system.

Loads also differ in the type of electric power they can use. Most motors are designed for alternating current (a.c.) at a specific frequency and voltage, although there are also direct current (d.c.) motors. Fluorescent lamps require a.c. because the radiation discharge depends on the continuing reversal of the electric field across the tube. Pure resistors are the most forgiving loads, being tolerant of low voltage and completely indifferent to the direction of current flow; any resistor will operate interchangeably on a.c. or d.c. Electronic devices operate on d.c. internally, but are manufactured to interface with the standard a.c. grid power. Conversely, many

---

Electric Power Systems: A Conceptual Introduction, by Alexandra von Meier
Copyright © 2006 John Wiley & Sons, Inc.

TABLE 5.1 Types of Loads

Pure resistive loads	Incandescent lamps
	Heaters (range, toaster, iron, radiant heater)
Motors	Compressors (air conditioner, refrigerator)
	Pumps (well, pool)
	Fans
	Household appliances (washer, mixer, vacuum cleaner)
	Power tools
	Large commercial three-phase motors (grocery store chiller)
Electronics	Power supplies for computers and other electronics
	Transformers (adapters, battery chargers)
	Microwave ovens
	Fluorescent lamps (ballast)

common appliances are sold in d.c. versions for use in camper vehicles or remote homes. Table 5.1 shows some examples of different types of loads. In the following sections, we will discuss the characteristics of the most common loads: resistive loads, motors, and electronic equipment. Section 5.4 examines aggregated loads from the utility or system perspective, and Section 5.5 deals with the connection between loads and the power distribution system.

5.1 RESISTIVE LOADS

The simplest type of load is a purely resistive load, that is, one without capacitive or inductive reactance. The power factor of such a load is 1.0 (see Section 3.3). Many familiar appliances fall into this category: incandescent light bulbs as well as all kinds of resistive heaters, from toasters to electric blankets, space heaters, and electric ranges. In each case, the heating element consists simply of a conductor that dissipates power according to the relationship $P = I^2 R$, which can be rewritten as $P = IV$ or $P = V^2/R$ by substituting Ohm's law.

As can be recognized from $P = V^2/R$, the amount of power dissipated increases with decreasing resistance, given that a constant voltage is supplied. Therefore the strongest heaters are those with the lowest resistance. Although it is the resistance that is the fixed physical property of a given device, the resistance in ohms is not usually stated on the package or explicitly referred to in the rating of the device, as most consumers would probably have no idea what to do with that information. What people care about is how much power, in watts, a device consumes. To determine a power rating, it is necessary to assume a voltage at which this device is to be operated. This assumption, usually 115 or 120 volts, is generally stated on the label along with the power rating. Note that the same light bulb connected to a different voltage supply will deliver a different number of watts.

In general, resistive loads are the simplest to operate and the most tolerant of variations in power quality, meaning variations in the voltage level above or below the nominal 120 V, or departures of a.c. frequency from the nominal 50 or 60 Hz. A resistive heating element can be damaged by excessively high voltage, which will cause it to overheat or wear out prematurely. The filaments of incandescent light bulbs, for example, will burn out sooner if they are operated at a higher voltage. Low voltage, on the other hand, causes no physical damage whatsoever in a resistive device, though the heat or light output will be reduced. Resistive loads are essentially indifferent to the frequency of a.c., or whether it alternates at all. For example, the performance of an incandescent lamp at 120-V a.c. and 120-V d.c. would be indistinguishable to the human eye.

As for voltage magnitude, despite the efforts of utilities to maintain it at a constant level, the actual voltage as seen by an appliance plugged into an outlet varies from one time and location to another. The tolerance for voltage supplied by U.S. utilities is typically $\pm 5\%$, which corresponds to an actual range of 114–126 V based on a nominal 120 V for residential customers. What happens to loads and their power consumption when voltage changes? Unlike motor loads, resistive loads are easily predictable in this regard. It follows from the equation $P = V^2/R$ that if the voltage is increased, the power drawn by resistive loads increases as the square of the voltage.

Example

Compare the power consumption of a 100-W light bulb at 114 versus 126 V.

We may assume that the bulb was rated for a nominal voltage of 120 V; thus, its resistance is $(120 \text{ V})^2/100 \text{ W} = 144 \text{ }\Omega$. At 114 V, this bulb draws a power of $(114)^2/144 \text{ }\Omega = 90.25 \text{ W}$. At 126 V, it draws $(126 \text{ V})^2/144 \text{ }\Omega = 110.25 \text{ W}$.

Another way to approach this calculation is to say that if the voltage is increased or decreased by 5%, it is changed by a factor of 1.05 or 0.95, respectively. To find the change in the amount of power, we can simply square these factors, since power is proportional to voltage squared. Thus, the increase in power is 1.1025 and the decrease is 0.9025. The new factors, multiplied by 100 W, give the correct results.

Voltage response is also relevant to dimmer circuits, where the brightness of a light is controlled by way of reducing the effective root-mean-square (rms) voltage across the lamp. A common dimmer switch contains solid-state circuitry that "clips" or selectively turns the voltage on and off for a certain fraction of each cycle, depending on the setting. Because resistive loads are indifferent to a.c. waveform, it does not matter that such a dimmer produces a rather jagged voltage curve; even to the extent that a light bulb filament cools off during the flat part of the cycle, the human eye cannot discern any flicker. What we do see is that the overall brightness is diminished in relation to the average amount of power dissipated by the filament, which in turn is given by the average or rms voltage over the course of each cycle. This average can be visualized in terms of the area left under the voltage curve after it has been clipped.

It is important to note that such dimmer switches work only with incandescent lights and with specially labeled (and more expensive) *dimmable fluorescents*. This is because fluorescent lamps are not resistive loads; rather, their various types of ballasts (the part that interfaces with the power supply) resemble inductors, electronic loads, or some combination thereof. The majority of fluorescent lamps and ballasts on the market today do not take well to having their supply voltage lowered or their waveform brutalized in the preceding manner and are therefore *not* dimmable. In general, reducing the voltage for appliances other than simple resistors may damage them, or they may not work at all. Dimmer switches should therefore only be installed in appropriate circuits.

The reader may wonder why it should be necessary to go through all the trouble of electronic clipping. Could not the voltage be reduced simply by inserting another resistance in series with the light bulb? A variable resistance used for this purpose is called a *rheostat*, and is mostly found on circuits with small amounts of power. The problem with the rheostat is its own power dissipation—that is, heat—which implies not only waste but, in the case of power circuits, a risk of meltdown and fire.

As illustrated by the following example, the rheostat's series resistance is added to the resistance already present in the light bulb, which reduces the total current (because it must now flow through both resistances; see Section 2.2.1). Because of the lower current, the light bulb will dissipate less power and appear dimmer. Also, the overall circuit voltage (say, 120 V), which remains unaffected, is now split between the light bulb and the rheostat according to the relative proportion of their resistances. The greater the resistance of the rheostat, the greater the fraction of the overall voltage it will sustain. This effect, too, dims the light bulb, which sees a lower voltage. At the same time, though, and despite the lower current, a significant amount of power is now dissipated by the rheostat, which can easily get too hot in a confined space. By contrast, the waste heat generated by an electronic dimmer is small enough as to be insignificant both in terms of safety and losses when compared to the savings from reduced lighting energy use.

Example

A 100-W incandescent light bulb in a 120-V circuit is dimmed to half its power output using an old-fashioned rheostat, or variable resistor, as a dimmer. What is the value of the resistance in the dimmer at this setting, and how much power is being dissipated by the rheostat itself?

First, let us determine the resistance of the light bulb by considering its normal operating condition, when the power is 100 W at 120 V. Rewriting $P = V^2/R$ as $R = V^2/P$, we obtain $R = (120 \text{ V})^2/100 = 14{,}400/100 = 144 \text{ }\Omega$. We are given the information that with the dimmer in series, the power drops from 100 W to 50 W. In this situation, we do not know the voltage drop across the light, because now $120 \text{ V} = V_{\text{light}} + V_{\text{dimmer}}$ according to Kirchhoff's voltage law. But since we know the light's resistance and its power at the new operating condition, we can determine the current using $P = I^2 R$, which gives $I = \sqrt{50 \text{ W}/144 \text{ }\Omega} = 0.59 \text{ A}$ (for comparison, the current at 100 W was 0.83 A). Knowing the current, we can now infer the total resistance in the circuit including the dimmer by using Ohm's law $V = IR$, or $R = V/I = 120$ V$/0.59$ A $= 204$ Ω. Thus, the rheostat's resistance at this setting is $R_{dim} = 204 - 144 = 60$ Ω. The power dissipated within the rheostat is $P = I^2 R = (0.59$ A$)^2 \cdot 60$ $\Omega = 21$ W. The light–dimmer combination consumes $50 + 21 = 71$ W. This is less than the 100-W light by itself, but considerably more than the 50 watts we might have expected if we failed to consider what happens inside the dimmer. Note that continually dissipating 21 W inside an electrical box in the wall would easily pose a fire hazard—do not try this at home!

When $R_{dim} \approx 0$, the dimmer has no effect, acting simply as a piece of conducting wire that, ideally, dissipates no power at all. As R_{dim} gets very large, the current becomes very small until the light goes out. The fraction of power dissipated within the dimmer switch increases with higher resistance setting, although the total power (switch plus light) becomes less.[1]

5.2 MOTORS

Electric motors represent an important fraction of residential, commercial, and industrial loads; in the neighborhood of 60% of the electric energy in the United States is consumed by motors of some kind. Motor loads comprise fans, pumps of all kinds including refrigerators and air conditioners, power tools from hand drill to lawn mower, and even electric streetcars—basically, anything electric that moves.

A motor is essentially the same thing as a generator operated backwards; electrical and mechanical energy are converted into one another by means of a magnetic field that interacts with both the rotating part of the machine and the electrons inside the conductor windings. The mechanical power output of a motor is conventionally expressed in units of *horsepower* (hp), where 1 hp $= 0.746$ kW, to distinguish it from the electrical power expressed in kilowatts. The essential physical properties of motors are analogous to those of generators discussed in Chapter 4.

Aside from differences in size and power, there are three distinct types of motors that correspond to the three main types of generators: induction, synchronous, and d.c. In each case, the motor is similar to its generator counterpart. Induction motors are the least expensive and by far the most common; they account for roughly two-thirds of motors in use with about 90% of the motor energy consumed in the United States.[2]

---

[1] The reader may entertain himself by finding the setting, expressed as a fraction of the light bulb's resistance, at which the rheostat itself dissipates the most heat (a first-year calculus problem whose solution might be guessed intuitively).

[2] Steven Nadel, R. Neal Elliott, Michael Shepard, Steve Greenberg, Gail Katz, and Anibal T. de Almeida, *Energy-Efficient Motor Systems: A Handbook on Technology, Program, and Policy Opportunities*, Second Edition (Washington, DC: American Council for an Energy-Efficient Economy, 2002).

Like the induction generator, the induction motor's rotational speed varies with the *torque*[3] applied to the rotor, and thus the amount of power transferred. In order to produce a torque by magnetic force, the induction motor requires *slip*, or a difference between the motor's mechanical rotation speed and the a.c. frequency. The rotor magnetic field comes from an induced current in the rotor, which has no independent electrical source, but receives electromagnetic induction from the stator windings (supplied by the a.c. source) as a result of the motor's internal geometry.[4] When an induction motor is started from rest, the rotor does not yet have any current circulating inside it, and thus no magnetic field; the a.c. current supplied to the stator windings thus encounters very little impedance for the first fraction of a second. The resulting phenomenon characteristic of induction motors is called the *inrush current*. The inrush current accounts for the familiar flicker of lights when a heavy motor load is starting up, as the local line voltage is momentarily reduced by the voltage drop associated with the high current flow.

Even after the rotor magnetic field has been established, any motor consumes additional power and current as it mechanically accelerates to its operating speed; this is the *starting current*. A typical motor's starting current may be five to seven times greater than the current under full load. Larger and more sophisticated commercial motor systems include starting controls designed to soften the impact of motor loads on the local electrical system while gradually ramping up their rotational speed.

Because of the geometric relationship between the rotor and stator and the need to induce a rotor current, an induction machine always consumes reactive power (VAR; see Section 3.3). This is true regardless of whether the machine is operating as a motor or generator (i.e., consuming or generating real power). Therefore, induction motors all have power factors less than unity; depending on the motor design and rating, typical power factors range from below 60 to the low 90s of percent. Induction motors are chiefly responsible for lowering a system's overall power factor, which is a combination of all connected loads. Fluorescent lamp ballasts, the other important reactive loads, make a much smaller contribution.

Synchronous motors, by contrast, have an independent source of magnetization for their rotor, which may be either a permanent magnet or an electromagnet created by an external current (*excitation*). This independent magnetization allows the machine to operate at synchronous a.c. speed—some fraction of 3600 rpm for 60 Hz, depending on the number of magnetic poles—regardless of load. Like a synchronous generator, it is possible to operate a synchronous motor at different

---

[3] "Torque" implies the application of force to rotate something. Technically, it is the product of force (in units of pounds or newtons) and the distance from the rotational axis (in feet or meters) where the force is applied; thus, the units of *foot-pounds* is used for torque. As we know from direct experience, it is easier (requiring less force) to turn something (apply a certain torque) by pushing at a point farther away from the center of rotation—say, turning a nut with a longer wrench, or using a screwdriver with a thicker handle. Oddly, in terms of physical dimension, units of torque are the same as units of energy. This makes sense when we note that to produce units of power, we multiply torque by the rotational or angular frequency, which has units of inverse time. Thus, power equals torque times revolutions per minute (rpm).

[4] The rotor is the rotating part and the stator the stationary part of the machine; see Section 4.2.1.

power factors. Synchronous motors are more complicated and expensive than induction motors of comparable size. They are characteristically used in industrial applications, especially those requiring high horsepower and constant speed.

D.c. motors also have independent magnetization. Most important, they require *commutation* of the d.c. in order to produce the correct torque. Commutator rings or brushes involve moving electrical contacts that are inherently prone to mechanical wear; d.c. machines therefore tend to require more maintenance than a.c. machines. D.c. motors are distinguished by a high starting torque, which is useful in applications such as accelerating vehicles from rest. They also afford convenient speed control, since the rotational speed varies directly with voltage, such as in a model electric train, where the locomotive speed is controlled quite precisely with a simple rheostat dial that varies the voltage between the tracks. While many smaller d.c. motors are used in off-grid applications, the most important d.c. motor loads in power systems throughout the last century have been electric trains and streetcars. Other types of motors—for example, the switched reluctance motor, which is a variation on the d.c. motor with electronic commutation—are under development, but do not yet account for a significant market share.

Besides motor type, another important distinction is between single- and three-phase motors. Like a generator, a motor benefits from the constant torque afforded by three separate windings, staggered in space and time, that in combination produce a rotating magnetic field of constant strength (this applies only to a.c., not d.c. machines). Three-phase motors therefore operate more smoothly and much more efficiently than single-phase motors, though they are also more expensive. They are commonly used for large industrial and commercial applications where high performance, including high horsepower output and high efficiency, is essential, and where three-phase utility service is standard. Most smaller commercial and almost all residential customers receive only single-phase service because of the significant cost difference in distribution.[5] Accordingly, there is not a significant market for smaller three-phase motors to date. It is technically possible, however, to operate a three-phase motor on single-phase service by inserting an electronic phase-shifting device that effectively splits the voltage and current along several circuits and changes their relative timing to produce three staggered sine waves.[6]

Because motors account for such a large portion of our society's energy consumption, they represent a significant opportunity for efficiency improvements. Typical energy conversion efficiencies range from an average of 65% for small (fractional horsepower) to over 95% for very large (over 500 hp) motors. The operating efficiency depends both on the motor design and the operating condition, with most motors running most efficiently near their rated capacity (the load for which they were presumably designed). A variety of approaches exist to increase motor efficiency, ranging from specific motor designs to systemic aspects such as speed controls and motor sizing in relation to the load. Short of replacing a motor with a more

---

[5] The standard combined 120–240-V service, though it affords different options for plugging in appliances, is still a single phase; see Section 5.5.
[6] For example, the PhaseAble device (Otto J. Smith, personal communication; www.phaseable.com).

efficient model, some of the standard options include rewinding existing motors with conductors of lower resistance and adding adjustable or variable speed drives (known as ASD or VSD) to allow for energy savings when less power is required.[7]

Today's standard ASDs control speed by rectifying the a.c. supply to d.c. and inverting it again to a.c. of variable frequency. The associated conversion losses are easily outweighed by the significant savings in mechanical energy. When a fan or pump motor moves a fluid (air or water),[8] the volume moved and the kinetic energy imparted to the fluid both increase with motor speed (the former linearly and the latter with the square of velocity), making the power theoretically proportional to the cube of rotational speed. Friction may have an additional impact. A doubling of motor speed thus implies roughly an eightfold increase in power, or, in practical terms, a speed reduction of 20% can yield 50% energy savings.

It is interesting to note that motors can have a very long life, on the order of 100,000 operating hours. Depending on how heavily a motor is used, the lifetime cost of supplying its energy can be much greater than the motor's initial capital cost; in fact, it is not unusual for a commercial motor's annual electric bill to exceed its purchase price by an order of magnitude. Therefore, the additional cost of a more efficient motor system may be recovered in a reasonable time even if the percentage efficiency gain appears small.

Unlike resistive loads, electric motors are sensitive to power quality, including voltage, frequency, harmonic content and, in the case of three-phase machines, phase imbalance. One of the key problems that tend to afflict motors is unequal and excessive heating of the windings, which leads to energy and performance losses, degradation of the insulating material, and possibly short-circuiting. Such heating can be caused by voltage levels that are either too high, too low, or too uneven across phases. Voltage also affects the power factor at which a motor operates. Excess heating and energy losses can further result from harmonic content in the a.c. sine wave, in addition to undesirable vibration. Finally, transient disturbances (brief spikes or notches) in voltage may impact motor controls, the commutation mechanism, or protective circuit breakers. Because of these sensitivities, it is not uncommon for owners of expensive and sophisticated motor systems to install their own protective or conditioning equipment, so as to guarantee power quality beyond the standard provided by the local utility.

5.3 ELECTRONIC DEVICES

Consumer electronics—basically, anything that has little buttons on it—are powered by low-voltage d.c. They may be operated either by batteries or through a *power*

---

[7]Nadel et al. (*Energy-Efficient Motor Systems*) offer an excellent review of strategies for increasing motor efficiency.

[8]This most common type of load is called a *variable torque load*, as opposed to *constant torque loads* (where torque and speed are independent) and *constant power loads* (where torque and speed are inversely related). The categorization alone illustrates some of the subtlety and complexity of electric motor systems.

supply that delivers lower-voltage d.c. by way of a step-down transformer and rectifier. Power supplies may be external and plug directly into the outlet in the form of the familiar adapter (for example, with a telephone answering machine), or they may be hidden inside the larger appliance (for example, a computer). While individual electronic loads tend not to be large power consumers, their proliferation in number is turning them into a significant load category.

Unlike appliances designed to do real "work" in the physical sense, such as heating or moving physical objects, the intended function of electronics is to move information, encoded by way of many tiny circuits switched on or off in some pattern. Moving information does not in and of itself require physical work, except for the inadvertent heating of circuit elements when current flows through them. The power consumption of any electronic appliance, from answering machine to calculator, can be gauged by how warm it gets during operation. This type of heat is very much like the waste heat resulting from mechanical friction, which can be reduced by clever design but never completely eliminated.

Since the main job of an electronic circuit component is not to deliver power but to relay the information whether a particular circuit is "on" or "off," this job can be performed with a very small current. Indeed, an important aspect of the continuing evolution of electronics, from vacuum tubes to transistors to integrated circuit chips, has been to reduce the operating currents to ever smaller amounts while fitting them into ever smaller spaces without overheating.

"Pure" electronic devices, such as pocket calculators or digital wristwatches, use very little energy, as evidenced by the fact that they can operate for a long time on a small battery—if not on a few square centimeters of solar cells in indoor light. Then there are those appliances whose external power supplies supply low-voltage d.c. through a rather skinny electrical cord, which tells us that the power consumption had better be small. Many of these devices require a modicum of mechanical work, such as spinning a compact disk or moving a loudspeaker membrane, in addition to their "brains."

Even when electronic appliances require little physical work to operate, they also contribute to energy use by way of *standby power*, which is consumed whenever the machine is plugged in and ready to respond. For example, a television set that can be switched on by remote control needs to keep some internal circuits activated to enable it to recognize the infrared signal from the remote. This standby power, which is typically on the order of several watts (comparable to a night light), can sometimes be recognized as warmth on the back of the appliance. Similarly, power supplies that remain plugged into an outlet continually dissipate heat from their transformer coils. Though it does not sound like much, a constant drain of 4 watts, for example, adds up to 35 kWh of waste heat over the course of a year. For a household full of electronic gadgets, an annual consumption of hundreds of kilowatt-hours for standby appliances is not unusual. A simple power strip with a switch allows energy-conscious consumers to avoid this phenomenon without too much hassle.

Many of today's conventional electronic loads are in fact combinations of electronic circuits and something that does relevant physical work. A microwave oven, for example, may have an impressive array of buttons, beeps, and displays, but its real job is to deliver hundreds of watts in the form of electromagnetic radiation (which it does by running a current through coils at high frequency), to be picked up by resonating water molecules in our food. Television and computer monitors continually shoot electrons at a phosphorescent screen, which produces heat on the order of hundreds of watts. Liquid crystal display (LCD) screens use considerably less energy. Most computers need a fan for cooling their central processing unit (thus the sound of the computer being "on"), and a laser printer uses the bulk of its electrical energy to heat the drum. Again, it is the amount of heating that gives us a direct measure of the electric power consumption involved, as all of the energy entering an appliance from the outlet must eventually exit and be dissipated as heat.[9]

To the power circuit, mixed electronic and power appliances appear as resistive or inductive loads, depending on the feature that dominates energy consumption. From the power system perspective, they differ from the plain variety of resistive and inductive loads mainly in their sensitivity to power quality. Arguably, one of the major cultural impacts of LED clocks has been to vastly increase consumer awareness of momentary power fluctuations in the grid, with the blinking 12:00 on the video recorder display an indelible icon of late 20th-century technological society.

5.4 LOAD FROM THE SYSTEM PERSPECTIVE

From the point of view of the power grid, individual customers and their appliances are small, numerous, and hardly discernible as distinct loads. This section thus deals with *aggregate* load, that is, the combined effect of many customers both in terms of the magnitude and timing of electric demand.

While consumers typically think of their electricity usage in terms of a quantity of *energy* (in kilowatt-hours) consumed over the course of a billing period, the quantity of interest to system operators and planners is the *power* (in kilowatts or megawatts, measuring the instantaneous rate of energy flow) demanded at any given time. The term *demand* thus refers to a physical quantity of power, not energy. Serving that instantaneous demand under diverse circumstances is the central challenge in designing and operating power systems, and the one that calls for the majority of investment and effort.

It is interesting to reflect upon the historical service philosophy that considers demand as the independent variable that is to be met by supply at any costs. The assumption embedded in both the hardware design and the operating culture of electric power systems is that customers freely determine how much power they want,

---

[9]Especially in commercial environments, this waste heat is often masked by aggressive air conditioning, not only for human comfort but to protect the machines themselves from overheating. In this sense, heat-producing electronics and computing equipment, in particular, represent a double energy load, and efficiency improvements to such equipment—including "energy saver" features for turning devices off when not in use—afford the additional benefit of reduced air-conditioning needs.

and that it is the job of power system designers and operators to bend over backwards if necessary to accommodate this demand. In the same vein, power engineering texts refer to "load" as an externally given quantity, a variable beyond control, in a completely unselfconscious manner. This assumption is codified in the social contract between utilities as regulated monopolies—or even just as transmission and distribution service companies—and the public, which insists upon the utilities' fundamental obligation to serve the load.

More recently, though, with power system economics being scrutinized and reconsidered from every angle, the philosophy by which demand rules the game is being challenged in some respects. We are now beginning to expect customers to vary their demand according to electricity prices (in turn driven by supply), and research and development efforts are directed toward technological approaches to make demand more responsive, including scenarios with remote-controlled and automated devices. The profound nature of this conceptual shift cannot be overstated; it is probably the most significant change to the mission of electric power systems since their inception. How the transition from a service-driven to a market-driven system will eventually play with consumers remains to be seen. For the time being, though, the modus operandi of power systems both in terms of day-to-day operating decisions and hardware investments continues to be based on demand as the key independent variable.

Consequently, an entire discipline is dedicated to predicting what that demand might be at a given future time. Drawing on detailed statistics of past demand behavior, meteorology, and any other factors (from school holidays to new technologies) that might impact electricity use, *load forecasting* is both a science and an art.

5.4.1 Coincident and Noncoincident Demand

In the utility context, analysts distinguish *coincident* and *noncoincident* demand. Coincident demand refers to the amount of combined power demand that could normally be expected from a given set of customers, say, a residential block on one distribution feeder. By contrast, the noncoincident demand is the total power that would be drawn by these customers if all their appliances were operating at the same time. It is called noncoincident because all these demands do not usually coincide. Coincident demand reflects the statistical expectation regarding how much of these individual demands will actually overlap at any one time.

For example, suppose each of 10 residences had a 600-W refrigerator. The noncoincident demand associated with these refrigerators would be 6000 W. Under ordinary circumstances, however, the compressor in each refrigerator goes on and off on a duty cycle, and so is operating only part of the time, let us say 20%. The duty cycles of these 10 refrigerators will usually be at random in relation to each other, so we could expect that at any given time, only one in five is operating. The coincident demand in this case would be 20% of 6000 W, or 1200 W. Clearly, this sort of statistical prediction becomes more reliable when greater numbers of customers are involved.

Although ordinarily the utility observes only the coincident demand, it must be prepared to face noncoincident demand under certain circumstances. Suppose, for example, that there is an outage that lasts for a sufficient time period— an hour or so—to let all the refrigerator compartments warm up above their thermostat settings. Now power is restored. What happens? All 10 compressors will kick in simultaneously, and the 6000-W noncoincident demand suddenly coincides!

What makes this particular scenario most troublesome, actually, is not just the simultaneous operation of normal loads; rather, it is the split-second inrush current of electric motors as they turn on and establish their internal magnetic field. The sum of these inrush currents from refrigeration and air-conditioning units can overload distribution transformers and even cause them to explode the moment that power is restored after an outage. For this reason utilities often request their customers to switch most appliances off during an outage until they know the service is back.

5.4.2 Load Profiles and Load Duration Curve

Instantaneous demand, as it varies over the course of a day, is represented in a *load profile*. A load profile may be drawn at any level of aggregation: for an individual electricity user, a distribution feeder, or an entire grid. It may represent an actual day, or a statistical average over typical days in a given month or season. The maximum demand, which tends to be of greatest interest to the service provider, is termed the *peak load*, *peak demand*, or simply the *peak*.

From the power system perspective, it is sometimes relevant to compare periods of higher and lower demand over the course of a year. Thus, one might compile the highest demand for each month and plot these 12 points, indicating the seasonal as opposed to the diurnal rhythm. In warmer climates where air conditioning dominates electric usage, demand will tend to be *summer-peaking*; conversely, heating-dominated regions will see *winter-peaking* demand.

A different way to represent a load profile for this purpose is by way of a *load duration curve*. The load duration curve still depicts instantaneous demand at various times (generally in one-hour intervals), except that the hours are sorted not in temporal sequence but according to the demand in each hour. Thus, the highest demand hour of the year appears to be the first hour, followed by the second highest demand hour (which may well have occurred on a different day), and so on. Each of the 8760 hours of the year then appears somewhere on the graph, with the night hours mostly at the low-demand end on the right-hand side.

Figures 5.1 and 5.2 illustrate a daily load profile and a load duration curve for the state of California.[10] The vertical scale is the same in both graphs, but the

---

[10] The daily load profile is published live on the internet by the California Independent System Operator (CAISO), along with the load forecast for the day and available generation resources (not shown here). While this particular day turned out to be slightly cooler than expected, the close proximity of forecast and actual demand offer a glimpse into the sophistication of load forecasting.

5.4 LOAD FROM THE SYSTEM PERSPECTIVE

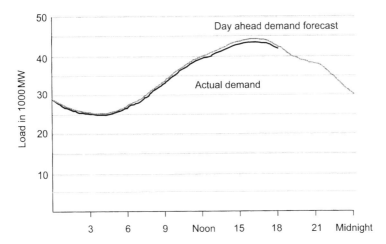

Figure 5.1 Load profile for an August day in California. [From California ISO (www.caiso.com); August 15, 2005, 6 P.M.]

time axis has a different meaning in each. If the data were for the same year, we might cross-reference the graphs as follows: at 5 P.M. on August 5 (a warm but not excruciatingly hot day), the load was about 43,000 MW. By inspection of the load duration curve, we see that this would put 5 P.M. August 5 within about the top 250 or 300 demand hours of the year.

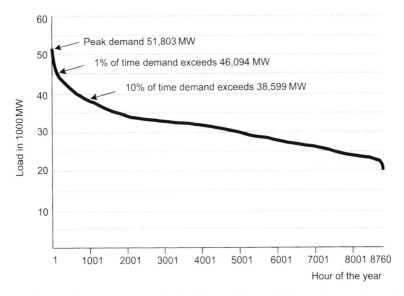

Figure 5.2 Load duration curve for California. [From California Energy Commission (www.energy.ca.gov); California statewide demand, 2002.]

The shape of the load duration curve's peak, which is obvious at a glance, is a useful way to characterize the pattern of demand. Quantitatively, the ratio between peak and average demand is defined as the *load factor*. From the standpoint of economics as well as logistics, a relatively flat load duration curve with a high load factor is clearly desirable for utilities. This is because the cost of providing service consists in large part of investments related to peak capacity, whereas revenues are generally related to total energy consumed (i.e., average demand). A pronounced peak indicates a considerable effort that the service provider must undertake to meet demand on just a few occasions, although the assets required to accomplish this will tend not to be utilized much during the remainder of the year. For example, in the case illustrated in the last section, note that the resources required to meet the top 5000 MW of demand—roughly 10% of the total capacity investment—are called upon for less than 1% of the year, or all of three days.

The load factor obviously depends on climate, but it also depends on the diversity within the customer base, or *load diversity*. For example, commercial loads that operate during the day may be complemented by residential loads before and after work hours. Improvement of the load factor through increased load diversity was a major factor in the historical expansion of power systems, as was the ability to share resources for meeting the peak.

5.5 SINGLE- AND MULTIPHASE CONNECTIONS

Taking full advantage of three-phase transmission, certain loads connect to all three phases. These are almost exclusively large motors, such as those in heavy machinery or commercial ventilation and refrigeration equipment, where efficiency gains from smooth three-phase operation are worth the extra cost. Three-phase motors contain three windings that make up three distinct but balanced circuits. A three-phase machine plugs into its appropriate power outlet with three phase terminals and one neutral to handle any current resulting from phase imbalance. Most utility customers, however, do not have three-phase service. All the familiar loads from residential and small commercial settings represent a single circuit with only two terminals to connect.

The standard, 120-V nominal outlet thus has two terminals (Figure 5.3), a "hot" or *phase* (black wire, small slot) and a *neutral* (white wire, large slot), in addition to a safety *ground* (bare or green wire, round hole). In older buildings, ungrounded outlets that predate electrical-code revisions can still be found.

The phase supplies an alternating voltage with an rms value of 120 V $\pm$ 5% between it and the neutral terminal. The neutral terminal is ostensibly at zero volts, but its voltage will tend to float in the range of a few volts or so, depending on how well the loads in the neighborhood are balanced among the three phases, and on the distance to the point where the neutral terminal is physically grounded.

The differently sized slots in the outlet accommodate the polarized plugs that are standard for lighting appliances. The purpose is to reduce the risk of shock for people changing a light bulb with the circuit energized: the small prong delivers

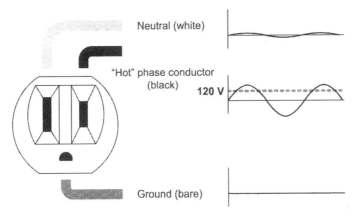

Figure 5.3 Standard electrical outlet.

the phase voltage to the back of the fixture where it is less likely to be touched accidentally.[11]

The ground is not part of the power circuit except during malfunction. It should connect to the earth nearby—typically, through a building's water pipes—and serves to protect against shock and fire hazards from appliances due to faulty wiring, exposure to water, or other untoward events. Removing the ground prong from a plug to fit it into an ungrounded outlet is not just contrary to the advice of appliance manufacturers, but probably a bad idea. (The appropriate way to use an adapter plug made for ungrounded outlets is to fashion an external ground with a wire connecting to the small metal loop on the plug.)

Most utility customers also have wiring for higher voltage appliances, which may provide 240 V or 208 V. This is why the utility service enters the house with three wires, which are not the same as the three phases. Instead, they are one neutral and two phase conductors.

In the 120/240 case, which is standard for residential service in the United States, the two phase conductors tap a distribution transformer at different points. The transformer has the correct turns ratio so that the secondary coil provides 240 V. By tapping the secondary coil at the halfway point, another wire can supply half the voltage, or 120 V. In this case, both the 120 and 240 are coming from the same phase (A, B, or C). Figure 5.4 illustrates the situation. Because the primary and secondary circuits are linked only magnetically, not electrically, and because voltage is inherently a relative, not an absolute quantity, the neutral wire on the secondary side can be connected to any arbitrary part of the winding, forcing that point of the winding to be at or near ground potential while the voltage of other points is simply measured relative to this neutral terminal. For example, if one end of the secondary winding were neutral, the two hot wires would carry 120 V and 240 V, respectively. The typical arrangement, which

---

[11] This paragraph seems to beg for some screw-in-a-lightbulb joke, but I can't quite put my finger on it.

142 LOADS

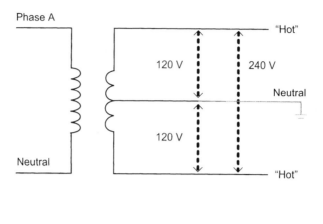

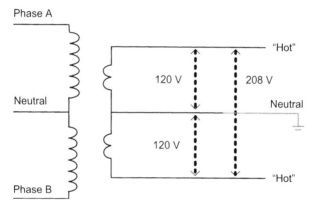

Figure 5.4 Transformer taps and multiphase service.

allows for maximum use of the three wires, is to connect the neutral wire to the center tap, as shown in the diagram. In this case, each hot or phase conductor carries 120 V with respect to the neutral wire, and thus can serve any of the customer's 120-V circuits. Because these 120 volts are measured in opposite directions, however—that is, they are 180° apart in phase—a load connected between these two will see their sum, or 240 V.

In the 120/208 case, two different phase combinations are tapped. The 120 V corresponds to the *phase-to-ground voltage* between one phase (say, A) and the neutral terminal. The 208 V corresponds to the *phase-to-phase voltage* between two different phases (say, A and B). Mathematically, this phase-to-phase voltage corresponds to the difference between two sine curves of equal magnitude, shifted by 120 degrees. As may be obvious only to connoisseurs of trigonometry, this difference between two sine curves is itself a sine curve, but with a magnitude that exceeds the phase-to-ground voltage by a factor of $\sqrt{3}$, or about 1.732. (Note that $208 \approx 120\sqrt{3}$.)

An arrangement where three loads are each connected between one phase and ground is also called a *wye connection*, because the schematic diagram for all three phases resembles the letter Y. By contrast, an arrangement where three loads are each connected between one pair of phases is called a *delta connection*, as in the Greek letter Δ. The term "load" here is usually understood in the aggregate sense, as in a transformer. Delta and wye connections are discussed further in Section 6.2.3.

CHAPTER 6
Transmission and Distribution

6.1 SYSTEM STRUCTURE

6.1.1 Historical Notes

Since the beginnings of commercial electric power in the 1880s, the systems for its delivery from production sites to end users have become increasingly large and interconnected. In the early days, the standard "power system" consisted of an individual generator connected to an appropriately matched load, such as Edison's famous Pearl Street Sation in New York city that served a number of factories, residences, and street lighting. The trend since the early 1900s has been to interconnect these isolated systems with each other, in addition to expanding them geographically to capture an increasing number of customers. Owing to a considerable investment in Public Works projects for rural electrification, most U.S. citizens had electricity by World War II. The process of interconnecting regional systems into an expansive synchronous grid has continued throughout the postwar era, leaving us today with only three electrically separate alternating current (a.c.) systems in the United States: the Western United States, Eastern United States, and Texas.[1] Similarly, the Western European system is completely interconnected and synchronous from Portugal to Denmark, Austria, and Italy.

The continuing geographical expansion and interconnection of power systems over the course of the last century has been motivated by a variety of technical, social, and economic factors. For example, early drivers included a sense of cultural progress associated with a connected grid; in recent years, economies of exchange, or opportunities for sales of electricity, have been a key motivator for strengthening transmission interconnections or *interties* between regions. The main technical justifications for expansion and interconnection are threefold: *economies of scale*, improvement of the *load factor*, and enhancement of reliability by pooling generating *reserves*.

---

[1]The state of Texas opted to forgo connectedness in the interest of avoiding interstate commerce regulations and federal oversight.

Electric Power Systems: A Conceptual Introduction, by Alexandra von Meier
Copyright © 2006 John Wiley & Sons, Inc.

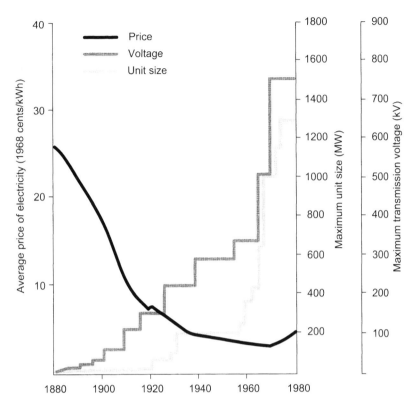

Figure 6.1 Historical growth of generation unit size and transmission voltage. (Adapted from Economic Regulatory Administration, 1981.)

An economy of scale simply means that it tends to be less expensive to build and operate one large generator than several smaller ones. This makes sense because much of the construction process of conventional power plants—from design and licensing to pouring concrete and bringing in the crane that lifts the generator—tends to involve fixed rather than variable costs, not depending very much on the unit's megawatt capacity. Even operating costs, though dominated by fuel, have economies of scale in aspects such as labor, maintenance, and operational support. Figure 6.1 shows the historical increase in the size of generation units as a result. The graph also reflects the limit that appears to have been reached in the late 1960s in terms of the maximum practicable and efficient unit size, marking the end of an era of declining costs.[2] In the early days of power systems, however, the advantage of increased unit size was substantial and provided an

---

[2]This point is carefully documented by Richard Hirsh in *Technology and Transformation in the American Electric Utility Industry* (Cambridge, England: Cambridge University Press, 1989). Hirsh argues that unexpectedly diminishing economies of scale played an important role in precipitating the economic crisis that U.S. utilities experienced in the 1970s, independent of the widely recognized factors of increasing fuel costs and unprofitable investments in nuclear power.

important incentive for utilities to connect enough customers so as to take full advantage of economies of scale.

The load factor relates to the ratio of a load's actual energy consumption over a period of time to the maximum amount of power it demands at any one instant. This is a key criterion for the economic viability of providing electric service, since the cost of building the supply infrastructure is related to the maximum amount of power (i.e., the capacity of generators and transmission lines), whereas the revenues from electricity sales are related to the amount of energy (kilowatt-hours) consumed. Thus, from the supply standpoint, the ideal customer would be demanding a constant amount of power 24 hours a day. Of course, this does not match the actual usage profile of real customers; nevertheless, a smoother consumption profile can be accomplished by *aggregating* loads, that is, combining a larger number and different types of customers within the same supply system whose times of power demand do not coincide.

For example, an individual refrigerator cycles on and off, using a certain amount of power during the time interval when it is on, and none the rest of the time. But if a number of refrigerators are considered together, their cycles will not all coincide; rather, they will tend to be randomly distributed over time. The larger the number of individual loads thus combined, the greater the driving force of statistics to level out the sum of power demand. By expanding their customer base to include both a larger number of customers and customers with different types of needs (such as machinery that operates during business hours versus lighting that is needed at night) so as to deliberately combine complementary loads, utilities improved their load factor and increased their revenues in relation to the infrastructure investment. Accordingly, electric power systems grew from the scale of city neighborhoods to cover entire counties and states.[3]

The third main factor driving geographical expansion and interconnection has to do with the ability to provide greater service reliability in relation to cost. The basic idea is that when a generator is unavailable for whatever reason, the load can be served from another generator elsewhere. To allow for unexpected losses of generation power or *outages*, utilities or independent system operators (ISOs) maintain a *reserve margin* of generation, standing by in case of need (see Section 8.1). Considering a larger combined service area of several utilities, though, the probability of their reserves being needed simultaneously is comparatively small. If neighboring utilities interconnect their transmission systems in a way that enables them to draw on each other's generation reserves, they can effectively share their reserves, each requiring a smaller percentage reserve margin at a given level of reliability.

More extensive interconnection of power systems also provides for more options in choosing the least expensive generators to dispatch, or, conversely, for utilities with a surplus of inexpensive generating capacity to sell their electricity. For example, the north–south interconnection along the west coast of the United

---

[3]This argument is developed in detail with ample historical illustration by Thomas P. Hughes in *Networks of Power: Electrification in Western Society, 1880–1930* (Baltimore: Johns Hopkins University Press, 1983).

States allows the import of hydropower from the Columbia River system down through California. In general, over the course of the development of electric power systems during the 20th century, larger and more interconnected systems have expanded the options for managing and utilizing resources for electric supply in the most economic way.

As the distance spanned by transmission lines has increased, so has the significance of energy losses due to resistive heating. Recall from Section 1.4.2 that high voltage is desirable for power transmission in order to reduce current flow and therefore resistive losses in the lines. Therefore, as systems have grown in geographical extent, there has been an increasing incentive to operate transmission lines at higher voltages. The maximum voltages used for transmission have thus steadily increased over the decades, as shown in Figure 6.1 (note that the first scale on the left indicates unit size in megawatts, the scale on the far right transmission voltage in kilovolts).

There are also liabilities associated with larger size and interconnection of power systems. Long transmission lines introduce the problem of stability (see Section 8.3). More interdependence among areas also means greater vulnerability to disturbances far away, including voltage and frequency fluctuations (see Section 8.4). Still, conventional wisdom in the electric power industry holds that the benefits of interconnection outweigh the drawbacks, at least up to the scale at which power systems are currently operated. Thus, while individual utilities retained their geographically delimited service territories throughout the era of regulated monopolies, the strength and importance of their interties has increased to the point where, when considering power systems from an operational standpoint, it is often more meaningful to speak of a group of interconnected utilities than of an individual corporate entity. Because the success of system operation depends on a level of technical cooperation, such groups are administratively organized under the North American Electric Reliability Council (NERC), shown in Figure 6.2, where utilities in each regional council collaborate in planning and sharing resources. The three major networks in the United States are synchronous internally and connected to each other only by direct current (d.c.) links; not shown in the figure are synchronous interties to Canada and Mexico.

6.1.2 Structural Features

High voltages are crucial for power transmission over long distances. Closer to the end users of electricity, however, safety prohibits the use of equipment at excessively high voltages lest people start electrical fires or electrocute themselves (a point that every parent of a curious toddler can especially appreciate). The sheer expense of properly insulating high-voltage equipment is also a consideration. In the design of power delivery systems, the greater energy efficiency of high voltage and low current must therefore be weighed against safety and capital cost. Rather than having to settle for some intermediate voltage as a compromise, the use of *transformers* (see Section 6.3) makes it possible to operate different parts

TRANSMISSION AND DISTRIBUTION

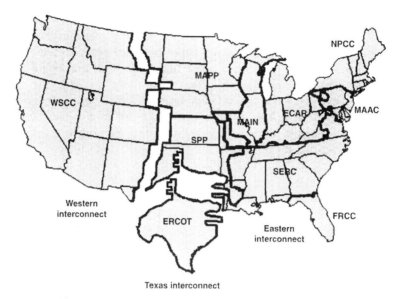

Figure 6.2 Main interconnections of the U.S. electric grid.

of the system at different voltages, retaining the respective advantages of higher and lower voltages in those places where they are most consequential.

Power delivery systems are therefore divided into two general tiers: a *transmission system* that spans long distances at high voltages on the order of hundreds of kilovolts (kV), usually between 60 and 500 kV, and a more local *distribution system* at intermediate voltages in the low tens of kV. The latter is more specifically referred to as *primary distribution*, in contrast to the *secondary distribution system*, which consists of the wires that directly connect most domestic and small commercial customers, at voltages in the 100-V range (nominally 120 V in the United States). Larger commercial and industrial customers often receive their service at higher voltages, connected directly into the primary distribution system. The transmission system is also further subdivided into *subtransmission*, operated in the neighborhood of 100 kV, and longer-distance transmission at several hundred kV.[4] Collectively, the entire power delivery system is referred to as the *transmission and distribution (T&D) system*.

The division between transmission and distribution is defined in terms of voltage level, though individual utilities have different customs for where, exactly, to draw the line. In general, "distribution" means below 60 or 70 kV. Physically, the boundary between transmission and distribution systems is demarcated by transformers,

---

[4]In European usage, the terms translate as follows: subtransmission voltage = high voltage; primary voltage = medium voltage; and secondary voltage = low voltage. Owing to the smaller distances in Europe (where line losses play a less important role), "high voltage" is generally limited to 380 kV, whereas 500 kV and 750 kV are used in the United States. Worldwide, the highest voltages used to date are 1200 kV a.c. for some very long transmission lines across Siberia and ± 600 kV d.c. in Brazil.

grouped at *distribution substations* along with other equipment such as circuit breakers and monitoring instrumentation. Organizationally, most utility companies have separate corporate divisions responsible for the operation and maintenance of transmission versus distribution systems; the substations themselves may be considered to lie on either side, but most often fall under the jurisdiction of power distribution.

6.1.3 Sample Diagram

Figure 6.3 shows the basic structure and components of a transmission and distribution system. First note that the illustration is not drawn to scale. Note also that it is a *one-line* diagram, which does not show the three phases for each circuit (see Section 6.2). The vertical lines represent *buses*, or common connections at key points in the system, especially power plants and substations. With power flowing from left to right, the diagram indicates the hierarchical relationship among the important subsystems, along with some typical voltage values.

On the far left side of the diagram, two generators deliver power at 21 kV. They connect to the transmission system through a transformer (indicated by the wavy symbol reminiscent of wire coils) that steps up the voltage to 230 kV. The squares on either side of the step-up transformer indicate circuit breakers that can be opened to isolate the generator from the system. Circuit breakers also appear elsewhere in the diagram; their functions will be discussed in more detail in Section 6.7 on protection.

The system shown includes both high-voltage transmission at 230 kV and subtransmission at 60 kV. The transmission and subtransmission systems meet in a transformer at a transmission substation. At the distribution substation, the voltage is stepped down further to the primary distribution voltage, in this case 12 kV. The primary distribution lines or *feeders* branch out from the substation to serve local areas. These main feeders carry all three phases (see Section 6.2).

From the main feeders, *lateral* feeders (*laterals* for short) carry one or two phases for a shorter distance, a few city blocks, for example. From the lateral, several

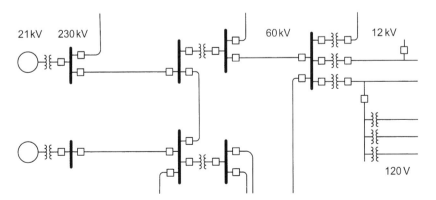

Figure 6.3 One-line diagram showing basic power system structure.

distribution transformers step the voltage down again to the secondary level at which most customers are served, generally 120 V. Typically, one distribution transformer serves several residences, up to one city block. Larger commercial or industrial consumers (not shown here) usually have three-phase service directly from the primary distribution level through their own dedicated transformer.

6.1.4 Topology

An important characteristic of transmission and distribution systems is their topology, or how their lines are connected. The most important distinction is between a *radial* configuration where lines branch out sequentially and power flows strictly in one direction, and a *network* configuration that is more interconnected. In a network, any two points are usually connected by more than one path, meaning that some lines form loops within the system.

Transmission systems are generally networks. Local portions of a transmission system can be radial in structure—for example, the simplified section shown in Figure 6.3, with all the power being fed from only one side. Since generating plants are more likely to be scattered about the service territory, though, the system must be designed so that power can be injected at various locations and power can flow in different directions along the major transmission lines, as necessitated by area loads and plant availability. Thus, high-voltage transmission systems consist of interconnected lines without a hierarchy that would distinguish a "front" or "back" end.

It is true, of course, that due to the geography of generation and major load centers, power will often tend to flow in one direction and not the other. Nevertheless, this kind of directionality is not built into the transmission hardware. For example, in the state of New York, power tends to flow from north to south into New York City, but as far as the transmission system is concerned, the power could just as easily be sent from south to north. The system structure does finally become hierarchical at the interfaces with the lower-voltage subsystems (subtransmission or distribution), where power flows only from high to low voltage.

The network character of the transmission system allows for different operating conditions in which power may flow in different directions. It also offers the crucial advantage of *redundancy*. Because there are multiple paths for power to flow, if one transmission line is lost for any reason, all the load can still be served (as long as the remaining lines can carry the additional load). Indeed, a standard design criterion for transmission systems requires that the entire system must continue to function if any one link is interrupted (see Section 8.1 on reliability).

The basic radial design is that shown in Figure 6.4. The radial system has a strict hierarchy: power flows only in one direction; there is always an "upstream" and a "downstream." The distribution lines or feeders extend and branch out in all directions from a substation somewhat like spokes from a hub. Owing to this hierarchy, any given line or component can only be energized from one direction. This property is crucial in the context of circuit *protection*, which means the interruption of circuits or isolation of sections in the event of a problem or *fault*

6.1 SYSTEM STRUCTURE   151

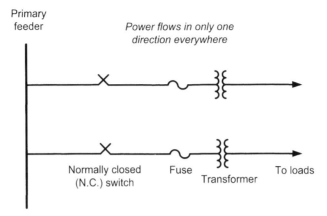

Figure 6.4 Radial distribution system.

(see Section 6.7). In a radial system, circuit breakers can readily be located so as to isolate a fault—for example, a downed line—immediately upstream of the problem, interrupting service to all downstream components. Economically, radial systems also have the advantage that smaller conductor sizes can be used toward the ends of the feeders, as the remaining load connected downstream diminishes.

Figure 6.5 illustrates a *loop system*. One switch near the midpoint of the loop is open (labeled N.O. for normally open) and effectively separates the loop into two radial feeders, one fed by each transformer. Thus, under normal operation, the

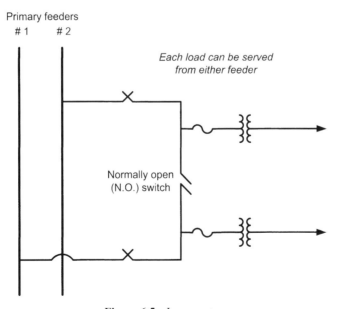

Figure 6.5 Loop system.

sections are not connected at that point, so that the system operates as a radial system. But under certain conditions—for example, a failure of one of the two substation transformers—the N.O. switch can be closed and one section of the distribution system energized through the other. By choosing which one of the other switches to open, sections of the loop can be alternatively energized from the left or right side. This has the advantage of enabling one transformer to pick up additional load if the other is overloaded or out of service, and of restoring service to customers on both sides of a fault somewhere on the loop. While loops are operated as radial systems at any given time, that is, with power flowing only outward from the substation transformer, the hardware including protective devices must be designed for power flow in either direction.

Finally, Figure 6.6 shows an example of a network distribution system. A networked system is generally more reliable because of the built-in redundancy: if one line fails, there is another path for the power to flow. The cost of a network system to serve a given area is higher than a simple radial system, owing not only to the number of lines but also the necessary equipment for switching and protection. Networks are often used in downtown metropolitan areas where reliability is considered extremely important, and where the load density justifies the capital expense.

From the standpoint of circuit protection, a network is much more challenging because there is no intrinsic upstream or downstream direction, meaning that a given point in the system could be energized or receiving power from either side. This means that any problem must be isolated on *both* sides, rather than just on the upstream side. However, the objective is still to make the separation as close to the fault as possible so as to minimize the number of customers affected by service interruptions. As a result, the problem of coordinating the operation of multiple circuit breakers becomes much more complex. This is the main reason why most distribution systems are radial or looped.

A special case of power system topology is the *power island*, or an energized section of circuits separate from the larger system. An island would be sustained by one or more generators supplying a local load, at whatever scale. For example,

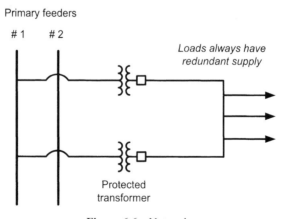

Figure 6.6 Network.

in the event of a downed transmission line to a remote area in the mountains, a hydroelectric plant in this area might stay on-line and serve customers in its vicinity, an occurrence described by operators as "not by the book." Similarly, small-scale distributed generation such as rooftop photovoltaics could in theory sustain local loads as a small island during a service interruption, if it were not for standard requirements that they be disconnected from the grid in the event of an outage.

Islanding is not routinely practiced or condoned by U.S. utilities, except temporarily while restoring loads after a widespread outage, where individual power islands are reconnected as quickly as possible in a centrally orchestrated effort. Reasons to avoid routine islanding include both safety and liability. First and foremost, the safety of line crews could be jeopardized if they encountered a power island while expecting to find a de-energized circuit. Second, the ability of the generators in the island to maintain power quality (including voltage and frequency) is not guaranteed, potentially causing problems for some customers with sensitive equipment for which the utility may then be held liable. However, with an increasing prevalence of small generators such as photovoltaics and fuel cells throughout distribution systems, and considering also the control capabilities introduced by distribution automation technology, the issue of islanding seems likely to be revisited and perhaps become the subject of some controversy.

6.1.5 Loop Flow

In addition to protection, an important operational complexity introduced by a network structure is *loop flow* (Figure 6.7). Loop flow can arise whenever there is more than one path for the current to travel between two points in the system. The basic problem is that current flow cannot be directed along any particular branch in the network, but is determined by Kirchhoff's laws (see Section 2.3)

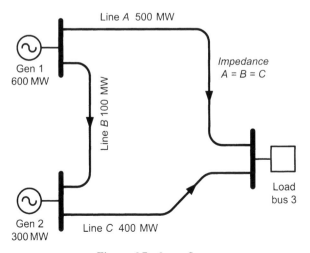

Figure 6.7 Loop flow.

and the relative impedances of the various branches. The concept is best explained by example, as shown in the simple case below. This example is intended to illustrate that actual line flows through network loops are not immediately obvious, even in apparently simple cases. Calculating these amounts generally requires *power flow analysis*, which is performed by computer.

How the power flows through a network tends to be of little interest until there is *congestion* or overloading of transmission lines, at which point it suddenly becomes critically important. In order for local transmission overloads to be relieved, operators need to know which generators could have their output adjusted so as to most effectively achieve a reduction in line flows. In a competitive market, it may be desirable to allocate the available transmission capacity in some economic fashion. If an individual generator is to be allocated the rights to a certain power transfer along a particular link, it is vital to know how this generator's output contributes to the total flow on any given transmission path. While this contribution cannot readily be measured, it can be calculated or at least estimated by power flow analysis (see Chapter 7). In the simplified situation shown in the Figure 6.7, there is only one load, located at Bus 3, and two generators, Gen 1 and Gen 2. Suppose the load is 900 MW, of which Gen 1 supplies 600 MW and Gen 2 300 MW. What are the flows on the transmission lines A, B, and C? Although line A is the most direct path from 1 to 3, not all of the 600 MW will flow along A. Some will flow along the combination $B-C$, which constitutes another path from 1 to 3. The magnitude of this portion depends on the relative impedance of the path $B-C$ as compared to A (see Section 2.2 about parallel circuits).

To make this example most transparent, let us suppose that the impedances of all three links A, B, and C are exactly the same (a highly idealized situation). The impedance of path $B-C$ (in series) is therefore exactly twice that of path A, and the current flowing through $B-C$ is half that flowing through A (see Section 1.2). To a very good approximation (assuming the voltages at 1, 2, and 3 have reasonable values), the power (megawatts) transmitted along each path will be directly proportional to the current (amperes) flowing though it; thus, twice as much power flows through A as through $B-C$. If these are the only two paths from this generator to the load, and their total is 600 MW, then 400 MW flow through A and 200 MW through $B-C$.

Now there is an additional 300 MW supplied to the load at 3 from Gen 2. Again, while line C is the most direct path, some of the current will flow around the loop $B-A$. Given our previous assumption about the impedances, twice as much current (or power) flows through C as through $B-A$; thus, we have 200 MW through C and 100 MW through $B-A$, for a total of 300 MW.

The total flows on each line can now be calculated quite simply by taking advantage of the *superposition principle* (see Section 2.3.4), which allows us to consider each power source individually at first and then add the currents due to each source in each link. We must only be careful about the direction and whether the currents in fact add or subtract. On line A, power from Gen 1 and Gen 2 flow in the same direction, from 1 to 3. We can therefore add the currents (or power), and the total flow on A is 500 MW. Similarly, on line C, we have 200 MW from Gen 1 and 200 MW from

Gen 2, each flowing in the direction from 2 to 3, for a total of 400 MW. But on line B, 200 MW from Gen 1 flow to Gen 2, whereas 100 MW from Gen 2 flow from 2 to 1. These line flows subtract, and we have a net flow of 100 MW on line B from 1 to 2. We can do a reality check by confirming that the total power arriving at the load is indeed $500 + 400 = 900$ MW.

This example shows that line flows in the presence of loops are not obvious. If there were more possible paths connecting the buses, the result would be accordingly more complicated and could hardly be calculated by hand. The example also illustrates that the effect of one generator's output may be to reduce rather than increase the flow on a given link. For example, suppose line B is overloaded and can only handle 90 MW, whereas there is plenty of capacity on lines A and C. If both load at Bus 3 and generation at Gen 2 are now increased by 30 MW, the result is that the flow on B is reduced by 10 MW, saving the day. (The reader can verify that the new flows on A and C are 510 MW and 420 MW, respectively.)

Under some conditions, there may be a circulating current around a loop within the network that does not actually serve a load but contributes to energy losses. These cases show that designing and operating a power network for safety as well as efficiency requires modeling the power flow under a diverse set of operating conditions so as to avoid unpleasant surprises.

INTERNATIONAL DIFFERENCES IN DISTRIBUTION SYSTEM DESIGN

In Europe, as well as in many countries formerly colonized by Europeans, power distribution systems have a distinctively different look. Unlike the United States, where there is usually a distribution transformer (typically single phase) for every few customers, connected by short service drops, in Europe there are fewer and larger transformers—usually hidden in vaults rather than mounted on poles—from which a more extensive system of secondary lines branches out. Because secondary (low-voltage) lines are less expensive to underground than the primary (medium-voltage) lines that make up the larger part of the U.S.-style distribution system, distribution systems are more often undergrounded and thus altogether less visible in Europe. These differences in design are consistent with differences in geography, population and load density, and the historical expansion of power systems.

Except when compared with downtown urban areas, the population density in Europe is generally higher and individual loads smaller, making it more feasible to extend secondary lines to many customers. Extension of secondary lines is also more feasible because the standard secondary voltage is higher in Europe (nominally 220 V). This design approach makes it relatively easy to geographically extend service areas by adding another secondary line to an existing transformer, though it becomes more difficult once the transformer capacity is insufficient. The layout of the system is generally optimized around load level.

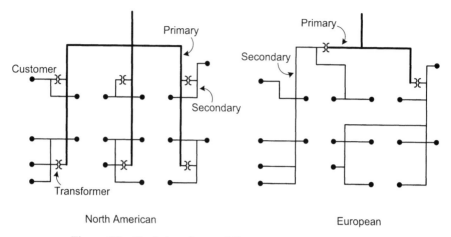

Figure 6.8 North American and European distribution systems.

In North America, particularly in rural areas, the location of customers and the size of the area to be covered are generally more important considerations in determining system layout than load level. Because customers tend to be much farther apart, higher voltage (primary) lines are needed to reach them. Systems here tend to have a higher loading capacity per mile of circuit, and load growth within the existing service area can usually be quite readily accommodated by adding small transformers.[5] Figure 6.8 illustrates the two different types of layout (adapted from Carr and McCall, 1992), where the heavier lines graphically represent primary and the lighter lines secondary distribution wires. Note that the line weight does not reflect physical wire diameter, since a lower-voltage line actually carries a greater current to transmit a given amount of power.

6.1.6 Stations and Substations

Transmission and distribution stations exist at various scales throughout a power system. In general, they represent an interface between different levels or sections of the power system, with the capability to switch or reconfigure the connections among various transmission and distribution lines. On the largest scale, a transmission substation would be the meeting place for different high-voltage transmission circuits. At the intermediate scale, a large distribution station would receive high-voltage transmission on one side and provide power to a set of primary distribution circuits. Depending on the territory, the number of circuits may vary from just a few to a dozen or so. The major stations include a control

---

[5] J. Carr and L.V. McCall, "Divergent Evolution and Resulting Characteristics among the World's Distribution Systems," *IEEE Transactions on Power Delivery* **7**(3), July 1992.

room from which operations are coordinated. Smaller distribution substations follow the same principle of receiving power at higher voltage on one side and sending out a number of distribution feeders at lower voltage on the other, but they serve a more limited local area and are generally unstaffed.

The central component of the substation is the *transformer*, as it provides the effective interface between the high- and low-voltage parts of the system. Other crucial components are *circuit breakers* and *switches*. Breakers serve as protective devices that open automatically in the event of a fault, that is, when a protective *relay* indicates excessive current due to some abnormal condition. Switches are control devices that can be opened or closed deliberately to establish or break a connection. An important difference between circuit breakers and switches is that breakers are designed to interrupt abnormally high currents (as they occur only in those very situations for which circuit protection is needed), whereas regular switches are designed to be operable under normal currents. Breakers are placed on both the high- and low-voltage side of transformers. Finally, substations may also include capacitor banks to provide voltage support. Because three phases are used, all equipment comes in sets of three, referred to as a "bank"—for example, a transformer bank is a set of three transformers, one for each phase.

A sample layout for a distribution substation is shown in Figure 6.9. Note that schematic diagrams like this are not drawn to scale, nor do they necessarily

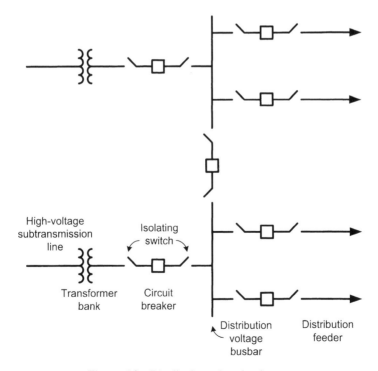

Figure 6.9 Distribution substation layout.

Figure 6.10 Distribution substation. (Courtesy of Marshal F. Merriam.)

provide a good sense of the actual spatial arrangement of the switchyard for those unfamiliar with the physical equipment represented here. Figure 6.10 shows a photograph of a similar substation.

6.1.7 Reconfiguring the System

Operating a transmission and distribution system involves switching connections among high-voltage power circuits. Switching operations can be carried out remotely from the computer screen at a switching center through a *supervisory control and data-acquisition* (*SCADA*) system. In many distribution systems, however, switching is still performed manually by field crews per telephone instructions from operators at a distribution switching center.

Reasons for switching include contingencies, work clearances, service restoration following an *outage*, managing overloads, and enhancing system efficiency. For example, a contingency might be a fault on one line that has to be isolated from the system, and other connections need to be rearranged so as to redistribute the load of the lost line among them. Similarly, in the case of maintenance work or replacement of a line or system component, this component is electrically isolated (*cleared*) and the system "patched" around it. Clearances may also be issued for work other than on the power system components themselves, including any construction work that is sufficiently close to power lines that workers may accidentally contact them.

In the event of an outage, or service interruption to some number of customers, care is taken to follow a sequential procedure of *restoring load*. The idea is to reconnect sections of load one at a time in a given order so that each new load added does not jeopardize the stability of the remainder of the connected system. This restoration process involves opening and closing switches in order to divide loads into appropriately small sections, to connect them, and sometimes to transfer sections temporarily. Restoration after a widespread outage is usually the only time that utilities operate parts of the system as power islands. In this situation, local areas with generation and load are brought back on line and then synchronized with each other and reconnected.

At the transmission level, because the system is networked, and because outages are less frequent, there tends to be less switching during normal operation. In radial distribution systems, by contrast, well-defined local blocks of load can readily and deliberately be shifted from one circuit to another. For example, if one particular substation transformer threatens to overheat on a day of heavy demand, part of the load of the corresponding distribution feeder can be switched over to the adjacent transformer. This is easily done in a loop system by shifting the position of the "gap" or open switch between the two ends of the loop (see Figure 6.5.)

Under extreme overload conditions, where there is either no available distribution capacity to reroute power, or if there is a shortfall in power generation, load may be *shed* or selectively disconnected. For example, if the systemwide generation is insufficient to meet demand, and consequently the system frequency drops below a specified limit, a number of customers will be disconnected for the sake of keeping the remainder of the system operational, as opposed to risking a more extensive failure of potentially much longer duration (see Section 6.7). In order to assure fairness in this procedure, customers are assigned *rotating outage block* numbers that appear on the electric bill for taking turns in being shed.

Finally, it is possible to redistribute loads in distribution systems for the purpose of *load balancing*, or equalizing the loads on distribution feeders or transformers in order to increase operating efficiency by minimizing losses. Because resistive power losses vary with the square of electric current ($P = I^2 R$), they are minimized when the currents are evenly balanced among alternative lines and transformers.[6] To achieve this, some load may be switched over from one to another, less heavily loaded circuit. This procedure is not part of standard operations in the industry today, but has been proposed as an economically viable procedure in the context of automating distribution switching operations.

6.2 THREE-PHASE TRANSMISSION

In the preceding transmission and distribution system diagrams, each circuit was indicated by a single line in what is known as a *one-line diagram*. In reality, however, a standard a.c. circuit includes three electrically separate parts or

---

[6]Readers to whom this result is not obvious may wish to work through a numerical example: Consider two currents that add up to 10. In an unbalanced combination—say, 9 and 1—the squares of the currents add up to 82. In a balanced combination—that is, 5 and 5—the squares add up to only 50.

160 TRANSMISSION AND DISTRIBUTION

phases. For example, transmission lines are generally seen to have three wires or conductors. In this section we describe the three-phase concept—how it works, why it was chosen, and basic aspects of analyzing three-phase circuits.

6.2.1 Rationale for Three Phases

In Chapter 4, we described power generators as having three sets of outgoing windings or *phases*, each carrying an a.c. of the same magnitude, but separated in timing by one-third of a cycle or 120 degrees from the other two. We also stated one rationale for this three-phase system; namely, that a three-phase generator experiences a constant torque on its rotor as opposed to the pulsating torque that appears in a single- or two-phase machine, which is obviously preferable from a mechanical engineering standpoint.[7]

In the system context, another benefit of supplying a.c. in multiple phases lies in its economy of transmission. As we show in this section, three-phase transmission permits the use of fewer wires than common sense might suggest, and indeed less conductor capacity than would be required to transmit an equivalent amount of power using only a single phase of a.c.

In general, the completion of an electric circuit requires two conductors between the power source and the load: one for the current to flow out, and one for it to return. In principle, this is true regardless of whether the current is direct or alternating. In practice, d.c. transmission can sometimes do away with one of the conductors and allow the current to return through the ground, in what is called *ground return*. For a.c., ground return is rarely feasible because the soil in most places presents too high an impedance. In any case, multiphase transmission does not rely on ground return to work.

Let us begin by imagining a.c. power being supplied from a source to a load in a single phase. Two conductors between the source and the load form one complete circuit. At any instant, the current everywhere along this circuit is the same. (This implies that if the conductors to and from the load are placed side by side, their currents would go in opposite directions.)

Suppose now that we want to supply three separate phases, one from each set of generator windings, to three individual loads. Each of these circuits will again have two conductors, one leading from one end of the generator armature winding to the load, and one connecting back from the load to the other end of the winding. Thus, we have a total of six conductor wires.

The secret of multiphase transmission is that the return conductors of each of these three circuits can be combined into a single hypothetical return wire, and ultimately that this combined return can be eliminated altogether, leaving us with only three wires for three circuits. To appreciate how this can work, we must consider the relative timing and magnitudes of the three individual phase currents, as shown in

---

[7]The constant torque results from the increasing and decreasing magnetic field components along the direction of each phase's coil that add up in space so as to emulate a rotating field of constant magnitude, as illustrated in Section 4.2.1. This addition is geometrically impossible with fewer than three phases.

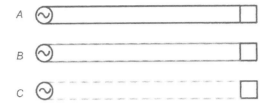

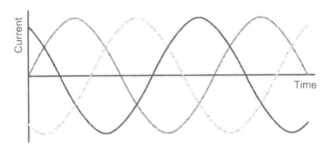

$I_A(t) + I_B(t) + I_C(t) = 0$ always

Figure 6.11 Three balanced single-phase a.c. circuits.

Figure 6.11. Specifically, consider the sum of the three currents at any particular instant. This is the same as asking, if we were to physically combine all three return wires into one, what would be the combined current in that wire?

In some places, it is obvious from inspection that the three curves add up to zero. For example, at the instant when current A is zero, current B is exactly the negative of current C. Or, when current A is at its maximum, B and C are both at one-half their negative maximum. Elsewhere, it is not as easy to see, but the math holds up: the sum of currents A, B, and C is always zero, anywhere along the graph. This is true as long as two conditions are met: (1) that the three phases are exactly 120° apart, and (2) that the respective amplitudes or maxima are the same. What this means in physical terms is that if we bundled together the return wires of the three circuits, the current through this combined conductor would be zero. This situation is shown in Figure 6.12.

Let us now think about the picture in terms of voltage or potential. The three windings in the generator are producing an alternating potential across them. This alternating potential is also seen by each load, which is connected to the two ends of a winding. But now all three generator windings share a point of common potential, point G, as do all three loads at point L. To determine what the actual potential is at point G (with respect to ground), we can add up a set of three sinusoidal waves just as we did for the current, only now indicating the voltage generated in each phase over time, and see that it remains zero at all times.

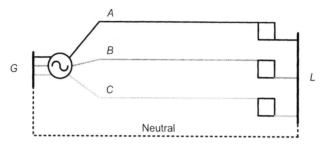

Figure 6.12 Three phases with common return.

The same should be true for point L, assuming that all three loads (their impedances, really) are identical. While the generator controls the voltages, the load impedances determine the currents. To assume equal loads on the three phases is to assume that both currents and voltages in the three phases are properly timed and remain of equal magnitude. Thus, with equal loads, the currents will be balanced and the potential with respect to ground at point L will also remain at a constant zero.

From this observation, we can conclude that there will be no current flowing through the return conductor or neutral connecting G and L, since there is no potential difference between G and L. Therefore, we can simply eliminate the wire. In reality, because loads may not be balanced, some residual net return current may need to flow. When the neutral points G and L are both grounded, this relatively small residual current can flow through the soil. The situation is illustrated in Figure 6.13.

In summary, then, we are transmitting power to three circuits, but, owing to the clever combination of phases staggered in time, we can do away with three of the six wires we would expect to need. Some readers may note that this result is not unique to three phases. In fact, any number of phases, spaced equally around the cycle and with equal amplitudes, will have the same property of always adding up to zero. For example, we could supply four circuits with four wires for four phases, spaced 90° apart; five phases 72° apart; or even just two opposite phases, 180° apart. Why, then, choose three?

First, we recall the advantage of using at least three phases for generator design. In addition, transmitting power with only two phases would mean a greater

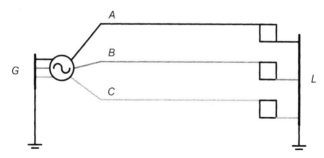

Figure 6.13 Three phases with neutral removed.

vulnerability to imbalances in the loading of the two circuits: if one circuit were loaded more heavily than the other, the effects on voltage and current would be more readily noticeable than they are with more phases to "absorb" the difference.

More than three phases, on the other hand, would become increasingly complicated and expensive. If all the power on the system were supplied in four phases, for instance, this would mean that all transmission and major distribution lines consist of four separate conductors. In addition, each bank of transformers at every substation would have to comprise four instead of three transformers, and the same for circuit breakers. Although for the same amount of total power the capacity of each of these four components could be proportionately smaller, their total cost would be greater due to economies of scale (i.e., a transformer of twice the capacity costs less than twice as much). Thus, three is the most practical number of phases, and its use in a.c. power systems is a global standard.

Because the three phases are so interdependent, they are described as a single circuit (in contrast to the nomenclature we used before, where we referred to each individual phase as a circuit). In drawings, each three-phase circuit can be represented by a single line. This perspective is generally applied in situations that assume balanced and trouble-free operation. It is the more complicated, less-than-ideal situations involving faults and unbalanced loads that call for a separate analysis for each phase.

6.2.2 Balancing Loads

In deriving the result that a circuit of three a.c. phases does not require a ground return conductor, we made the assumption that the impedance or total load connected to each phase is identical, making the amplitudes of the three currents equal. But how is this accomplished in reality? Certain loads such as large commercial motors are connected to all three phases and draw power equally from all of them. But, as we noted earlier, electric service to most residential and small commercial customers generally consists of only one phase. The challenge for the distribution planner is to distribute these customers among the phases as evenly as possible so that the combined load on each phase is about the same. Where single-phase laterals branch out from the main distribution feeders, distribution engineers take care that the local areas served by these feeders comprise similar loads. If a three-phase feeder runs down a street, transformers that serve several houses each will tend to alternate among the phases, and when new customers are connected, the phase is chosen by considering the balance with other loads in the vicinity.

As the reader can imagine, balancing loads[8] is a rather approximate procedure, unlike the idealized, perfectly balanced phases encountered in engineering texts. For one, the physical equipment limits the precision with which loads can be balanced, because loads come in chunks and not in arbitrarily small increments. The balance will also fluctuate in real-time as customers connected to different

---

[8]This notion of balancing loads among individual phases is not to be confused with a similar term, *load balancing*, that refers to the practice of equalizing loads among different distribution circuits or feeders so as to avoid overloading one of them, or for the purpose of minimizing resistive losses (see Section 6.1.5).

phases turn their appliances on and off. In practice, the imbalance among current or power delivered by each phase will be in the range of several percent, sometimes as high as 10% or more. However, this variation will be greatest near the customers, and will diminish at higher levels of aggregation (such as the transmission system) by virtue of statistics, since the local imbalances are more likely to cancel each other out once many loads are added together.

When the loads on all phases are not equal, the currents no longer add to zero. This means that some return flow of current is needed. If only three conductors are provided for three phases, the return current must travel through the ground. Along distribution feeders, where more pronounced imbalances are to be expected, a fourth *neutral* wire is often provided for just this purpose. In Figure 6.12, we would observe a phase imbalance as a potential between points L and G that causes a current to flow in the neutral conductor. The magnitude of this current will tend to be on the order of a few percent of the phase current, as the voltage at L and G with respect to ground will be a few percent of the regular line voltage, depending on the extent of the imbalance. The conductor and insulator size of the neutral wire therefore may be noticeably smaller than the three phase conductors.

6.2.3 Delta and Wye Connections

An individual load generally requires two conductors to connect to, as this constitutes a closed circuit (an exception are three-phase motors that connect to all three conductors). When power is delivered in three phases, the designer has two distinct choices: a load can be connected either between one phase and ground, or between one phase and another phase. In each case, the three phases will supply three separate loads. The first arrangement, where three loads are connected between one phase each and ground, is called a *wye connection* because the schematic diagram resembles the letter Y, as in Figure 6.14b. The second arrangement, in which three loads are connected between one pair of phases each, is called a *delta connection* (Figure 6.14a) for the same reason.

For the wye connection, it is obvious that the voltages (known as *phase-to-ground voltages*) and currents for each load will be spaced 120° apart. The same is true for the delta connection, but it is less obvious. The key is to remember

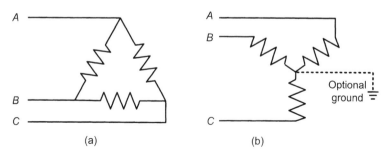

Figure 6.14 (a) Delta and (b) wye connections.

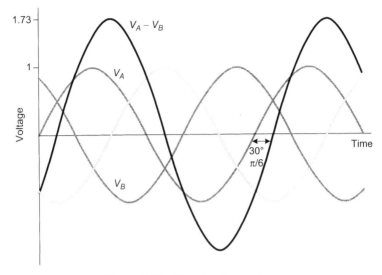

Figure 6.15 Phase-to-phase voltage.

that the voltage "across" a load is in fact the *difference* between the voltage on one side and the other. Thus, mathematically, we can simply plot a curve that shows the difference between each pair of phase voltages at any moment in time, which gives the *phase-to-phase voltage* as a function of time (as shown in Figure 6.15). Each of these three possible curves ($A-B$, $A-C$, or $B-C$) is still sinusoidal (as we might derive from trigonometry), but is shifted in time (phase) from the original phases A, B, and C. It also has a different amplitude: as we could further derive from trigonometry, the magnitude of the phase-to-phase voltage is greater than the phase-to-ground voltage by a factor of $\sqrt{3}$ (about 1.73).

The choice of phase-to-phase versus phase-to-ground offers one way to select different voltage levels for loads from the same three-phase power supply. This is, in fact, how some utility customers (usually in the industrial or commercial sector) obtain both 120 V and the so-called 220 V from two phases and a neutral conductor. (The alternative approach for supplying dual voltages involves transformer taps and is discussed in Section 6.3.)

Example

A service line to a grocery store comprises three conductors—Phase A, Phase B, and neutral—where each phase-to-ground voltage is 120 V. The 120-V loads around the store, such as lights and regular outlets, may be alternatively connected to Phase A or B. Heavier loads, such as big refrigerators, would be connected phase-to-phase between A and B. The voltage they see is $\sqrt{3} \cdot 120\,\text{V} = 208\,\text{V}$.

In general, generating equipment and loads are connected to power systems through *transformers*. Because the wire coils in the transformer always involve a pair of conductors that constitute one circuit, everything has to have either a delta or a wye connection, and both are common.

Aside from the difference in voltage level, the choice of delta or wye connection has some ramifications for reliability, in case there should be a short circuit. The delta configuration as a whole is ungrounded or *floating*, meaning that no point on the circuit is connected to ground or to any point that has a particular, known *potential*. The (time-varying) potential differences between and among the points on the circuit are all that matters. The entire delta system can thus float at any arbitrary, absolute value of potential. Consequently, if any part of the circuit accidentally gets grounded—which would otherwise cause an outage—the delta circuit can continue to operate on an emergency basis. (It is crucial, though, to locate and fix the problem before a second one occurs elsewhere on the circuit, which could now cause a very damaging fault current.) Because of this property, the delta configuration is used where reliability is most important, such as auxiliary equipment in power plants, or on smaller transformers.

The wye configuration, by contrast, is normally grounded at the center or neutral point. Here, a single ground anywhere else in the system will immediately cause a fault, and ground relay protection (see Section 6.7) is always used to open the circuit breakers in such an event to protect the lines and equipment. In this case, the risk of damage to equipment overrides the short-term reliability concern. The wye connection is typically used on generators, main transformer banks, and transmission lines. It is also possible to switch between delta and wye connections, which is frequently done by transformers that are wired in a delta configuration on one side and wye on the other (see Section 6.3.3).

6.2.4 Per-Phase Analysis

Based on the assumption that the voltages and currents in the three phases are equal, the analysis of a three-phase circuit can be simplified to a single phase that is then representative of what happens in all three. Whereas phase imbalances are often encountered in the operating context, they are not easily dealt with in the engineering context since the differences among loads that give rise to the different currents and voltages are not by design and cannot readily be predicted. At the transmission level, phases are generally assumed to be balanced. At the distribution level, there are single-phase feeders, but main feeders are also assumed to be balanced among their three phases. An important exception is the analysis of what happens during a fault—say, if a single phase is accidentally connected to ground, or two phases are short-circuited together. For the most part, though, there is no need to explicitly keep track of the three phases individually. Therefore, *one-line diagrams* are often used that are understood to represent a complete three-phase circuit.

6.2.5 Three-Phase Power

In order to calculate the power transmitted on a three-phase transmission or distribution line, one essentially multiplies the voltage and current together for each individual phase and multiplies by three. However, we must be careful to take into account a factor of $\sqrt{3}$. It arises in the case of a wye connection because the effective voltage seen by the load is the line-to-ground, not the line-to-line voltage. In the case of a delta connection, the current through each conductor contributes to two phase pairs, so that the applicable current that can be "credited" to each load is only $1/\sqrt{3}$ of the total current measured in the line (it is more than half because the timing of the current is shifted by 120° for each phase pair). Thus, in the wye case, we have for apparent power

$$S = 3 I_{\text{rms}} \frac{V_{\text{rms}}}{\sqrt{3}}$$

where V_{rms} refers to the phase-to-phase voltage (according to which line voltages are conventionally labeled) and I_{rms} to the actual current in each conductor. In the delta case

$$S = 3 \frac{I_{\text{rms}}}{\sqrt{3}} V_{\text{rms}}$$

As we can readily see, both come out to the same formula,

$$S = \sqrt{3} I_{\text{rms}} V_{\text{rms}}$$

regardless of how the load is connected.

For real and reactive power, it is only necessary to include the factors of $\cos \phi$ and $\sin \phi$, respectively, depending on the load power factor (p.f.).

6.2.6 D.C. Transmission

In some situations it is beneficial to use d.c. transmission. This is true despite the fact that the transmission system is otherwise a.c., and the power has to be converted both at the beginning and the tail end of the d.c. line through relatively costly solid-state devices called *thyristors*. The reason this effort is worthwhile is that d.c. eliminates the problem of a *stability limit* (see Section 6.5.2), which poses a power transmission constraint on long lines with a significant *inductance*. Because its effect depends on a.c. frequency, inductance is irrelevant for d.c., and therefore only the thermal limit (determined by the *resistance* of the line) applies. An overhead d.c. transmission line can be easily identified by the fact that it has two, not three, conductors. D.c. is also used on undersea cables in order to avoid the effect of *capacitance* on coaxial cables.

The use of d.c. in modern power systems is often confusing to those who remember that d.c. was historically associated with excessive line losses, which motivated the selection of a.c. as a national and international standard. The crucial difference is that early d.c. systems were operated at low voltage—hence the losses—because there was no practical way to step d.c. voltage up and down between transmission and end users (since transformers do not work with d.c.). By contrast, modern d.c. lines, connected to the high-voltage a.c. system with advanced solid-state technology, carry very high voltages and are therefore highly efficient.

168 TRANSMISSION AND DISTRIBUTION

Besides transmitting bulk power over long distances, another application for a.c.-to-d.c.-and-back-to-a.c. conversion in modern power systems is to provide an *intertie* between two a.c. systems that are not synchronous. In this way, power can be shared between these systems even though their a.c. frequency and phase may differ, and the systems are therefore neither subject to stability constraints across the intertie nor vulnerable to frequency-related disturbances from each other. The eastern and western United States are connected by such a d.c. intertie.[9]

6.3 TRANSFORMERS

6.3.1 General Properties

A transformer is a device for changing the voltage in an a.c. circuit. It basically consists of two conductor coils that are connected not electrically but through magnetic flux. As a result of electromagnetic induction, an alternating current in one coil will set up an alternating current in the other. However, the comparative magnitude of the current and voltage on each side will differ according to the geometry, that is, the number of turns or loops in each coil.

Consider the diagram in Figure 6.16 of a very simple transformer. The coil on the left-hand side, which we will label as the *primary* side, might be connected to a power source such as a generator, while the right-hand or *secondary* coil would supply a load. Looking at the electrical connections, we see two separate circuits in this diagram: the circuit between the generator and the transformer (with

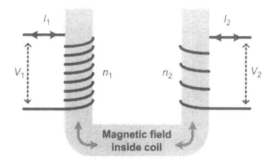

Figure 6.16 Transformer schematic.

---

[9]Another interesting example is the German power system. For four decades, East and West German systems were operated completely independent of each other—one interconnected with eastern Europe and the U.S.S.R., and the other with western Europe. Although much of the hardware infrastructure is very similar, reconnecting these systems after the reunification in 1990 was complicated by different operating standards: while the nominal frequency on both sides was 50 Hz, the tolerance for frequency deviations was much less on the West German side, and the systems could not be kept synchronous. In order to share generation resources without compromising stability, the first step was to install d.c. interties at key points along the former border.

current I_1 and voltage V_1) and the circuit between the transformer and the load (with current I_2 and voltage V_2). Yet electrical power is somehow transmitted *across* the transformer from the generator to the load. In reality, a small percentage of the power will be dissipated in the form of heat from the transformer, just as some resistive heating will occur along the wires. But in the idealized case, where there is no heating in the transformer, all the power is transmitted across. In other words, the same amount of power goes into the transformer as goes out.

Recalling that electric power is given by the product of current and voltage, we can state that the product of current and voltage on each side of the transformer must be the same. Referring to the labels in the diagram, we would write

$$I_1 \cdot V_1 = I_2 \cdot V_2$$

However, based on examining only the electrical connections, we cannot say anything more about how the voltage and current relate. For this, we must look to what happens with the magnetic flux and induction inside the transformer.

First, in the primary coil, the a.c. supplied by a generator produces a magnetic field, or flux inside the core of the coil. Like the conductors inside a generator, transformer coils are wound around a core of magnetically susceptible material, generally some type of iron, to enhance the magnetic field.[10] The magnetic flux resulting from the current is proportional to both the current's magnitude and the number of turns in the coil.

Using the terminology introduced in Section 2.4, we can state that the product of the current I and the number of turns n in the coil gives the *magnetomotive force* (*mmf*), which produces the *magnetic flux* (denoted by Φ, the Greek phi) inside its core. The flux is enhanced because the iron core has a high magnetic permeability, or a low magnetic *reluctance* (denoted by $\mathscr{R}$). In equation form,

$$\text{mmf} = nI$$

$$\Phi = \frac{\text{mmf}}{\mathscr{R}}$$

Because the reluctance of the transformer core is significantly lower than that of the surrounding air, the flux is in effect "captured" inside the core, meaning that the leakage flux outside the core is small.

As the current on the primary transformer winding alternates, the flux inside the core changes direction back and forth 60 times per second (for 60-Hz a.c.). But this same changing flux links the secondary winding around the same core. Proportional to the rate of change of this flux, an electromotive force (emf) is induced in the secondary winding. A current then flows in this winding as determined by the

---

[10] The choice of material for a transformer core affects its behavior under different a.c. frequencies, which is relevant in contexts other than power systems. At high frequencies, it becomes more difficult for iron to sustain the rapid reversal of the magnetic field, causing increased heat losses. Transformers do work in principle even with nothing but air in the core.

impedance connected to it. The emf is also proportional to the number of turns, denoted by n, of this winding around the core. This makes sense in that an emf is induced in each turn, and the successive turns of the same conductor are effectively in series, so that the voltages or emfs are additive.

With regard to changing the voltage from the primary to the secondary side, the important measure is the *turns ratio*, or ratio of the number of turns in the primary and secondary winding. If both primary and secondary windings had the same number of turns, the voltage would be the same on either side of the transformer. In order to increase or step up the voltage, the number of turns must be greater on the secondary side; conversely, to step down voltage, the number of turns must be less. For example, the primary winding in Figure 6.16 has eight turns and the secondary winding four, so that the voltage is stepped down by a factor of 2. A real transformer would have hundreds of turns in each winding so as to make the magnetic induction process more efficient, but as long as the ratio of primary to secondary turns remains the same the effect on voltage is the same. Because power is conserved (to a first approximation), and since power equals voltage times current, the currents through either winding are inversely proportional to the number of turns. In equation form,

$$\frac{V_1}{V_2} = \frac{n_1}{n_2} = \frac{I_2}{I_1}$$

The transformer in this example would be called a *step-down* transformer if the greater number of turns is on the primary side, which is defined as the side to which the power source is connected. An identical transformer, if connected to a generator on the right and a load on the left, would be called a *step-up* transformer.

The voltage on the secondary side can be deliberately changed if there is a moveable connection between the winding and the conductor. Such a connection is called a *transformer tap*. Depending on where the conductor taps the secondary winding, this circuit will effectively "see" a different number of turns, and the transformer will have a different effective turns ratio. By moving the tap up or down along the winding, the voltage can be adjusted. Distribution transformers, especially at the substation level, generally have *load tap changers* (*LTCs*) to adjust the connection in a number of steps. These LTCs are moved to different settings in order to compensate for the changes in voltage level that are associated with changes in load (see Section 6.6).

6.3.2 Transformer Heating

In a real transformer, some power is dissipated in the form of heat. A portion of these power losses occur in the conductor windings due to electrical resistance and are referred to as *copper losses*. However, so-called *iron losses* from the transformer core are also important. The latter result from the rapid change of direction of the magnetic field, which means that the microscopic iron particles must continually realign themselves—technically, their *magnetic moment*—in the direction of the field (or flux). Just as with the flow of charge, this realignment encounters friction on the microscopic level and therefore dissipates energy, which becomes tangible as heating of the material.

Taking account of both iron and copper losses, the efficiency (or ratio of electrical power out to electrical power in) of real transformers can be in the high 90% range. Still, even a small percentage of losses in a large transformer corresponds to a significant amount of heat that must be dealt with. In the case of small transformers inside typical household adaptors for low-voltage d.c. appliances, we know that they are warm to the touch. Yet they transfer such small quantities of power that the heat is easily dissipated into the ambient air (bothering only conservation-minded analysts, who note the energy waste that could be avoided by unplugging all these adaptors when not in use). By contrast, suppose a 10-MVA transformer at a distribution substation operates at an efficiency of 99%: A 1% loss here corresponds to a staggering 100 kW.

In general, smaller transformers like those on distribution poles are passively cooled by simply radiating heat away to their surroundings, sometimes assisted by radiator vanes that maximize the available surface area for removing the heat.

Figure 6.17 Transformer. (Courtesy of Pacific Gas & Electric.)

Large transformers like those at substations or power plants require the heat to be removed from the core and windings by active cooling, generally through circulating oil that simultaneously functions as an electrical insulator.

The capacity limit of a transformer is dictated by the rate of heat dissipation. Thus, as is true for power lines, the ability to load a transformer depends in part on ambient conditions including temperature, wind, and rain. For example, if a transformer appears to be reaching its thermal limit on a hot day, one way to salvage the situation is to hose down its exterior with cold water—a procedure that is not "by the book," but has been reported to work in emergencies. When transformers are operated near their capacity limit, the key variable to monitor is the internal or oil temperature. This task is complicated by the problem that the temperature may not be uniform throughout the inside of the transformer, and damage can be done by just a local *hot spot*. Under extreme heat, the oil can break down, sustain an electric arc, or even burn, and a transformer may explode.

A cooling and insulating fluid for transformers has to meet criteria similar to those for other high-voltage equipment, such as circuit breakers and capacitors: it must conduct heat but not electricity; it must not be chemically reactive; and it must not be easily ionized, which would allow arcs to form. Mineral oil meets these criteria fairly well, since the long, nonpolar molecules do not readily break apart under an electric field.

Another class of compounds that performs very well and has been in widespread use for transformers and other equipment is polychlorinated biphenyls, commonly known as PCBs. Because PCBs and the dioxins that contaminate them were found to be carcinogenic and ecologically toxic and persistent, they are no longer manufactured in the United States; the installation of new PCB-containing utility equipment has been banned since 1977.[11] However, much of the extant hardware predates this phase-out and is therefore subject to careful maintenance and disposal procedures (somewhat analogous to asbestos in buildings).

Introduced in the 1960s, sulfur hexafluoride (SF_6) is another very effective arc-extinguishing fluid for high-voltage equipment. SF_6 has the advantage of being reasonably nontoxic as well as chemically inert, and it has a superior ability to withstand electric fields without ionizing. While the size of transformers and capacitors is constrained by other factors, circuit breakers can be made much smaller with SF_6 than traditional oil-filled breakers. However, it turns out that SF_6 absorbs thermal infrared radiation and thus acts as a greenhouse gas when it escapes into the atmosphere; it is included among regulated substances in the Kyoto Protocol on global climate change. SF_6 in the atmosphere also appears to form another compound by the name of trifluoromethyl sulfur pentafluoride (SF_5CF_3), an even more potent greenhouse gas whose atmospheric concentration is rapidly increasing.[12]

---

[11] See http://www.epa.gov/history/topics/pcbs/01.htm (accessed December 2004).

[12] SF_5CF_3 is the most potent greenhouse gas currently known in terms of radiative forcing per molecule; it also has a very long residence time in the atmosphere. See W.T. Sturges et al., "A Potent Greenhouse Gas Identified in the Atmosphere: SF_5CF_3," *Science* **289**: 611–613 (July 28, 2000); http://cdiac.esd.ornl.gov/trends/otheratg/sturges/sturges.html; http://www.eia.doe.gov/oiaf/1605/gg01rpt/emission.html (accessed December 2004).

This surprising and unfortunate characteristic may motivate future restriction of SF$_6$ use.

6.3.3 Delta and Wye Transformers

In the explanation given in the preceding section, each side of a transformer has one winding, or a single conductor with two ends to connect to a circuit. What happened to the three phases? Transforming three-phase power actually requires three transformers, one for each phase. These three may be enclosed in a single casing, labeled to contain a single three-phase transformer, or they may be three separate units standing next to each other, called a *transformer bank*. In either case, the three transformers are magnetically separate, meaning that they should not share magnetic flux among their cores, because the magnetic field or flux in each one oscillates in a different phase (i.e., with different timing). Figure 6.17 shows a large, three-phase transformer at a utility substation.

The two distinct ways of connecting a set of single-phase loads to a three-phase system, the delta and the wye connections (see Section 6.2.3), also apply to connecting a set of transformers. In the wye connection, each transformer winding connects between one phase and ground, or a common neutral point shared by all three windings. In the delta connection, each winding connects between one phase and another. However, because the primary and secondary side of a transformer are electrically separate, the type of connection on either side need not be the same. This provides for four distinct possibilities for a three-phase transformer connection: $\Delta-\Delta$, $Y-Y$, $\Delta-Y$, and $Y-\Delta$, all of which are commonly used. One of these four possibilities, the $\Delta-Y$ connection, is illustrated in Figure 6.18, showing both the shorthand symbolic representation and a schematic of the actual wiring. The letter n in the diagram stands for neutral.

The choice of delta or wye connection has implications for the voltage. Specifically, if the connection is different on either side, this will in itself affect the apparent amount of voltage increase or decrease, in addition to the effect of the turns ratio.

For the case of the delta–delta and the wye–wye connections, the change in voltage from the primary to the secondary side is simply given by the turns ratio. Here, delta and wye connections are relevant only with respect to grounding, which is important in case of a fault. If the connections on the primary and secondary sides differ, though, the voltage is modified, because the transformer is in effect converting a phase-to-phase to a phase-to-ground voltage or vice versa.

Suppose the delta–wye transformer in Figure 6.18 is connected on its left-hand (primary) side to a transmission line with 115-kV phase-to-phase and 66.4-kV phase-to-ground voltage. This transformer has a turns ratio of 1:1, so that the voltage across any pair of windings is identical. On the primary side with the delta connection, this voltage corresponds to the phase-to-phase voltage, 115 kV. This same voltage on the secondary side with the wye connection, however, is now made to correspond to the phase-to-ground voltage. Because the relationship

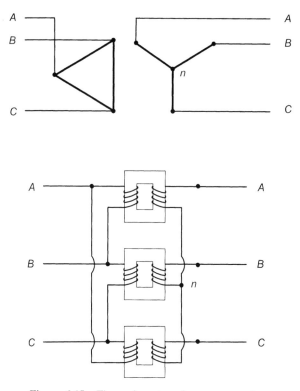

Figure 6.18 Three-phase transformer connections.

between phase-to-phase and phase-to-ground is mathematically fixed, the phase-to-phase voltage on the secondary side will be raised by a factor of $\sqrt{3}$, to 199 kV. Comparing the phase-to-phase voltages on the primary and secondary side, we find that the delta–wye transformer has effectively increased voltage by a factor of $\sqrt{3}$. A wye–delta transformer has the same effect in the opposite direction.

Note that the $\sqrt{3}$ voltage change effected by a wye–delta or delta–wye transformer involves no difference in the number of turns in the primary and secondary coils. For a step-down or step-up transformer that requires a voltage change by more than a factor of $\sqrt{3}$, the turns ratio will be chosen such that the combined effect of turns ratio and $\sqrt{3}$ equals the desired voltage ratio.

Example

A wye–delta transformer with a 10:1 turns ratio steps voltage down from a 230-kV transmission line to a distribution circuit. What is the voltage on the secondary side?

Transmission-line voltages are labeled on the basis of phase-to-phase measurements. The wye connection on the primary side indicates that on its primary side, the transformer "sees" the phase-to-ground voltage, $230/\sqrt{3} = 133$ kV. Due to the turns ratio, the voltage across the secondary transformer winding is less by a factor of 10, or 13.3 kV. Because of the delta connection on the secondary side, this is the phase-to-phase voltage, and 13.3 kV is the correct label for the secondary voltage. All in all, the voltage has been stepped down by a factor of $10 \cdot \sqrt{3} = 17.32$.

Beyond the change in the voltage magnitude, the delta–wye conversion in transformers also affects the phase angle. From the curves in Figure 6.15, we can appreciate that there is a change in the timing when we compare the phase-to-ground with the phase-to-phase voltage. The difference, as can be determined through trigonometry or by inspection of the graph, is $30°$ ($\pi/6$ radians). This can pose a problem if two parallel paths in a network have a different phase shift. In this case, two voltages would meet again after having gone through their respective transformers, one shifted $30°$ relative to the other. Such an addition within a network loop creates a circulating current, which is wasteful (not to mention terribly messy for the engineering calculations). Therefore, power system designers try to ensure that any pair of parallel paths in a network have equal transformer gains in terms of both voltage and phase. A system for which this is true is termed a *normal* system.

6.4 CHARACTERISTICS OF POWER LINES

6.4.1 Conductors

Conductors of overhead transmission and distribution lines typically consist of aluminum, which is lightweight and relatively inexpensive, and are often reinforced with steel for strength. *Stranded* cable is often used, which, as the name suggests, is twisted from many individual strands. At the same diameter or *gauge*, stranded cable is much easier to bend and manipulate. For underground lines, *cables* with insulation are used. Here heat dissipation is more of an issue, whereas weight is not. Copper is the material of choice for underground cables because, while it is more expensive, it has a lower resistance than aluminum. Low resistance is generally desirable for power lines to minimize energy losses, but also because heating limits the conductor's ability to carry current.

Recall from Section 1.2 that resistance is given by $R = \rho l/A$, where A is the cross-sectional area, l is the length of the conductor, and ρ (rho) is the resistivity (inverse of conductivity). The electrical resistance of a power line thus increases linearly with distance and decreases with the conductor cross section (which, in turn, is proportional to the square of the radius or wire diameter). For the purpose of minimizing resistance, then, conductors should be chosen large. However, resistance must be weighed against other factors, including the cost of the conductor cable itself and its weight that needs to be supported by the towers. Because even aluminum conducts so well, this trade-off comes out in favor of surprisingly slender lines considering the amount of current and power transferred.

Note that while resistance of lines is critical in the context of line losses, it is less important in the context of power flow and stability. This is because the overall impedance of a line tends to be dominated in practice by its inductive reactance, to such an extent that it is sometimes appropriate in calculation to make an approximation where a line has zero resistance and only reactance—the *lossless* line. It is perhaps surprising that transmission and distribution lines should have any inductance at all, even though they do not resemble wire coils. Recall that inductance is based on magnetic flux lines linking a loop of wire (see Section 2.4). This notion extends to a straight wire, which can be considered an infinitely large loop, and the magnetic flux around the wire does link it. Since there is only a fraction of a turn in a straight line, this magnetic effect is quite weak. But it is cumulative on a per–unit-length basis, and with a conductor that extends over tens or hundreds of miles, it does eventually add up. Indeed, there are two contributions to line inductance: the self-inductance, which is just a property of the individual conductor, and the mutual inductance, which occurs between the conductors of the three different phases.

Transmission lines have capacitance, too. It is a bit easier to see how two lines traveling next to each other would vaguely resemble opposing plates with a gap in between. In fact, there is also capacitance between a conductor and the ground. Because the lines are small and the gap wide, the capacitance tends to be fairly small. Capacitance is especially important, though, for *coaxial* cables where one conductor surrounds another with insulation in between. Coaxial cables are used where simplicity and compactness matters; for example, on underground or undersea cables for d.c. transmission.

In describing transmission-line parameters, the inductance is generally considered to be in series and the capacitance in parallel. Figure 6.19 illustrates the modeling of a transmission line. Without delving into the details, we can appreciate qualitatively that a line can be characterized in terms of an equivalent resistance,

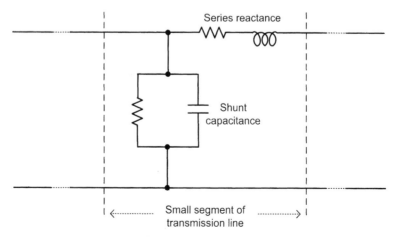

Figure 6.19 Transmission-line modeling.

inductance, and capacitance on a per-meter or per-mile basis. Readers familiar with calculus will recognize the logic of summing up the varying contributions from many small segments of a transmission line over its length to obtain its equivalent properties for that entire length. As the length increases, inductance and capacitance both increase, as does resistance.

The ratio of series impedance (the combination of resistance and reactance) and shunt admittance (the inverse of impedance) determines a quantity called the *characteristic impedance*. In the case where resistance is negligible, this reduces to the ratio of inductance to capacitance (actually, the square root of this ratio) and is called the *surge impedance* of a line. The significance of the surge impedance in telecommunications is that a voltage signal can be transmitted with minimal loss if the resistance connected to the end of a line is equal in magnitude to the line's surge impedance. For power transmission, it is more common to speak of a line's *surge impedance loading (SIL)*, an amount of real power in MW that is given by the square of transmission voltage divided by the surge impedance. The SIL does not measure a line's power carrying capacity, but rather states the amount of real power transmission in the situation where the line's inductive and capacitive properties are completely balanced. To system operators, this provides a benchmark: if the power transmitted along a line (at unity power factor) is less than the SIL, the line appears as a capacitance that injects reactive power (VARs) into the system; if transmitted power exceeds the SIL (the more common situation), the line appears as an inductance that consumes VARs, and thus contributes to reactive losses in the system.

The fact that the inductive property dominates at higher loading makes sense because the line's reactive power consumption is a function of the line current, whereas the capacitive property of injecting reactive power is a function of the voltage at which the line is energized. In equation form, $Q_{loss} = I^2 X_L$ and $Q_{prod} = V^2 / X_C$. From these equations we can also see that $Q_{loss} = Q_{prod}$ when $X_L X_C = V^2 / I^2$. Substituting $X_L = \omega L$ and $X_C = 1/\omega C$ (see Section 3.2), this gives us the impedance (the ratio of voltage to current, as in Ohm's law) of $V/I = \sqrt{L/C}$, which is the surge impedance. The surge impedance loading is then given by $SIL = V^2 / \sqrt{L/C}$.

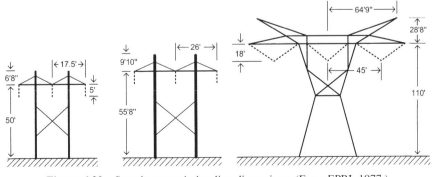

Figure 6.20 Sample transmission-line dimensions. (From EPRI, 1977.)

TABLE 6.1 Sample Transmission Line Data

	Line Voltage (kV)		
	138	345	765
Conductors per phase	1	2	4
Number of strands aluminum/steel	54/7	45/7	54/19
Diameter (in.)	0.977	1.165	1.424
Conductor geometric mean radius (ft)	0.0329	0.0386	0.0479
Current-carrying capacity per conductor (A)	770	1010	1250
Geometric mean diameter phase spacing (ft)	22.05	32.76	56.7
Inductance (H/m $\times$ 10^{-7})	13.02	9.83	8.81
Inductive reactance X_L (Ohms/mi)	0.789	0.596	0.535
Capacitance (F/m $\times$ 10^{-12})	8.84	11.59	12.78
Capacitive reactance X_C (MOhms/mi)	0.186	0.142	0.129
Resistance (Ohms/mi)	0.1688	0.0564	0.0201
Surge impedance loading (MVA)	50	415	2268

Source: Bergen, Arthur R.; Vittal, Vijay, *Power Systems Analysis*, 2nd Edition, © 2000, p. 85. Adapted by permission of Pearson Education, Inc., Upper Saddle River, NJ.

Figure 6.20 and Table 6.1 give some examples for physical dimensions and electrical properties of transmission lines. Note that the inductance, since it is modeled in series, corresponds to a relatively small impedance. The capacitance, by contrast, is modeled in parallel and therefore corresponds to a large impedance (since the larger a parallel impedance, the smaller its impact).

Conductors on transmission lines—especially high-voltage, high-capacity lines—are sometimes *bundled*, meaning that what is electrically a single conductor is actually composed of two, three, or four wires a few inches apart, held together every so often with connectors known as *conducting frames*. There are several reasons for bundling conductors: increasing heat dissipation, reducing *corona losses*, and reducing inductance.

The first of these reasons is straightforward: by dividing a conductor in two (or more), one can keep the cross-sectional area (and thus the resistance and the weight) the same while increasing the surface area. This larger surface area allows the conductor to more effectively radiate heat off into the surrounding environment. Thus, for the same amount of power dissipated (at the same current), the conductor's equilibrium temperature will be lower.

The second reason for bundling conductors is to reduce corona losses. The corona results from the electric field that surrounds the conductor at high voltage. As at the top of a *Tesla coil*[13] (though fortunately much less intense), microscopic arcs occur between the conductor surface at high potential and ionized air molecules in the vicinity. The high frequency in a Tesla coil results in an arc "boring" its way into the air much farther than one would expect based on the ionization potential of

---

[13] A device invented by Nikola Tesla (1856–1943) that produces very high voltages and lightning discharges by transforming a rapidly alternating current.

air (see Section 1.3). A frequency of 60 Hz is less potent for producing this effect, but still, the audible crackling sound around high-voltage a.c. equipment comes from the corona of tiny arcs that discharge into the air. Because the arcs are so small, they are not visible even at night. Yet there is a measurable energy loss associated with what is in fact a small electric current flowing to ground through the air. The power associated with this current is the corona loss. When the surface area of the conductor is increased, the electric potential or surface charge density is spread out more, reducing the electric field strength. This in turn reduces the formation of arcs, and thus reduces corona losses.

The third reason to bundle conductors is to reduce line inductance, which is the least intuitive. Careful analysis of the flux both outside and within a conductor shows that the inductance is less for a bigger wire diameter, and less if the three phases are closer together.[14] Changing a single wire into a bundled conductor makes it resemble, as far as the magnetic field is concerned, a wire of larger diameter, and reducing the magnetic field or flux linkage in turn reduces the line's inductance. The effective conductor size approximated by a set of bundled conductors is described in terms of a *geometric mean radius* (*GMR*).

Occasionally, one sees transmission towers where the positions of the three phases are changed or *transposed*. This is done to keep the mutual inductance about equal on all three phases, which is desirable for purposes of balancing the load (not to mention keeping up with the engineering analysis). Arranging the three conductors symmetrically as an equilateral triangle solves the problem, but is not always practical, depending on the tower or pole design. If three conductors are in a row, there is an asymmetry because the one in the middle is closest to both other conductors. Over a long distance, therefore, the wires are transposed every so often, allowing each individual phase to cover roughly the same distance in each of the three positions.

6.4.2 Towers, Insulators, and Other Components

The poles or towers that support overhead transmission and distribution lines are usually made of wood or, for the larger transmission towers, metal. Designs used by different utilities vary depending on line voltage, conductor size and weight, terrain, aesthetic preferences, and tradition. Figure 6.21 shows a 500 kV line; note the scale compared to the parked vehicle. Ideally, the distance between conductors is maximized while using a minimum amount of materials for tower construction. In any case, towers have to be made tall enough so that there is sufficient clearance between the hanging conductor and the ground or any other objects that may threaten to come into contact with an energized line. Of course, clearance also has to be maintained between the conductors and the towers. Depending on the line voltage, there are standard design criteria for minimum safe clearance, which have to take into account the *sagging* of lines due to thermal expansion

---

[14] A thorough explanation can be found in A. Bergen, *Power Systems Analysis* (Englewood Cliffs, NJ: Prentice Hall, 1986).

Figure 6.21 High-voltage transmission line. (Courtesy of Gene Rochlin.)

under high current. In order to achieve a given clearance, there is an engineering trade-off between making the towers tall versus using more towers at closer spacing. Sometimes the separation between towers is constrained; for example, when crossing a body of water. Here, the towers have to be made extra tall in order to allow for sufficient clearance.

Insulators serve to electrically separate the conductor from the tower. They are made in carefully designed, rounded shapes consisting of one or several *bells* made of a nonconducting ceramic or plastic. The surface of the bells is round and smooth to minimize the potential for arcs to form. Dividing a long insulator into a string of individual bells is more effective than a single cylindrical shape, because it provides more surface area. This area helps spread out surface charge and discourages a current from creeping along the surface, especially when the insulator is wet. The length of insulators, or number of bells, is roughly proportional to the line voltage, which corresponds to the potential difference that the insulator has to maintain.[15] As a rule of thumb, each one of the typical bells contributes to insulate against a voltage on the order of 10 kV. Thus, counting the number of bells on the insulators can give a rough idea of the voltage on a transmission or distribution line. Single bells are used for primary distribution below or around 10 kV.

Most often, transmission lines hang down from the tower on a single insulator, but sometimes one sees configurations with horizontally extended insulators and the conductor describing a semicircle underneath. These are the places where horizontal tension is applied to the conductors. In order to create a reasonable clearance and not let the wires droop down, the amount of tension is several times greater than the actual weight of the conductor.[16]

Owing to their shape and conductivity, metal transmission towers are a likely target for lightning strike. The electric current associated with lightning can usually travel through a metal tower into ground without doing any damage. However, one wants to prevent lightning from traveling along the actual conductors, where it would cause severe voltage fluctuations and potential equipment damage. The metal top of a transmission tower serves to attract lightning and direct it straight down into the tower to ground. In areas where lightning is common, wooden distribution poles are specifically equipped with metal lightning arresters. In addition, it is common to connect metal towers with a ground wire—usually a small diameter wire at the top of the towers—that is electrically separate from the power circuit. In case one tower experiences a change in potential due to lightning, the grounding wire provides a path for current to flow and the potential to equalize among neighboring towers. This minimizes the danger of an arc jumping across an insulator between a tower and conductor and lightning current flowing along the conductor.

A curious item one sees on transmission lines are the small metallic objects, usually with a round bulb on each end, attached to the conductors not far from the tower. Their function is to reduce the amount of swinging and vibration of the conductor in the wind, by changing its mechanical resonant frequency. Finally, there are the large red and white plastic balls which, as most people guess correctly, are intended for airplane and helicopter pilots to see.

---

[15]To be precise, the insulator must be able to sustain the maximum instantaneous line-to-ground voltage (given that the tower or pole is at ground potential), whereas the "line voltage" is conventionally quoted as the line-to-line rms value (see Section 3.1). The maximum is greater than the root mean squared (rms) value by a factor of $\sqrt{2}$, while the line-to-ground voltage is less than line-to-line by a factor of $\sqrt{3}$. The insulators on a 230-kV line, for example, must sustain an instantaneous potential difference between conductor and tower of $230 \text{ kV} \cdot \sqrt{2}/\sqrt{3} = 188 \text{ kV}$.

[16]Readers with some background in physics may enjoy the classic problem of deriving the parametric equations that specify the shape described by the conductor suspended under tension (no, it is not exactly a parabola). (*Hint*: Consider the angle relative to horizontal that the conductor makes at any given point. Note that any given small segment of conductor is not accelerating (the forces on it are equal in all directions), and that the force pulling it toward the bottom of the curve is determined by the mass below it.)

6.5 LOADING

6.5.1 Thermal Limits

Distribution lines and short to medium transmission lines are limited in their capacity to transmit power by resistive heating. It is thus the magnitude of the current, continuing over time and increasingly heating the conductor, that limits the loading; this is the *thermal limit*. As the conductor heats up, it stretches from thermal expansion, and the line sags. If it sags too far, the distortion becomes irreversible. This is bad for the conductor and may violate the clearance requirement. Also, the resistance of the conductor will increase with temperature. This is generally a small effect, but it does eventually become noticeable, especially since increasing resistance in turn will increase heating, not to mention losses. In the worst-case scenario—say, if a fault did not get cleared—a conductor can actually melt off the pole.

Because conductor temperature is the real limiting factor, the *rating* of lines by the amount of current they can safely carry is an approximation based on assumptions about the weather. If it is cold and windy, the line can dissipate more power while remaining at the same temperature. General practice among transmission and distribution engineers has been to err on the safe side and rate lines conservatively for hot weather with no wind. However, in the interest of improved asset utilization, it has become common to adopt variable ratings that take into account ambient conditions. A relatively crude approach is to have a summer and a winter rating. Of course, this does not help very much if load peaks and transmission congestion occur in the summer. *Dynamic ratings* are ones that are updated more frequently with current weather information.

Besides the ambient conditions, it is also important to consider the loading history of the line, since heating occurs gradually. In this vein, lines often have a *normal* and an *emergency rating*. The emergency rating may be good for a number of minutes or hours, which may be just long enough to sustain the load in case of a contingency (such as the loss of another line).

Because current is the key variable for heating, line ratings are generally expressed in terms of *ampacity*, or current-carrying capacity in amperes. This ampacity is independent of voltage. Thus, the amount of power that can actually be transmitted by a given conductor depends on the operating voltage. Note that when the current is translated into power, this means apparent power in volt-amperes (VA), not real power in watts, since only the total magnitude and not the phase of the current matters for heating purposes.

Heating also limits the operation of transformers at high current. The situation is slightly more complicated, because the amount of energy dissipated within the transformer core depends to some extent on voltage (and on a.c. frequency, though this is unimportant here), but the key concern is again temperature. Although transformers are not as exposed to the elements as power lines, environmental factors such as ambient temperature, wind, and rain do play a role, as does the circulation of oil or coolant. Transformer ratings are given in terms of apparent power (kVA or MVA).

Because of the importance of environmental factors as well as the time dimension in overheating equipment, any thermal rating is inherently approximate. As mentioned in the context of generators, there has been some historical and cultural trend from conservative toward more exact ratings, a trend that surely is not unique to the electric power industry. Depending on the vintage and design of an individual piece of utility equipment, its official nameplate rating may or may not coincide with an engineer's or operator's judgment of what loading it can actually withstand at a given time without damage. Variable ratings can be understood as an effort to formalize this judgment while deriving maximum economic benefit from the extant hardware.

6.5.2 Stability Limit

Aside from heat, another type of limit on power transmission is sometimes important for longer transmission lines: the *stability limit*. Stability refers to maintaining the feedback between generators on either end of the line that keeps them locked in synchronicity, by making each generator push harder if it tries to speed up and less hard if it tries to slow down (see Section 4.2). Casually speaking, transmitting power along a line requires that a generator on the sending end pushes harder than one on the receiving end. Accordingly, the sending generator has a *power angle* δ that is somewhat ahead of the receiving generator. This power angle indicates the exact timing of the generator emf or voltage pulse in relation to the voltage maximum of the system.

As further discussed in Section 8.3, the amount of real power transmitted on an ideal, *lossless* line (with no resistance, only reactance) is given by the equation

$$P = \frac{V_1 V_2}{X} \sin \delta_{12}$$

where δ_{12} is the difference in power angles between the sending and receiving end of the line,[17] V_1 and V_2 are the voltage magnitudes at either end, and X is the reactance of the line in between. In order to transmit a large amount of power on a given line, it is necessary to increase the angle difference δ_{12}. However, this difference cannot be made arbitrarily great, because the feedback between the generators becomes less effective for large δ_{12}, and there is an increasing risk of loss of synchronism.

For short lines, the reactance X in the denominator is usually small, so that a small δ_{12} still results in a large amount of power transmitted. Consequently, if one were to apply the maximum permissible δ_{12} to such a line, the power transmitted could easily exceed the line's thermal capacity. For long lines, however, the reactance becomes more significant. In this case, a dangerous δ_{12} well may be reached before the thermal limit of the line. Hence, a more stringent limit on power transmission called the *stability limit* is imposed. While the thermal limit is expressed in terms of either current (amps) or apparent power (MVA), the stability limit has units of real power (MW).

---

[17]Note that a δ can be assigned to any location within a power system, not necessarily just an individual generator; it simply describes the time at which the voltage peaks at that location. In Chapter 7, in keeping with the convention for power flow analysis, the same angle is called θ.

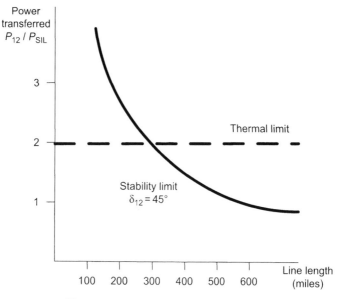

Figure 6.22 Thermal and stability limits.

Figure 6.22 shows the stability limit and thermal limit as a function of length for a hypothetical transmission line. The label P_{12}/P_{SIL} is a measure of the real power transmitted between the two ends of the line, expressed as a ratio of the actual power in watts and the surge impedance loading, which is a characteristic of a given line.[18] The curve shows the amount of power transmitted at a particular value of δ_{12}, 45° in this case, that represents the maximum permissible phase separation between the two ends of the line before synchronicity is lost. The diagram shows that for a short line the thermal limit applies, whereas beyond a certain length the stability limit becomes more constraining.

6.6 VOLTAGE CONTROL

Voltage in power systems is controlled both at generators and on location throughout the transmission and distribution system. As described in Section 4.3.2, the voltage at a generator terminal or *bus* is controlled by the excitation or rotor field current, which determines the rotor magnetic field strength and thus the magnitude of the induced emf in the armature. Generator voltage is directly linked to reactive power generation; the two variables cannot be controlled independently of each other. Since various generators may be producing different amounts of real and reactive power, their bus voltages vary slightly from the nominal voltage that is the same for all.

This variation is part of a subtle profile of voltage levels that rise and fall on the order of several percent throughout any interconnected power system. Such a profile

---

[18] The SIL represents the amount of power transferred when the impedance connected at the end of the line matches the characteristic impedance of the line itself; see Section 6.4.1.

exists separately from the order-of-magnitude voltage changes introduced by transformers. For example, within a network that is nominally operating at 230 kV, the actual voltage at different locations may vary by thousands of volts (1% being 2.3 kV or 2300 V).[19] The exact voltage level at each location depends mainly on two factors: the amount of reactive power generated or consumed in the vicinity, and the amount of voltage drop associated with resistive losses.

In radial distribution systems, the voltage-drop effect dominates. Here the voltage simply decreases as one moves from the substation (the power source, in effect) out toward the end of a distribution feeder. This change in voltage is known as the *line drop*. The line drop is described by Ohm's law, $V = IZ$, where I is the current flowing through the line, Z is the line's impedance, and V is the voltage difference between the two ends. Ohm's law also shows us that the line drop depends on the connected load, since a greater power demand implies a greater current. While the line impedance stays the same, the voltage drop varies in proportion to the load.

In practice, the voltage drops in distribution systems are quite significant, especially for long feeders. Recognizing that it is physically impossible to maintain a perfectly flat profile, operational guidelines in the United States generally prescribe a tolerance of $\pm 5\%$ of the nominal voltage. This range applies throughout transmission and distribution systems, down to the customer level. For example, a customer nominally receiving 120 V should expect to measure anywhere between 114 and 126 V at their service drop.

Figure 6.23 illustrates the problem of voltage drop along a feeder. If the feeder is very long, the voltage drop may exceed the window of tolerance, so that if the first customer is receiving no more than 126 V, the last would receive less than 114 V. In order to maintain a permissible voltage level along the entire length of a feeder, it may therefore be necessary to intervene and boost the voltage somewhere along the way. Furthermore, because the voltage drop varies with load, this boost may need to be adjusted at different times.

The two methods for controlling or supporting voltage in the transmission and distribution system are transformer taps and reactive power injection (usually capacitance). Transformer taps provide for a variable turns ratio, and thus a variable amount of voltage change. The tap is simply where the conductor connects to the transformer coil on the secondary or load side. By moving the tap up or down, the effective number of turns of that transformer winding is changed. This mechanism is called a load tap changer (LTC). An LTC typically has some number of discrete settings to be adjusted by distribution operators according to conditions on the circuit.

The same basic device, when installed midway on an individual feeder rather than at the substation transformer, is called a *voltage regulator*. Voltage regulators look like tall transformers on distribution poles, often with large fins for heat

---

[19]Birds sitting on power lines are famously unharmed because their entire bodies are at the same voltage as the line, with no current flowing through them. In fact, this is only because their feet are so close together. If a giant bird could straddle across many miles to touch two ends of the same line, it would instantly get shocked.

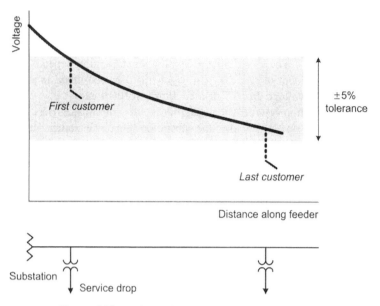

Figure 6.23 Voltage drop along a distribution feeder.

dissipation, but with no secondary lines going out to customers; they are simply transformers between two segments of the same line. Their turns ratio is adjusted to boost the voltage just enough to compensate for line drop.

Reactive power affects voltage in a very different manner. It is provided by *capacitors*, *static VAR compensators* (*SVC*s), or *synchronous condensors*, which are simply synchronous generators operating at zero real power output. All of these devices provide capacitive reactance, meaning that they inject reactive power into the system in order to boost the local voltage level. While the VAR output of SVCs and synchronous condensors can be continually adjusted, capacitor banks are controlled by being either switched in or out of the system. The latter are the simplest and are most common at the distribution level. Capacitors can be automatically controlled, either by sensing local variables such as voltage or current or—very simply—by the time of day, which may be sufficiently well correlated with load. Any of these capacitive devices are usually connected in parallel with the load; a parallel capacitance is also known as a *shunt capacitance*.[20] Series capacitance is used in some specific applications, mostly on transmission lines.

---

[20] This makes sense if we consider that one would generally wish to make only a small adjustment, and ask what size capacitor would be needed. In the series case, we can think of the capacitor as replacing a short circuit, so in order to add only a small impedance, the capacitor needs to be huge. In the parallel case, the capacitor is effectively replacing an open circuit (two previously unconnected parts of the circuit), and adding a small capacitor means adding a small amount of admittance (or lowering the impedance by a small amount).

The effect of voltage control through capacitance can be understood in terms of a reduction in the line drop as a result of injecting reactive power. Compensating for the inductive load with nearby capacitance brings the power factor of the area load closer to unity. This means a reduction in current, since now a smaller apparent power is needed to deliver the real power demanded by the load, which in turn causes a reduction in the voltage drop along the line according to Ohm's law.

A more detailed way to see this is in terms of the circulating current required to serve a reactive load, which has to travel between wherever reactive power is "generated" and "consumed," effectively swapping stored energy between electric or magnetic fields during certain portions of the cycle. If reactive power is injected at a generator bus to match the reactive demand, this circulating current must travel throughout the transmission system to reach any given load. If reactive power is injected locally instead, this circulating current does not need to accompany the real power on its way from the power plant or along the distribution feeder. Besides saving line losses, this means that the feeder current up to the point of the reactive power injection is reduced, with an associated reduction in voltage drop.

The effect of voltage regulation on a distribution feeder can be seen in Figures 6.24 and 6.25, which show the voltage profile on a distribution feeder as a function of distance, with and without intervention.

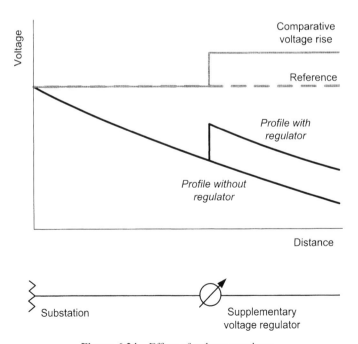

Figure 6.24 Effect of voltage regulator.

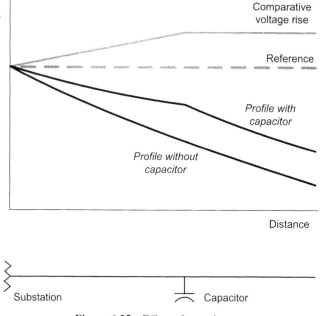

Figure 6.25 Effect of capacitor.

6.7 PROTECTION

6.7.1 Basics of Protection and Protective Devices

Circuit *protection* refers to a scheme for disconnecting sections or components of an electric circuit in the event of a *fault*. A fault means that an inadvertent electrical connection is made between an energized component and something at a different potential. If two conductors are touching directly, this makes a short circuit, or connection between two points that were initially at different potentials with essentially zero resistance in between.

The basic types of faults in power systems are *phase-to-ground* and *phase-to-phase* faults. A phase-to-ground fault means that one or more conductors make electrical contact with the ground, or point of zero-volt potential, such as a line coming in contact with a tree (which, owing to its moisture, will conduct a current to ground). A phase-to-phase fault means that two different phases (or, rarely, all three) come into direct or indirect contact with each other, for example, if a bird with a large wingspan touches two conductors simultaneously.

When analyzing what would happen during any conceivable fault, the main quantity of interest is the *fault current*. The fault current is determined by the *fault impedance*—that is, the impedance of whatever it is between the two points that are inadvertently connected—and by the ability of the power source to sustain the voltage while an abnormally high current is flowing.

A fault is always something to be avoided, not only because it implies a wasteful flow of electric current, but because there is a risk of fire or electrocution when current flows where it was not intended to go. The object of circuit protection is to reliably *detect* a fault when it happens and interrupt the power flow to it, *clearing* the fault. A fault is generally detected by the magnitude of its associated current, though it may also be sensed by way of a phase imbalance or other unusual voltage differences between circuit components.

The simplest protective device that can detect an overcurrent and interrupt a circuit is the *fuse*. It consists of a thin wire that simply melts when the current is too high. While fuses are very reliable, they have practical drawbacks. First, it takes a certain minimum amount of time before the wire heats up enough to melt. Once installed, it is not possible to change the sensitivity of a fuse, or how much current it will take to melt it. Then, once the wire has melted, it has to be physically replaced before the connection can be reestablished; it cannot be reset. This usually means a time delay for restoring the connection. Fuses are used for radial feeders in distribution systems, generally for a lateral feeder where it connects to the main. In these situations, the desired sensitivity of the fuse is fixed, and the time delay for restoring service is considered acceptable because only a small number of customers are affected. Fuses were installed in homes earlier in the 20th century (and are often still in use) before circuit breakers became standard.

Circuit breakers differ from fuses in that they have movable contacts that can open or close the circuit. This means that a circuit breaker can be reset and reused after it opens. The mechanical opening or *tripping* of the breaker is actuated by a *relay* that measures the current and, if the measurement is above a determined value, sends the signal that opens the breaker. Such a relay can have multiple settings, depending on the desired sensitivity.

While a circuit breaker can move more quickly than a melting fuse, it does take a certain amount of time for a current to persist before the relay will actuate. This time is inversely related to the magnitude of the current. At the same setting, the relay could be tripped either by a very large current for a very short amount of time or by a smaller current for a longer duration. The sensitivity of relays and fuses is thus characterized by a *time–current* curve that indicates the combination of current and duration that will cause a trip. A sample curve for a certain relay is shown in Figure 6.26. Note that both current and time are plotted on a logarithmic scale. For a large fault current, the *fault clearing time* should be a fraction of a second, on the order of several or tens of cycles. In some situations it may be problematic to distinguish between what is a fault current and what is just a high load current. This is especially true for *high-impedance faults*, where whatever is making the improper connection between a conductor and ground does not happen to conduct very well, and the fault current is therefore small. This problem is circumvented by another method of fault detection that compares the currents on the two or three different phases, or between a phase and return flow. Even a small fault current from one of the phases to ground will result in a difference between the currents in each conductor. This difference is detected by a *differential relay*, which sends a signal to an actuator that opens the circuit.

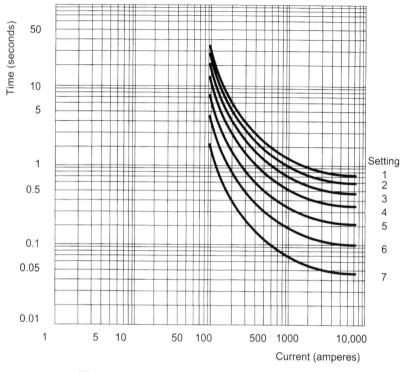

Figure 6.26 Time-current characteristic of a relay.

Differential relays are used in transmission and distribution systems, but also in the familiar *ground-fault circuit interrupters* (*GFCIs*) in residential settings where outlets are close to water sources and there is a danger of appliances coming in contact with water (which could cause persons who are also in contact with the water to be electrocuted, the worst-case scenario being an appliance dropped into the bathtub). Electrical code in the United States requires the installation of GFCIs in kitchens and bathrooms of new homes. GFCIs not only can detect a smaller fault current than a conventional circuit breaker, they can also operate more quickly in situations where any fraction of a second may be crucial.

Switches and circuit breakers in power transmission and distribution systems are collectively referred to as *switchgear*; they serve the purposes of deliberately isolating individual pieces of equipment (say, for maintenance) and to automatically isolate portions of the system (in case of a fault). An important distinction between a circuit breaker and a regular switch is that the breaker can safely interrupt a fault current, which may be much larger than a normal load current. In order to do so, circuit breakers must be specifically designed to control and extinguish the arc of *plasma* (see Section 1.1.5) drawn as the contacts separate. For simple air switches, the design crux is in the shape of the contacts. To be able to operate at higher voltages and currents, the contacts can be immersed in a tank of nonconducting fluid such as transformer oil or sulfur hexafluoride (SF_6; see Section 6.3.2) that is also difficult to ionize and rapidly quenches the arc. (The key is to have a symmetric molecule without free electrons that does not readily break apart under the influence of an electric field.) More effective yet, the contacts can be placed in a vacuum. Finally, there are *air-blast circuit breakers* and *puffer-type arc interrupters* in which a burst of compressed gas such as SF_6 or cleaned air is precisely directed at the arc to quench it.

In the operation of a circuit breaker, time is of the essence, both in terms of when the breaker first actuates and in terms of the physical movement of the contacts. The key problem is that the ionization of the medium (air or otherwise) between the contacts, which forms and sustains the arc, depends on both the voltage across the gap and its width (for example, the ionization potential of air is on the order of a million volts per meter). When the metallic contacts initially move apart, they will necessarily draw an arc between them, just like the arc that can be seen when an operating appliance is unplugged from its outlet, only bigger. But within half a cycle, the alternating voltage and current become zero, and the arc will naturally extinguish. (This, of course, is peculiar to a.c. and helps explain why high-voltage d.c. circuit breakers present an even greater engineering challenge.) By the time the voltage rises again in the opposite direction—and we are talking about mere milliseconds—the breaker contacts may or may not have already moved far enough apart to prevent the formation of a new arc, or *restrike*. Restriking is undesirable because it wears out the breaker contacts, in addition to prolonging the time before the fault is cleared. Ideally, the contacts can be physically moved fast enough so that the growing distance between them outpaces the sinusoidally increasing voltage. The properties of the circuit are relevant here, too, because any reactance will affect the relative timing between current and voltage. Considering these complexities along with the fact that arc plasma temperatures are on the order of tens of thousands of degrees Celsius, that large mechanical components are accelerated to high speeds within thousandths of a second, and that pressurized quenching gas flows at supersonic speeds, we can appreciate that the design of power switchgear is a serious business indeed.

Many times faults are *transient*, meaning that their cause disappears. For example, lightning strike may cause a fault current that will cease once the lightning is over; power lines may make contact momentarily in the wind; or a large bird may electrocute itself across two phases and drop to the ground, removing the connection. In these situations, it is desirable for the circuit to be restored to normal operation immediately after the fault disappears. For this purpose, *reclosing* breakers (or *reclosers* for short) are used. The idea is that the breaker opens when the fault is detected, but then, after some time has passed—the *reclosing time*—closes again to see if the fault is still there. If the current is back to normal, the breaker stays closed and everything is fine; customers have only suffered a very brief interruption. If the fault current is still there, the recloser opens again. This cycle may repeat another time or two, and if the fault persists on the last reclosing attempt, the breaker stays in the open or *lockout* position until it is reset.

The reclosing time and number of attempts can be adjusted as appropriate. In distribution systems, reclosing times tend to be much longer than in transmission systems—five seconds, perhaps, as compared to half a second. Primarily, this is because distribution equipment tends to be closer to the ground and more exposed to environmental factors that take a little longer to go away, including unfortunate incidents with animals. Another consideration is the number of customers exposed to the interruption, which is of course much greater for a transmission fault. This shifts the desired reclosing time in transmission systems toward the short side, especially if it can be made brief enough to escape customers' notice completely. Like time-current settings on a breaker, the choice of recloser settings illustrates the fact that any circuit protection inherently involves some trade-off between safety and convenience.

6.7.2 Protection Coordination

In order to cause the minimum interruption of service, power system protection is carefully designed to interrupt the circuit as close as possible to the fault location. There is also redundancy in protection, meaning that in the event one breaker fails to actuate, another one will. With both of these considerations in mind, protection throughout the system is *coordinated* so that for any given fault, the nearest breaker will trip first. Such a scheme is analyzed in terms of *protection zones*, or sections of the system that a given device is "responsible" for isolating. These zones are nested inside each other, as illustrated in Figures 6.27 to 6.30.

In such a scheme, any one protective device may simultaneously serve as the primary protection for its own zone and as backup for another. For example, Fuse 1 is the primary protection for the section of line between it and Fuse 2, while also

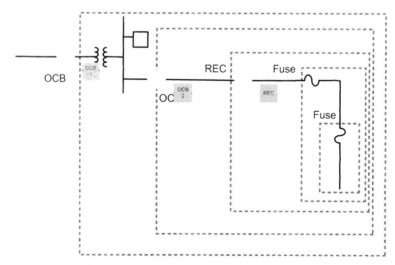

Figure 6.27 Example of protection zones. (OCB = oil circuit breaker, REC = recloser.)

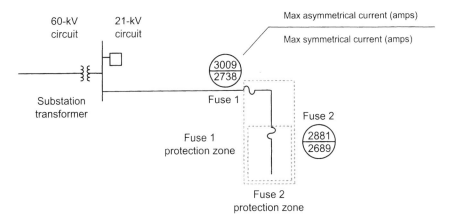

Figure 6.28 Protection zones—fuses.

serving as a backup in case Fuse 2 should fail to clear a fault in its own protection zone. Of course, we do not wish for Fuse 1 to melt and unnecessarily inconvenience customers in its own protection zone as the result of a problem beyond Fuse 2 on the circuit. For this reason, Fuse 1 ought to be less sensitive, that is, tolerate a greater current than Fuse 2. The illustrations show that the maximum current is greater for devices upstream in the circuit, which is also necessary simply because there is more load connected whose normal current must flow through them. The terms *symmetrical* and *asymmetrical* refer to fault currents that are symmetrical about the horizontal axis (i.e., the regular a.c.) versus those with a d.c. component. The sample specifications on the recloser and circuit breaker relay illustrate the crucial

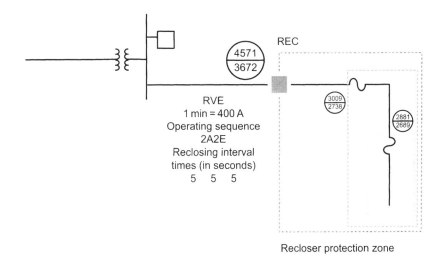

Figure 6.29 Protection zones—recloser.

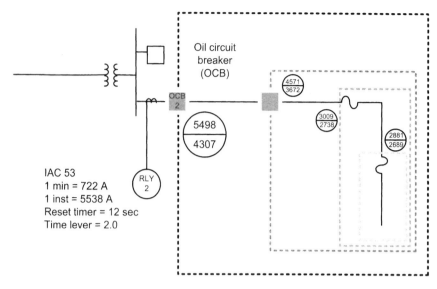

Figure 6.30   Protection zones—relay.

time dimension of the protection problem: it is important not only whether a device operates, but precisely *when*. With this we can begin to appreciate the multivariable character of protection coordination as involving location, current magnitude, individual phases, and time, all of which must be combined into a scheme that can be expected to perform safely and reliably—yet without causing nuisance interruptions—under any foreseeable circumstance.

Still, the example in the figure has a key simplicity: namely, that the distribution system layout here is *radial* and power flow unidirectional. In a network, protection coordination becomes even more challenging, because here the roles of primary and backup protection (i.e., which one trips first) must be reversed depending on what side the fault is on. Yet the only means of discriminating the distance to a fault is by the impedance of the line in between. This complexity alone is sufficient reason for the majority of power distribution systems to be laid out radially. It also explains why protection engineering is a subtle business usually carried out by highly specialized experts who draw not only upon mathematical analysis but also on experience and intuition to make it work in practice.

CHAPTER 7
Power Flow Analysis

7.1 INTRODUCTION

Power flow analysis is concerned with describing the operating state of an entire power system, by which we mean a network of generators, transmission lines, and loads that could represent an area as small as a municipality or as large as several states. Given certain known quantities—typically, the amount of power generated and consumed at different locations—power flow analysis allows one to determine other quantities. The most important of these quantities are the voltages at locations throughout the transmission system, which, for alternating current (a.c.), consist of both a magnitude and a time element or phase angle. Once the voltages are known, the currents flowing through every transmission link can be easily calculated. Thus the name *power flow* or *load flow*, as it is often called in the industry: given the amount of power delivered and where it comes from, power flow analysis tells us how it flows to its destination.

Owing mainly to the peculiarities of a.c., but also to the sheer size and complexity of a real power system—its elaborate topology with many nodes and links, and the large number of generators and loads—it turns out to be no mean feat to deduce what is happening in one part of the system from what is happening elsewhere, despite the fact that these happenings are intimately related through well-understood, deterministic laws of physics. Although we can readily calculate voltages and currents through the branches of small direct current (d.c.) circuits in terms of each other (as seen in Chapter 2), even a small network of a handful of a.c. power sources and loads defies our ability to write down formulas for the relationships among all the variables: as a mathematician would say, the system cannot be solved analytically; there is no *closed-form solution*. We can only get at a numerical answer through a process of successive approximation or *iteration*. In order to find out what the voltage or current at any given point will be, we must in effect simulate the entire system.

Historically, such simulations were accomplished through an actual miniature d.c. model of the power system in use. Generators were represented by small d.c.

Electric Power Systems: A Conceptual Introduction, by Alexandra von Meier
Copyright © 2006 John Wiley & Sons, Inc.

power supplies, loads by resistors, and transmission lines by appropriately sized wires. The voltages and currents could be found empirically by direct measurement. To find out how much the current on line A would increase, for example, due to Generator X taking over power production from Generator Y, one would simply adjust the values on X and Y and go read the ammeter on line A. The d.c. model does not exactly match the behavior of the a.c. system, but it gives an approximation that is close enough for most practical purposes. In the age of computers, we no longer need to physically build such models, but can create them mathematically. With plenty of computational power, we can not only represent a d.c. system, but the a.c. system itself in a way that accounts for the subtleties of a.c. Such a simulation constitutes *power flow analysis*.

Power flow answers the question, What is the present operating state of the system, given certain known quantities? To do this, it uses a mathematical algorithm of successive approximation by iteration, or the repeated application of calculation steps. These steps represent a process of trial and error that starts with assuming one array of numbers for the entire system, comparing the relationships among the numbers to the laws of physics, and then repeatedly adjusting the numbers until the entire array is consistent with both physical law and the conditions stipulated by the user. In practice, this looks like a computer program to which the operator gives certain input information about the power system, and which then provides output that completes the picture of what is happening in the system, that is, how the power is flowing.

There are variations on what types of information are chosen as input and output, and there are also different computational techniques used by different programs to produce the output. Beyond the straightforward power flow program that simply calculates the variables pertaining to a single, existing system condition, there are more involved programs that analyze a multitude of hypothetical situations or system conditions and rank them according to some desired criteria; such programs are known as *optimal power flow* (*OPF*), discussed in Section 7.5.

This chapter is intended to provide the reader with a general sense of what power flow analysis is, how it is useful, and what it can and cannot do. The idea is not necessarily to enable readers to perform power flow analysis on their own, but to carry on an intelligent conversation with someone who does. The level of mathematical detail here is intended to be just sufficient so the reader can appreciate the complexity of the problem.

Section 7.2 introduces the problem of power flow, showing how the power system is abstracted for the purpose of this analysis and how the known and unknown variables are defined. Section 7.3 discusses the interpretation of power flow results based on a sample case and points out some of the general features of power flow in large a.c. systems. For interested readers—though others may skip it without loss of continuity—Section 7.4 explicitly states the equations used in power flow analysis and outlines a basic mathematical algorithm used to solve the problem, showing also what is meant by *decoupled power flow*. Finally, Section 7.5 addresses applications of power flow analysis, particularly optimal power flow.

7.2 THE POWER FLOW PROBLEM

7.2.1 Network Representation

In order to analyze any circuit, we use as a reference those points that are electrically distinct, that is, there is some impedance between them, which can sustain a potential difference. These reference points are called *nodes*. When representing a power system on a large scale, the nodes are called *buses*, since they represent an actual physical *busbar* where different components of the system meet (see Section 4.3 and 6.1.3). A bus is electrically equivalent to a single point on a circuit, and it marks the location of one of two things: a generator that injects power, or a load that consumes power. At the degree of resolution generally desired on the larger scale of analysis, the load buses represent aggregations of loads (or very large individual industrial loads) at the location where they connect to the high-voltage transmission system. Such an aggregation may in reality be a transformer connection to a subtransmission system, which in turn branches out to a number of distribution substations; or it may be a single distribution substation from which originate a set of distribution feeders (see Section 6.1.3).[1] In any case, whatever lies behind the bus is taken as a single load for purposes of the power flow analysis.

The buses in the system are connected by transmission lines. At this scale, one does not generally distinguish among the three phases of an a.c. transmission line (see Section 6.2). Based on the assumption that, to a good approximation, the same thing is happening on each phase, the three are condensed by the model into a single line, making a so-called *one-line diagram*. Most power engineering textbooks provide a detailed justification of this important simplification, demonstrating that what we learn about the single "line" from our analysis can legitimately be extrapolated to all three phases. Indeed, a single line between two buses in the model may represent more than one three-phase circuit. Still, for this analysis, all the important characteristics of these conductors can be condensed into a single quantity, the *impedance* of the one line (see Section 3.2.3). Since the impedance is essentially determined by the physical characteristics of the conductors (such as their material composition, diameter, and length), it is taken to be constant.[2] Note that this obviates the need for geographical accuracy, since the distance between buses is already accounted for within the line impedance, and the lines are drawn in whatever way they best fit on the page.

---

[1] Of course, it is possible to run power flow analysis at different scales, including a smaller scale that explicitly incorporates more distribution system elements. In the present discussion we emphasize the largest transmission scale because of its political and economic importance. Also note that distribution systems usually succumb to simpler methods of analysis because of their radial structure (implying that power flows in only one direction). Power flow analysis is indispensable, however, for the networks that characterize the transmission system.

[2] Ambient conditions such as conductor temperature have a sufficiently small effect on line impedance that they are usually neglected.

198 POWER FLOW ANALYSIS

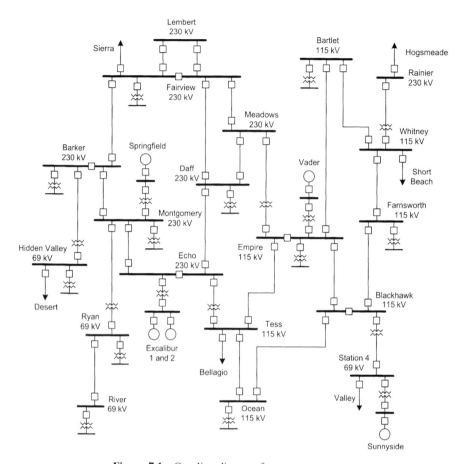

Figure 7.1 One-line diagram for a power system.

Thus, the model so far represents the existing hardware of the power system, drawn as a network of buses connected by lines. An example of such a one-line diagram is shown in Figure 7.1. This topology or characteristic connection of the network may in reality be changed by switching operations, whereby, for example, an individual transmission line can be taken out of service. Such changes, of course, must be reflected by redrawing the one-line diagram, where now some lines may be omitted or assigned a new impedance value. For a given analysis run, though, the network topology is taken to be fixed.

7.2.2 Choice of Variables

From the analysis of simple d.c. circuits in Chapter 2, we are familiar with the notion of organizing the descriptive variables of the circuit into categories of "knowns" and "unknowns," whose relationships can subsequently be expressed in terms of

7.2 THE POWER FLOW PROBLEM

multiple equations. Given sufficient information, these equations can then be manipulated with various techniques so as to yield numerical results for the hitherto unknowns. As some readers may recall from high school algebra, the conditions under which such a system of equations is solvable (meaning that it can yield unambiguous numerical answers) are straightforward: for each unknown quantity, there must be exactly one equation for each unknown. Each equation represents one statement relating one unknown variable to some set of known quantities. This set of equations must not be redundant: if any one equation duplicates information implied by the others, it does not tell us anything new and therefore does not count toward making the whole system solvable. If there are fewer equations than unknowns, we do not have enough information to decide which values the unknowns must take (in other words, the information given does not rule out a multiplicity of possibilities); if there are more equations than unknowns, the system is overspecified, meaning that some equations are either redundant or mutually contradictory. In order to determine whether an unambiguous, unique solution to a system of equations such as those describing an electric power system can be found, one must begin by taking an inventory of variables and information that translates into equations for those variables.

As introduced in Section 1.1, there are two basic quantities that describe the flow of electricity: voltage and current. Recognizing these quantities in simple d.c. circuits in Chapter 2, we saw that both voltage and current will vary from one location to another in a circuit, but they are everywhere related: the current through each circuit branch corresponds to the voltage or potential difference between the two nodes at either end, divided by the impedance of this branch (see Section 2.1). It is generally assumed that the impedances throughout the circuit are known, since these are more or less permanent properties of the hardware. Thus, if we are told the voltages at every node in the circuit, we can deduce from them the currents flowing through all the branches, and everything that is happening in the circuit is completely described. If one or more pieces of voltage information were missing, but we were given appropriate information about the current instead, we could still work backwards and solve the problem. In this sense, the number of variables in a circuit corresponds to the number of electrically distinct points in it: assuming we already know all the properties of the hardware, we need to be told one piece of information per node in order to figure out everything that's going on in a d.c. circuit.

For a.c. circuits, the situation is a bit more complicated, because we have introduced the dimension of *time*: unlike in d.c., where everything is essentially static (except for the instant at which a switch is thrown), with a.c. we are describing an ongoing oscillation or movement. Thus each of the two main variables, voltage and current, in an a.c. circuit really has *two* numerical components: a magnitude component and a time component. By convention, a.c. voltage and current magnitude are describes in terms of *root-mean-squared (rms) values* (see Section 3.1) and their timing in terms of a *phase angle*, which represents the shift of the wave with respect to a reference point in time (see Section 3.1). To fully describe the voltage at any given node in an a.c. circuit, we must therefore specify two numbers: a voltage magnitude and a voltage angle. Accordingly, when we solve for the currents in each branch, we will again obtain two numbers: a current magnitude and a current angle. And when we consider the amount of power transferred at any point of an a.c. circuit, we again have two numbers: a real and a reactive component (see Section 3.3). An a.c. circuit thus requires exactly two pieces of information per node in order to be completely determined. More than two, and they are either redundant or contradictory; fewer than two, and possibilities are left open so that the system cannot be solved.

A word of caution is necessary here: Owing to the nonlinear nature of the power flow problem, it may be impossible to find one unique solution because more than one answer is mathematically consistent with the given configuration.[3] However, it is usually straightforward in such cases to identify the "true" solution among the mathematical possibilities based on physical plausibility and common sense. Conversely, there may be no solution at all because the given information was hypothetical and does not correspond to any situation that is physically possible. Still, it is true in principle—and most important for a general conceptual understanding—that two variables per node are needed to determine everything that is happening in the system.

Having discussed voltage and current, each with magnitude and angle, as the basic electrical quantities, which are known and which are unknown? In practice, current is not known at all; the currents through the various circuit branches turn out to be the last thing that we calculate once we have completed the power flow analysis. Voltage, as we will see, is known explicitly for some buses but not for others. More typically, what is known is the amount of *power* going into or out of a bus. Power flow analysis consists of taking all the known real and reactive power flows at each bus, and those voltage magnitudes that are explicitly known, and from this information calculating the remaining voltage magnitudes and all the voltage angles. This is the hard part. The easy part, finally, is to calculate the current magnitudes and angles from the voltages.

From Section 3.3, we know how to calculate real and reactive power from voltage and current: power is basically the product of voltage and current, and the relative phase angle between voltage and current determines the respective contributions of real and reactive power. Conversely, one can deduce voltage or current magnitude and angle if real and reactive power are given, but it is far more difficult to work out mathematically in this direction. This is because each value of real and reactive power would be consistent with many different possible combinations of voltages and currents. In order to choose the correct ones, we have to check each node in relation to its neighboring nodes in the circuit and find a set of voltages and

---

[3]This can be understood as follows: Finding a solution to a set of equations like a power flow problem is in some sense equivalent to finding the point where a curve crosses an axis. For a straight line (a *linear* problem), there is one and only one crossing (except for the rare special case where the line is parallel to the axis and there are zero crossings). But a bent, nonlinear curve may cross the axis more than once: a parabola, for example, could cross it twice; a cubic function three times. In such a case, there would be more than one mathematically correct solution.

currents that are consistent all the way around the system. This is what power flow analysis does.

7.2.3 Types of Buses

Let us now articulate which variables will actually be given for each bus as inputs to the analysis. Here we must distinguish between different types of buses based on their actual, practical operating constraints. The two main types are *generator buses* and *load buses*, for each of which it is appropriate to specify different information. At the load bus, we assume that the power consumption is given—determined by the consumer—and we specify two numbers, real and reactive power, for each load bus.[4] Referring to the symbols P and Q for real and reactive power, load buses are referred to as *P,Q buses* in power flow analysis.

At the generator buses we could in principle also specify P and Q. Here we run into two problems, however: the first has to do with balancing the power needs of the system, and the second with the actual operational control of generators. As a result, it turns out to be convenient to specify P for all but one generator, the *slack bus*, and to use the generator bus voltage, V, instead of the reactive power, Q, as the second variable. Generator buses are therefore called *P,V buses*.

7.2.4 Variables for Balancing Real Power

Balancing the system means that all the generators in the system collectively must supply power in exactly the amount demanded by the load, plus the amount lost on transmission lines. This applies to both real and reactive power, but let us consider only real power first. If we tried to specify a system in which the sum of P generated did not match the P consumed, our analysis would yield no solution, reflecting the fact that in real life the system would lose synchronicity and crash. Therefore, for all situations corresponding to a stable operation of the system, and thus a viable solution of the power flow problem, we must require that real power generated and consumed matches up. Of course, we can vary the contributions from individual generators—that is, we can choose a different *dispatch*—so long as the *sum* of their P's matches the amount demanded by the system. As mentioned earlier, this total P must not only match the load demand, it must actually *exceed* that amount in order to make up for the transmission losses, which are the resistive I^2R energy losses (see Section 1.4).

Now we have a problem: How are we supposed to know ahead of time what the transmission losses are going to be? Once we have completed the power flow analysis, we will know what the current flows through all the transmission lines

---

[4]We make the assumption here that the load's power demand is independent of the voltage at that bus. The reader might object, recognizing that, for example, an appliance with a fixed impedance uses a different amount of power depending on the voltage with which it is supplied. However, it is fair to assume that the actual service voltage to customers is independent of bus voltage, owing to the transformers and voltage regulators in the distribution system that keep the service voltage constant. The power drawn at the bus then remains independent of bus voltage to very a good approximation.

are going to be, and combining this information with the known line impedances will give us the losses. But we cannot tell a priori the amount of losses. The exact amount will vary depending on the dispatch, or amount of power coming from each generator, because a different dispatch will result in a different distribution of current over the various transmission paths, and not all transmission lines are the same. Therefore, if we were given a total P demanded at the load buses and attempted now to set the correct sum of P for all the generators, we could not do it.

The way to deal with this situation mathematically reflects the way it would be handled in actual operation. Knowing the total P demanded by the load, we begin by assuming a typical percentage of losses, say, 5%. We now dispatch all the generators in the system in some way so that the sum of their output approximately matches what we expect the total real power demand (load plus losses) to be: in this case, 105% of load demand. But since we do not yet know the exact value of the line losses for this particular dispatch (seeing that we have barely begun our power flow calculation), we will probably be off by a small amount. A different dispatch might, for example, result in 4.7% or 5.3% instead of 5% losses overall. We now make the plausible assumption that this uncertainty in the losses constitutes a sufficiently small amount of power that a single generator could readily provide it. So we choose one generator whose output we allow to adjust, depending on the system's needs: we allow it to "take up the slack" and generate more power if system losses are greater than expected, or less if they are smaller. In power flow analysis, this one generator bus is appropriately labeled the *slack bus*, or sometimes *swing bus*.

Thus, as the input information to our power flow analysis, we specify P for *one less* than the total number of buses. What takes the place of this piece of information for the last bus is the requirement that the system remain balanced. This requirement will be built into the equations used to solve the power flow and will ultimately determine what the as yet unknown P of the slack bus has got to be. The blank space among the initial specifications for the slack bus, where P is not given, will be filled by another quantity, the voltage angle, which will be discussed later in this chapter, following the discussion of reactive power.

7.2.5 Variables for Balancing Reactive Power

Analogous to real power, the total amount of reactive power generated throughout the system must match the amount of reactive power consumed by the loads.[5] Whereas in the case of a mismatch of real power, the system loses synchronicity,

---

[5]Recall that in Section 3.3, we put quotation marks around the terms "supplied" and "consumed" as they apply to reactive power because this is a somewhat arbitrary nomenclature: averaged over an entire cycle, no net power is either produced or consumed in either generator or load. Rather, the reactive power distinguishes the exact *time* during the cycle at which a device takes in and releases the power that is transferred back and forth between the electric and magnetic fields within reactive devices. Balancing reactive power thus really means balancing the *instantaneous* (not average) real power. When instantaneous power is imbalanced, the difference is made up from the potential energy stored in the electric and magnetic fields, until this stored energy runs out and the voltage collapses.

a mismatch of reactive power leads to voltage collapse (see Section 8.3). Also analogous to real power transmission losses, there are *reactive power losses*. Reactive losses are defined simply as the difference between reactive power generated and reactive power consumed by the metered load.

Physically, these losses in Q reflect the fact that transmission lines have some reactance (see Section 3.2) and thus tend to "consume" reactive power; in analogy to I^2R, we could call them I^2X losses. The term "consumption," however, like the reactive power "consumption" by a load, does not directly imply an energy consumption in the sense of energy being withdrawn from the system. To be precise, the presence of reactive power does necessitate the shuttling around of additional current, which in turn is associated with some real I^2R losses "in transit" of a much smaller magnitude. But these second-order I^2R losses (i.e., the side effect of a side effect) are already captured in the analysis of real power for the system. The term "reactive losses" thus does not refer to any physical measure of something lost, but rather should be thought of as an accounting device. While real power losses represent physical heat lost to the environment and therefore always have to be positive,[6] reactive losses on a given transmission link can be positive or negative, depending on whether inductive or capacitive reactance plays a dominant role.

In any case, what matters for both operation and power flow analysis is that Q, just like P, needs to be balanced at all times. Thus, just as for real power, all the generators in the system must generate enough reactive power to satisfy the load demand *plus* the amount that vanishes into the transmission lines.

This leaves us with the analogous problem of figuring out how much total Q our generators should produce, not knowing ahead of time what the total reactive losses for the system will turn out to be: as with real losses, the exact amount of reactive losses will depend on the dispatch. Operationally, though, the problem of balancing reactive power is considered in very different terms. When an individual generator is instructed to provide its share of reactive power, in practice this is not usually done by telling it to generate a certain number of MVAR (see Section 3.3). Instead, the generator is instructed to maintain a certain voltage magnitude at its bus. The voltage is continually and automatically adjusted through the generator's field current (see Section 4.3), and is therefore a straightforward variable to control.

Their own bus voltage is in fact the one immediate measure available to the generators for determining whether the correct amount of reactive power is being generated: when the combined generation of reactive power by all the generators in the system matches the amount consumed, their bus voltage holds steady. Conversely, if there is a need to increase or decrease reactive power generation, adjusting the field current at one or more generators so as to return to the voltage set point will automatically accomplish this objective.[7] The new value of MVAR produced by each generator can then be read off the dial for accounting purposes.

---

[6]Lest we violate the second law of thermodynamics, which forbids heat from flowing spontaneously from the air into the wire and making electricity.
[7]See Section 4.3.4 on how generators share MVAR load.

Conveniently for power flow analysis, then, there is no need to know explicitly the total amount of Q required for the system. Specifying the voltage magnitude is essentially equivalent to requiring a balanced Q. In principle, we could specify P and Q for each generator bus, except for one slack bus assigned the voltage regulation (and thus the onus of taking up the slack of reactive power). For this "reactive slack" bus we would need to specify voltage magnitude V instead of Q, with the understanding that this generator would adjust its Q output as necessary to accommodate variations in reactive line losses. In practice, however, since voltage is already the explicit operational control variable, it is customary to specify V instead of Q for all generator buses, which are therefore called *P,V buses*.[8] In a sense, this assignment implies that all generators share the "reactive slack," in contrast to the real slack that is taken up by only a single generator.

7.2.6 The Slack Bus

We have now, for our power flow analysis, three categories of buses: P,Q buses, which are generally load buses, but could in principle also be generator buses; P,V buses, which are necessarily generator buses (since loads have no means of voltage control); and then there is the slack bus, for which we cannot specify P, only V. What takes the place of P for the slack bus?

As introduced in Section 4.3, *real* power balance manifests operationally as a steady frequency such as 60 Hz. A constant frequency is indicated by an unchanging voltage angle, which for this reason is also known as the *power angle*, at each generator. When more power is consumed than generated, the generators' rotation slows down: their electrical frequency drops, and their voltage angles fall farther and farther behind. Conversely, if excess power is generated, frequency increases and the voltage angles move forward. While generators are explicitly dispatched to produce a certain number of megawatts, the necessary small adjustments to balance real power in real-time are made (by at least one or more *load-following* generator) through holding the generator frequency steady at a specified value. Not allowing the frequency to depart from this reference value is equivalent to not letting the voltage angle increase or decrease over time.[9]

In power flow analysis, the slack bus is the one mathematically assigned to do the load following. Its instructions, as it were, are to do whatever is necessary to maintain real power balance in the system. Physically, this would mean holding the voltage angle constant. The place of P will therefore be taken by the *voltage angle*, which is the variable that in effect represents real power balance. We can think of the voltage angle here as analogous to the voltage magnitude in the context of reactive power, where balance is achieved operationally by maintaining

---

[8]It is important to remember that the V of the P,V bus represents only voltage magnitude, not angle. To avoid any confusion, the careful notation $P,|V|$ is sometimes used, where the vertical lines indicate magnitude.

[9]Mathematically speaking, the frequency departure from the reference value represents the time rate of change of the voltage angle.

a certain voltage (magnitude) set point at the generator bus. Specifying that the bus voltage magnitude should be kept constant effectively amounts to saying that whatever is necessary should be done to keep the system reactive power balanced. Similarly, specifying a constant voltage angle at the generator bus amounts to saying that this generator should do whatever it takes to keep real power balanced.

We thus assign to the slack bus a voltage angle, which, in keeping with the conventional notation for the context of power flow analysis, we will call θ (lowercase Greek theta). This θ can be interpreted as the relative position of the slack bus voltage at time zero. Note that this θ is exactly the same thing that is elsewhere called the *power angle* (see Section 4.3.3 and Section 8.3) and labeled as δ (delta).

What is important to understand here is that the actual numerical value of this angle has no physical meaning; what has physical meaning is the implication that this angle will not change as the system operates. The choice of a numerical value for θ is a matter of convenience. When we come to the output of the power flow analysis, we will discover a voltage angle θ for each of the other buses throughout the system, which is going to take on a different (constant) value for each bus depending on its relative contribution to real power. These numerical values only have meaning in relation to a reference: what matters is the *difference* between the voltage angle at one bus and another, which physically corresponds to the phase difference between the voltage curves, or the difference in the precise timing of the voltage maximum.[10]

We now conveniently take advantage of the slack bus to establish a systemwide reference for timing, and we might as well make things simple and call the reference point "zero." This could be interpreted to mean that the alternating voltage at the slack bus has its maximum at the precise instant that we depress the "start" button of an imaginary stopwatch, which starts counting the milliseconds (in units of degrees within a complete cycle of 1/60th second) from time zero. In principle, we could pick any number between 0 and 360 degrees as the voltage angle for the slack bus, but 0° is the simple and conventional choice.

7.2.7 Summary of Variables

To summarize, our three types of buses in power flow analysis are P,Q (load bus), P,V (generator bus), and θ,V (slack bus). Given these two input variables per bus, and knowing all the fixed properties of the system (i.e., the impedances of all the transmission links, as well as the a.c. frequency), we now have all the information required to completely and unambiguously determine the operating state of the system. This means that we can find values for all the variables that were not originally specified for each bus: θ and V for all the P,Q buses; θ and Q for the P,V buses; and P and Q for the slack bus. The known and unknown variables for each type of bus are tabulated later in the following paragraph for easy reference.

---

[10]Since the voltage continually alternates, it would be of little use to say that the voltage maximum at Bus A occurs at exactly 3:00:00 P.M. Rather, we would want to know that the voltage maximum at Bus A always occurs one-tenth of a cycle later than at Bus B.

TABLE 7.1 Variables in Power Flow Analysis

Type of Bus	Variables Given (Knowns)	Variables Found (Unknowns)
Generator	Real power (P)	Voltage angle (θ)
	Voltage magnitude (V)	Reactive power (Q)
Load or generator	Real power (P)	Voltage angle (θ)
	Reactive power (Q)	Voltage magnitude (V)
Slack	Voltage angle (θ)	Real power (P)
	Voltage magnitude (V)	Reactive power (Q)

Once we know θ and V, the voltage angle and magnitude, at every bus, we can very easily find the current through every transmission link; it becomes a simple matter of applying Ohm's law to each individual link. (In fact, these currents have to be found simultaneously in order to compute the line losses, so that by the time the program announces θ's and V's, all the hard work is done.) Depending on how the output of a power flow program is formatted, it may state only the basic output variables, as in Table 7.1, it may explicitly state the currents for all transmission links in amperes; or it may express the flow on each transmission link in terms of an amount of real and reactive power flowing, in megawatts (MW) and MVAR.

7.3 EXAMPLE WITH INTERPRETATION OF RESULTS

7.3.1 Six-Bus Example

Consider the six-bus example illustrated in Figure 7.2.[11] This example is simple enough for us to observe in detail, yet too complex to predict its behavior without numerical power flow analysis.

Each of the six buses has a load, and four of the buses also have generators. Bus 1, keeping with convention, is the slack bus. Buses 2, 3, and 4, which have both generation and loads, are modeled as P,V buses; the local load is simply subtracted from the real and reactive generation at each. Buses 5 and 6, which have only loads, are modeled as P,Q buses.

The distribution of loads and the generation dispatch, for both real and reactive power, are completely determined somewhere outside the power flow program, whether in the real world or the program user's fantasy. The one exception is the generator at the slack bus, whose real power output varies so as to accommodate systemwide losses. In addition to the MW and MVAR loads and the MW generation levels for every generator (except the slack), the user specifies the voltage magnitudes to be maintained at each generator bus; the program then computes

---
[11] This example is taken from PowerWorld, a power flow software application available from www.powerworld.com for both commercial and educational use; a short version is available as a free download. The case illustrated here is from the menu of the standard demonstration cases in PowerWorld, labeled "Contour 6-Bus" (accessed October 2004).

7.3 EXAMPLE WITH INTERPRETATION OF RESULTS

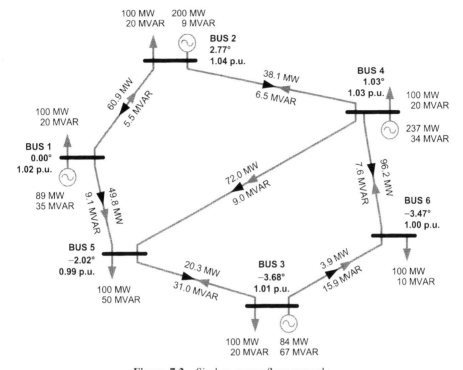

Figure 7.2 Six-bus power flow example.

the MVAR generation necessary to maintain this voltage at each bus. (It is also possible to specify MVAR generation and allow the program to determine the voltage magnitudes, but, as mentioned earlier, the former method better resembles real-life operations.)

By convention, the voltage angle at the slack bus is set to 0.00 degrees. The power flow program computes the voltage angle at each of the other five buses in relation to the slack bus. We may now begin to observe the relationship between real power and voltage angle: a more positive voltage angle generally corresponds to an injection of power into the system and a more negative voltage angle to a consumption of real power. Buses 2 and 4, which both have generation exceeding local load, have positive voltage angles of 2.77° and 1.03°, respectively. Bus 3, though it has a generator, is still a net consumer of real power, with 100 MW load and only 84 MW generated; its voltage angle is −3.68°. Buses 5 and 6 have loads only and voltage angles of −2.02° and −3.47°, respectively.

Note, however, that the voltage angles are not in hierarchical order depending on the amount of power injected or withdrawn at each individual bus. This is because we also must consider the location of each bus relative to the others in the system and the direction of power flow between them. For example, consider Buses 2 and 4. Net generation at Bus 4 is greater than at Bus 2 (137 MW compared to 100 MW), yet the voltage angle at Bus 2 is more positive. We can see that this is due to the location of these buses in the system: real power is generally flowing from north to south, that is, from Bus 2 to the neighborhood of Buses 5, 3, and 6 where there is more load and less generation. As indicated by the black arrow on the transmission link, real power is flowing from Bus 2 to Bus 4. As a rule, real power flows from a greater to a smaller voltage angle. This rule holds true for six of the seven links in this sample case; the exception is Link 3–6, where both the power flow and the difference in voltage angle are very small. The reader can verify that throughout this case, while power flow and voltage angle are not exactly proportional, a greater flow along a transmission link is associated with a greater angle difference.

We now turn to the relationship between reactive power and voltage magnitude, which is similar to that between real power and voltage angle. The nominal voltage of this hypothetical transmission system is 138 kV. However, just as the timing or angle of the voltage differs by a small fraction of a cycle at different locations in the grid, the magnitude, too, has a profile across the system with different areas a few percent higher or lower than the nominal value. Because it is this percentage, not the absolute value in volts, that is most telling about the relationship among different places in the grid, it is conventional to express voltage magnitude in *per-unit* terms. Per-unit (p.u.) notation simply indicates the local value as a multiple of the nominal value; in this case, 138 kV equals 1.00 p.u. The voltage magnitude at Bus 1 is given as 1.02 p.u., which translates into 141 kV; at Bus 5, the voltage magnitude of 0.99 p.u. means 137 kV.

As a rule, reactive power tends to flow in the direction from greater to smaller voltage magnitude. In our example, this rule holds true only for the larger flows of MVAR, along Links 1–5, 3–5, 4–5, and 3–6. The reactive power flows along Links 1–2, 4–2, and 6–4 do not follow the rule, but they are comparatively small.

Note that real and reactive power do not necessarily flow in the same direction on a given link. This should not be surprising, because the "direction" of reactive power flow is based on an arbitrary definition of the generation or consumption of VARs; there is in fact no net transfer of energy in the direction of the gray arrow for Q. Also, note that having Q flow opposite P does not imply any "relief" or reduction in current. For example, on Link 3–5, the real power flow P is 20.3 MW and reactive flow Q is 31.0 MVAR. In combination, this gives apparent power S of 37.1 MVA (using $S^2 = P^2 + Q^2$), regardless of the direction of Q. (Recall that MVA are the relevant units for thermal line loading limits, since total current depends on apparent power.)

From Figure 7.2, it is possible to evaluate the total real and reactive system losses, simply by observing the difference between total generation and total load. The four generators are supplying 89, 200, 84, and 237 MW, respectively, for a total of 610 MW of real power generated. Subtracting the six loads of 100 MW each, the total real power losses throughout the transmission system for this particular scenario are therefore 10 MW. On the reactive side, total generation is 145 MVAR, while total reactive load is 140 MVAR, and system reactive losses amount to roughly 5 MVAR.

The discerning reader may have noticed, however, that the stated line flows in Figure 7.2, which are average values for each link, cannot all be reconciled with the power balance at each bus. To account for losses in a consistent fashion, we must record both the power (real or reactive) entering and exiting each link. In Figure 7.3, these data are given for real power (MW) in black and reactive power (MVAR) in gray. The numbers in parentheses represent the losses, which are the difference between power flows at either end. Bus power, line flows, and losses are rounded to different decimal places, but the numbers do add up correctly for each bus and each link.

The most significant losses tend to occur on links with the greatest power flow. In this case, Link 4–6 has the greatest power flow with 96.2 MW real and 7.6 MVAR reactive, yielding 96.5 MVA apparent, and the greatest losses. While the real line losses are all positive, as they should be, the negative signs on some of the reactive losses indicate negative losses; we might consider them "gains," although nothing is actually gained. Reactive losses depend on operating conditions and impedance, where the model of a transmission link may incorporate reactive compensation such as capacitors. It is typical for system reactive losses to be positive overall, as they are in this example. Like real losses, reactive losses are related to the current and therefore apparent power flow. Thus, we also observe the greatest reactive

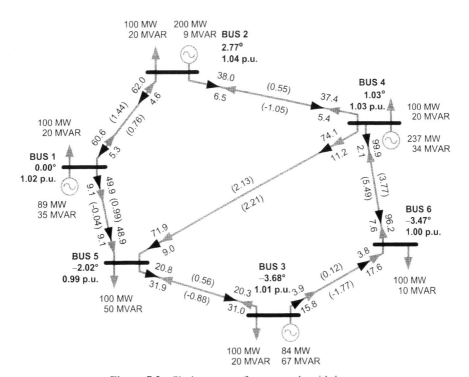

Figure 7.3 Six-bus power flow example with losses.

losses in our example on Link 4–6. The real and reactive losses for every link can be totaled to confirm the estimated system losses given earlier.

7.3.2 Tweaking the Case

To gain a better sense of a power system's behavior and the information provided by power flow modeling, let us now make a small change to the operating state in the six-bus example and observe how the model responds. We simply increase the load at Bus 5 by 20% while maintaining the same power factor, thus changing it from 100 MW real and 50 MVAR reactive to 120 MW real and 60 MVAR reactive. This change is small enough for the generator at the slack bus to absorb, so we need not specify increased generation elsewhere. Indeed, generation at Bus 1 increases from 89 to 110 MW. Note that the difference amounts to 21, not 20 MW, as the increased load also entails some additional losses in the system. The new scenario is illustrated in Figure 7.4.

As we would expect, the line flows to Bus 5 increase by a total of 20 MW. The bulk (about 14.5 MW) of this additional power comes from Bus 1, about 4 MW from Bus 4, and the balance appears as a reduction of about 1.5 MW in the flow to Bus 3.

The changes do not stop here, however; they have repercussions for the remainder of the system. Three of the other buses are defined as P,V buses, and therefore

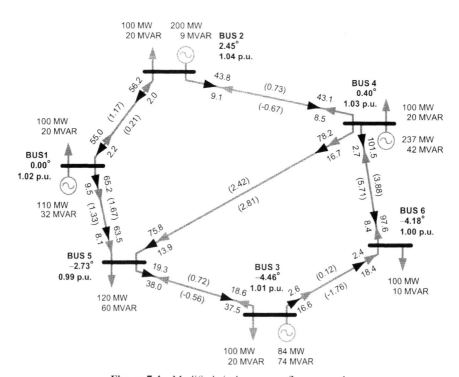

Figure 7.4 Modified six-bus power flow example.

have fixed voltage magnitudes. The voltage magnitude at Bus 6 (a P,Q bus) is affected slightly (from 0.9951 to 0.9950 p.u.), although the change does not show up after rounding. In order to maintain the preset voltages at the P,V buses, reactive power generation increases at Buses 4 and 6, as it does at the slack bus. Indeed, system reactive generation now totals 157 MVAR, which are needed to accommodate the additional reactive load introduced at Bus 5 as well as a substantial increase in system reactive losses of 2.33 MVAR (up almost 50% from 4.74 to 7.07 MVAR). While real power is fixed at all buses other than 1 and 5, their voltage angles change as a result of the changed power flow pattern.

The most serious repercussion in this example occurs on Link 4–6, which was fully loaded before the change to Bus 5 was made. Owing to the vagaries of network flow, the change at Bus 5 results in a slight increase in real power flow from Bus 4 to Bus 6. Link 4–6 now carries slightly more current and apparent power, going from 100 to 101.5 MVA. This is significant because, in this hypothetical scenario, each transmission link has a thermal limit of 100 MVA. The power flow program thus shows the line becoming overloaded as a result of the change at Bus 5, even though Buses 4 and 6 are located on the opposite geographic end of the system, and neither generation nor load levels there were affected. In reality, this violation would mean that the proposed change is inadmissible and other options would have to be pursued—specifically, a generator other than Bus 1 would be required to increase generation—in order to meet the additional load without violating any transmission constraints.

The reader is encouraged to further tinker with power flow scenarios, for which the PowerWorld^{TM} application is an invaluable tool. As this chapter illustrates, it is very difficult to produce an intuitive comprehension of power flow from the formal analytic description of the problem. Actual system operators, whose success hinges on just such an intuitive comprehension (see Section 9.3), develop it over the course of time by empirical observation of countless scenarios. Experimentation with a small network model like the one just discussed offers students of power systems an opportunity both to create some contextual meaning for the abstract mathematical power flow relationships, and to appreciate the complexity and challenge of the task faced by system operators and engineers.

7.3.3 Conceptualizing Power Flow

One interesting and conceptually somewhat difficult aspect of a.c. power systems has to do with time. The synchronous oscillation along with its profile of voltage angles across the network provides a consistent temporal reference frame for an entire synchronous grid, which may span half a continent. The 60-cycle voltage oscillation is like a pulse that pervades this grid. Indeed, we can imagine a universal clock for the entire system, whose ticking marks the synchronized rhythm of all the connected generators. Yet, as seen in the power flow analysis, a time difference emerges and the oscillations do not coincide perfectly among the various buses. After arbitrarily assigning one bus as the reference for the system clock (a voltage angle of 0.00), some buses will be slightly anticipating and others slightly dragging behind this pulse, like a group of musicians who fail to quite come together on the downbeat (except that each player remains perfectly consistent from one measure to the next). Converted into seconds, the time difference is minuscule: one degree is 1/360th of a cycle, which is 1/60th of a second, making one degree equal to about 46 microseconds (μs) (0.0000463 s).

It could be tempting to attribute this time differential to the propagation time of a signal, but that would be incorrect. As mentioned in Chapter 1, transmission lines can easily be long enough so that the time it would take for an electrical signal to propagate at the speed of light from one end to the other is not negligible. Thus we might wonder about a delay between the voltage maximum (which could be considered a signal) occurring at different buses. Are the power-producing buses senders and the power-consuming buses receivers of a signal that is delayed by the distance between them?

Not exactly. The steady-state behavior we are concerned with differs from a situation where a message originates from one distinct location and is being received by another. A signal in the sense of "traveling message" in an a.c. grid would actually consist of a disturbance or departure from the 60-cycle oscillation in the background. Such disturbances (caused by faults, for example) do in fact propagate from one distinct location to another at the speed of light in a conductor, and when studying them, their "travel time" on transmission and distribution lines has to be taken into account. These types of studies are complex and certainly beyond the scope of this text. They are distinct from power flow analysis, which deals with a steady operating state of the system. In power flow, rather than instabilities, we are describing the stable equilibrium itself. This equilibrium consists of a pulse that initially had to make its rounds through the system as a disturbance, while the grid was being energized by the first generator on-line. But once it has been echoed back from every direction and the transient effects have decayed, the pulse is established as an ambient, steady-state condition and appears to reside everywhere in the system at once.

The time differentials expressed in the voltage-angle profile arise not from long-distance communication, but from the subtleties of a.c. power transfer: the precise overlay of oscillating voltages and currents, each shifted by the influences of a vast multitude of nearby and remote devices, multiplied together at every instant to yield an energy transfer. Thus the voltage-angle difference between two ends of a link is not explicitly a function of length; it depends on the amount of power flow and the impedance of the link. Of course, the impedance of a given type of conductor varies with length, which is why *stability limits* (where more power flow would cause excessive separation of voltage angle) tend to be a concern only for long transmission lines. But the same time delay or voltage-angle profile will appear regardless of physical dimension for any network with the same power flows and impedances, even a very small circuit.

One way to visualize power flow in a network is by way of a mechanical analogy where transmission lines are represented by rubber bands tied together into a grid. Each bus is a place where the rubber bands are either suspended by hooks from the ceiling (generation) or have weights hanging from them (loads). The real power in megawatts injected or consumed at each node corresponds to the weight or amount of force pulling the node up or down. The voltage angle roughly corresponds to the elevation of each point in the rubber-band grid. The requirement that power injected equals power consumed corresponds to the requirement that the rubber-band grid be in balance, that is, neither fall down nor snap to the ceiling.

The rubber-band model helps visualize *dynamic stability* (Section 8.3.3), which is analogous to what happens when a weight suddenly falls off (a load is lost) or a hook pops out of the ceiling (a generator goes off-line). Even if we assume that the remaining hooks can accommodate the weight (i.e., generators compensate for the change in system load), the dynamic problem is that the network of rubber bands will bounce up and down following the sudden change. Thus, dynamic stability addresses the question of how much bouncing the hooks will tolerate before the web of rubber bands starts falling off.

We can also use the model to appreciate the role of stability limits. Clearly, any given rubber band can only be stretched so far before it breaks. This translates into the observation that any given transmission link can only sustain a limited difference in voltage angle between its two ends (*steady-state stability*; see Section 8.3.2). Once this limit is exceeded and synchronicity is lost, the link no longer transmits power, just as the broken rubber band no longer transmits a force from weight to ceiling hook.

The analogy becomes a bit awkward if we try to stretch it further (so to speak) and bring line impedance into it. We note that rubber bands may come in different elasticities and strengths, referring to the amount by which they stretch under a given tension and their ability to withstand tension without breaking. This could roughly translate into line impedance, but not in an obvious way. A line with a high impedance would have to correspond to a band that stretches farther under a given tension, either because it is longer or because it is more elastic. (We encounter a visualization problem here in that the dimension of linear distance in the rubber-band model relates to energy or voltage angle, not geographical distance.) To make the analogy work out, we must require that all rubber bands "break" (i.e., violate a stability limit) when elongated by a certain number of inches. It would then hold true that the stability limit is increasingly important for longer lines and those with higher impedance.

The thermal limit, by contrast, would be related to the amount of tension that can be sustained by each band, regardless of stretch. Physically, while the amount of heat produced on a transmission line is a function of its resistance and current, this heat is significant to power flow analysis only in terms of real power losses; the conductor temperature and the associated operating limit are derived outside the power flow calculation. In the analogy we would stipulate a certain tension or force deemed safe for each rubber band.

An ideal superconducting transmission line would translate into a perfectly strong and firm cord with no stretch at all. If all transmission links were superconducting, all the rubber bands would be replaced with perfect cords, nothing would bounce or stretch, and the subject of power flow analysis would become utterly boring.

This is about as far as the rubber-band analogy goes; incorporating reactive power and voltage magnitude into the model is not obvious and, if possible at all, too contrived to help develop intuition. A direct physical analogy to reactive power would be a rapid vibration in the weight (load), which would have to be met by a matching vibration in the suspension (generation). The mechanical system would then need a way of allocating this "vibration" (reactive power) deliberately among different hooks (generators) by way of some property corresponding to the bus voltage magnitude. Perhaps the most appropriate conclusion is that a.c. power systems have a certain complexity which, in its defiance of human intuition, is unmatched by any mechanical system.

7.4 POWER FLOW EQUATIONS AND SOLUTION METHODS

7.4.1 Derivation of Power Flow Equations

In Section 7.2, we stated the known and unknown variables for each of the different types of buses in power flow analysis. The *power flow equations* show explicitly how these variables are related to each other.

The complete set of power flow equations for a network consists of one equation for each node or branch point in this network, referred to as a *bus*, stating that the complex power injected or consumed at this bus is the product of the voltage at this bus and the current flowing into or out of the bus. Because each bus can have several transmission links connecting it to other buses, we must consider the sum of power entering or leaving by all possible routes. To help with the accounting, we will use a summation index i to keep track of the bus for which we are writing down the power equation, and a second index k to keep track of all the buses connected to i.

We express power in complex notation, which takes into account the two-dimensionality—magnitude and time—of current and voltage in an a.c. system. As shown in Section 3.3, complex power S can be written in shorthand notation as

$$S = VI^*$$

where all variables are complex quantities and the asterisk denotes the complex conjugate of the current.[12]

Recall that S represents the complex sum of real power P and reactive power Q, where P is the real and Q the imaginary component. While the real part represents a tangible physical measurement (the rate at which energy is transferred), the imaginary part can be thought of as the flippety component that oscillates. At different

---

[12] The complex conjugate of a complex number has the same real part but the opposite (negative) imaginary part. It is used here to produce the correct relationship between voltage and current angle—their difference, not their sum—for purposes of computing power (see Section 3.4.3).

7.4 POWER FLOW EQUATIONS AND SOLUTION METHODS

times it may be convenient to either refer to P and Q separately or simply to S as the combination.

In the most concise notation, the power flow equations can be stated as

$$S_i = V_i I_i^*$$

where the index i indicates the node of the network for which we are writing the equation. Thus, the full set of equations for a network with n nodes would look like

$$S_1 = V_1 I_1^*$$
$$S_2 = V_2 I_2^*$$
$$\ldots$$
$$S_n = V_n I_n^*$$

We can choose to define power as positive either going into or coming out of that node, as long as we are consistent. Thus, if the power at load buses is positive, that at generator buses is negative.

So far, these equations are not very helpful, since we have no idea what the I_i are. In order to mold the power flow equations into something we can actually work with, we must make use of the information we presumably have about the network itself. Specifically, we want to write down the impedances of all the transmission links between nodes. Then we can use Ohm's law to substitute known variables (voltages and impedances) for the unknowns (currents).

Written in the conventional form, Ohm's law is $V = IZ$ (where Z is the complex impedance). However, when solving for the current I, this would require putting Z in the denominator: $I = V/Z$. It is therefore tidier to use the inverse of the impedance, known as the *admittance* Y (where $Y = 1/Z$), so that Ohm's law becomes $I = VY$. Not only does this avoid the formatting issues of division, it also allows us to indicate the absence of a transmission link with a zero (for zero admittance) instead of infinity (for infinite impedance), which vastly improves the appearance of the imminent matrix algebra.

Recall from Section 3.2.4 the complex admittance $Y = G + jB$, where G is the *conductance* and B the *susceptance*. The admittances of all the links in the network are summarized by way of an *admittance matrix* **Y**, where the lowercase $y_{ik} = g_{ik} + b_{ik}$ indicates that matrix element that connects nodes i and k. (Essentially, the matrix notation serves the purpose of tidy bookkeeping.)

The relationship $I = VY$ is what we wish to write down and substitute for every I_i that appears in the power flow equations. But now we face the next complication: the total current I_i coming out of any given node is in fact the sum of many different currents going between node i and all other nodes that physically connect to i. We will indicate these connected nodes with the index k, where k could theoretically include all nodes other than i in the network from 1 to n. In practice, fortunately, only a few of these will actually have links connecting to node i. For any node k that is not connected to i, $y_{ik} = 0$.

For the current between nodes i and k, we can now write

$$I_{ik} = V_k y_{ik}$$

or, expanding the admittance into its conductance and susceptance components,

$$I_{ik} = V_k(g_{ik} + b_{ik})$$

Note that if this equation stood by itself, we would want V_k to mean "the voltage difference between nodes i and k" rather than simply "the voltage at node k." However, we will see the voltage differences between nodes naturally arise out of the complex conjugate once we expand the terms; specifically, the voltage differences will appear in terms of differences between phase angles at each node.

To complete the statement $S_i = V_i I_i^*$ for node i, we must now sum over all nodes k that could possibly be connected to i.[13] This summation over the index k means accounting for all the current that is entering or leaving this one particular node, i, by way of the various links it has to other nodes k. (For the complete system of power flow equations, we also have to consider every value of the index i so as to consider power flow for every node, but we do not need to write this out explicitly.) Thus,

$$S_i = V_i I_i^* = V_i \left(\sum_{k=1}^{n} y_{ik} V_k \right)^*$$

Now we expand the y's into g's and b's, noting that the complex conjugate gives us a minus sign in front of the jb:

$$S_i = V_i \sum_{k=1}^{n} (g_{ik} - jb_{ik}) V_k^*$$

After changing the order of terms to look more organized, we write the voltage phasors out in longhand, first as exponentials and then broken up into sines and cosines:

$$S_i = \sum_{k=1}^{n} |V_i||V_k| e^{j(\theta_i - \theta_k)} (g_{ik} - jb_{ik})$$

$$= \sum_{k=1}^{n} |V_i||V_k| [\cos(\theta_i - \theta_k) + j\sin(\theta_i - \theta_k)](g_{ik} - jb_{ik})$$

---

[13]The notation with the capital Greek sigma (for "sum") indicates the sum of indexed terms, with the index running from the value below the sigma ($k = 1$) to the value above it ($k = n$).

7.4 POWER FLOW EQUATIONS AND SOLUTION METHODS

The term ($\theta_i - \theta_k$) in this equation denote the *difference* in voltage angle between nodes i and k, where the minus sign came from having used the complex conjugate of I initially, which gave us the complex conjugate of V_k. This angle difference is not the same as the instantaneous difference in magnitude between the two voltages, but it is the correct measure of separation between the nodes for the purpose of calculating the power transferred.[14]

The equation we now have for S_i entails the product of two complex quantities, written out in terms of their real and imaginary components. By cross-multiplying all the real and imaginary terms, we can separate the real and imaginary parts of the result S, which will be the familiar P and Q. Remembering that j times j gives -1, we obtain:

$$P_i = \sum_{k=1}^{n} |V_i||V_k|[g_{ik}\cos(\theta_i - \theta_k) + b_{ik}\sin(\theta_i - \theta_k)]$$

$$Q_i = \sum_{k=1}^{n} |V_i||V_k|[g_{ik}\sin(\theta_i - \theta_k) - b_{ik}\cos(\theta_i - \theta_k)]$$

The complete set of power flow equations for a network of n nodes contains n such equations for S_i, or pairs of equations for P_i and Q_i. This complete set will account for every node and its interaction with every other node in the network.

7.4.2 Solution Methods

There is no analytical, closed-form solution for the set of power flow equations given in the preceding section. In order to solve the system of equations, we must proceed by a numerical approximation that is essentially a sophisticated form of trial and error.

To begin with, we assume certain values for the unknown variables. For clarity, let us suppose that these unknowns are the voltage angles and magnitudes at every bus except the slack, making them all P,Q buses (it turns out that having some P,V buses eases the computational volume in practice, but it does not help the theoretical presentation). In the absence of any better information, we would probably choose a *flat start*, where we assume the initial values of all voltage angles to be zero (the same as the slack bus) and the voltage magnitudes to be 100% of the nominal value, or 1 p.u. In other words, for lack of a better guess, we suppose that the voltage magnitude and angle profile across the system is completely flat.

We then plug these values into the power flow equations. Of course, we know they do not describe the actual state of the system, which was supposed to be consistent with the known variables (P's and Q's). Essentially, this will produce a contradiction: Based on the starting values, the power flow equations will predict a different set of P's and Q's than we stipulated at the beginning. Our objective is

---

[14] In effect, it gives us the "difference voltage" described in Section 4.3.3 as being associated with power transferred between generators that are out of phase with each other.

218 POWER FLOW ANALYSIS

to make this contradiction go away by repeatedly inserting a better set of voltage magnitudes and angles: As our voltage profile matches reality more and more closely, the discrepancy between the P's and Q's, known as the *mismatch*, will shrink. Depending on our patience and the degree of precision we require, we can continue this process to reach some arbitrarily close approximation. This type of process is known as an *iterative* solution method, where "to iterate" means to repeatedly perform the same manipulation.

The heart of the iterative method is to know how to modify each guess in the right direction and by the right amount with each round of computation (iteration), so as to arrive at the correct solution as quickly as possible. Specifically, we wish to glean information from our equations that tells us which value was too high, which was too low, and approximately by how much, so that we can prepare a well-informed next guess, rather than blindly groping around in the dark for a better set of numbers. There are several standard techniques for doing this; the ones most commonly used in power flow analysis are the Newton–Raphson, the Gauss, and the Gauss–Seidel iterations. While an elementary version of Newton's method is shown in the following Box, a comparison among these algorithms is beyond the scope of this text. Suffice it to say that the choice of algorithm for a particular situation involves a trade-off among the number of iterations required to arrive at the solution, the amount of computation required for each iteration, and the degree of certainty with which the solution is found.

Regardless of which method is used, we will need to press our power flow equations for the crucial information about the direction of the error. Some readers may recognize this as a kind of *sensitivity analysis*, which asks how sensitive one variable is with respect to another. We obtain this information by writing down the *partial derivatives* of the power flow equations, or their rates of change with respect to individual variables. Specifically, we need to know the rate of change of real and reactive power, each with respect to voltage angle or magnitude. This yields four possible combinations of partial derivatives.[15] For example, $\partial P/\partial \theta$ (read: "partial P partial theta") is the partial derivative of real power with respect to voltage angle, and similarly there are $\partial P/\partial V$, $\partial Q/\partial \theta$, and $\partial Q/\partial V$. Each of these combinations is in fact a matrix (known as a *Jacobian* matrix) that, in turn, includes every bus combined with every other bus. In expanded form, with three buses, $\partial P/\partial \theta$ would look like this:

$$\frac{\partial P}{\partial \theta} = \begin{cases} \dfrac{\partial P_1}{\partial \theta_1} & \dfrac{\partial P_1}{\partial \theta_2} & \dfrac{\partial P_1}{\partial \theta_3} \\ \dfrac{\partial P_1}{\partial \theta_1} & \dfrac{\partial P_2}{\partial \theta_2} & \dfrac{\partial P_2}{\partial \theta_3} \\ \dfrac{\partial P_3}{\partial \theta_1} & \dfrac{\partial P_3}{\partial \theta_2} & \dfrac{\partial P_3}{\partial \theta_3} \end{cases}$$

[15] The partial derivative means the rate of change of a function with respect to only one of several variables, and is conventionally indicated by the curly ∂ instead of plain d for "differential element."

7.4 POWER FLOW EQUATIONS AND SOLUTION METHODS   219

Each of these four types of partial derivatives ($\partial P/\partial \theta$, $\partial P/\partial V$, $\partial Q/\partial \theta$, and $\partial Q/\partial V$) constitutes one partition of the big Jacobian matrix **J**. Within each partition there are rows for the θ or V and columns for the P or Q belonging to every bus (except the slack bus, so that the dimensionality of each partition is one less than the number of buses in the system). Note that when we write V we mean the magnitude of V, which would be more properly designated by $|V|$ except that the notation is already cumbersome enough without the vertical lines.

We must now combine the system of equations and its partial derivatives with our guess for the unknown variables in such a way that it suggests to us a helpful modification of the unknowns, which will become our improved guess in the next iteration.

ITERATIVE COMPUTATION

Our task can be stated as trying to find an unknown value of a function whose explicit form we do not know, based on information from elsewhere along the function. This can be done with a *Taylor series expansion*, which may be familiar to readers who have studied calculus. The idea is that we can express the unknown value of the function $f(x)$ at some particular x in terms of two pieces of information: the function's value at a different, nearby x; and the rate of change of the function with respect to x—its slope—at the same nearby x. Suppose we already know the value f at location x, and we also know how steep the function is there. Now we want to learn the value of f at the nearby location that we call $x + \Delta x$, so that Δx represents the difference between the two x's. If the function $f(x)$ is a straight line, we can write

$$f(x + \Delta x) = f(x) + f'(x)\,\Delta x$$

where $f'(x)$ is the *first derivative* or slope of the line at location x.

In the more general case, where $f(x)$ is not a straight line, but some type of curve, this equation is incomplete; we would have to include additional ("higher order") terms that correct for the curvature. Specifically, we would include the second derivative (the rate of change of the rate of change, which describes the upward curvature) to correct the straight-line approximation, multiplied once again by the increment Δx. We may also need to include the third derivative or more, depending on how curvy the function is. Written out, the Taylor series looks like

$$f(x + \Delta x) = f(x) + f'(x)\,\Delta x + f''(x)\,\Delta x^2 + f'''(x)\,\Delta x^3 + \cdots$$

up to the point where the function's higher-order derivatives are zero.

Fortunately, we are in the business of making approximations rather than finding exact values. If the Δx is not too big and the function not too radically curved, then the higher-order terms ought to be quite small compared to the first derivative term. Thus, we can use the linear version for a general function,

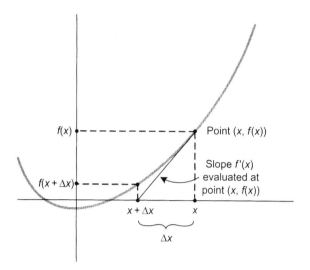

Figure 7.5 Newton's method.

with the understanding that it will not take us exactly to the new value $f(x + \Delta x)$, but a good bit of the way there and almost certainly closer than we were before.[16]

The reader may wonder why we do not we simply plug the new x-value, $x + \Delta x$, into the function $f(x)$ and be done. Usually, this is because we do not know how to write down $f(x)$ or $f'(x)$ in algebraic form, although we have a numerical value for both f and f' at one particular x. In our application, we do know what $f(x)$ looks like algebraically—the power flow equations—but we cannot solve it in the way that we would like. Specifically, we would like to go backwards to find the x that will yield a particular value of f. Even though we know how to get f from any given x, we cannot simply solve for x given the f because the equation does not allow itself to be turned inside out. This is precisely the situation for the power flow equations: given the θ's and V's, we can readily solve for the P's and Q's, yet we cannot go backwards from the P's and Q's to explicitly solve for the θ's and V's. Thus, we are forced to try out different sets of θ's and V's (which translate here into x's) until we hit the right P's and Q's (represented by the f).

The standard way to proceed is to rearrange the equations as necessary so that the target value is $f(x) = 0$. The problem can then be stated in the tidy format, "Find the x that makes $f(x) = 0$ a true statement." It is illustrated in Figure 7.5, where we start with a certain x and a known value $f(x)$ that is not zero, but wish to find the Δx for which the value of the function $f(x + \Delta x)$ is zero. Note that in this diagram Δx happens to be negative (i.e., measuring to the left), but it could go either way. After writing down the first terms of the Taylor series and declaring that $f(x + \Delta x) = 0$, it takes only a minor manipulation

---

[16]This is true unless the function is very badly behaved or we started in an awkward spot.

7.4 POWER FLOW EQUATIONS AND SOLUTION METHODS

to solve explicitly for the Δx that makes this true:

$$f(x + \Delta x) = f(x) + f'(x)\Delta x$$
$$0 = f(x) + f'(x)\Delta x$$
$$\Delta x = -f(x)/f'(x)$$

We have then found the Δx that must be added to the original x to obtain the new x-value, for which the function is zero.

But because the function is curved, not straight, our answer will not be exactly right. We have evaluated the derivative $f'(x)$ at the location of our original x, meaning that we have used the slope at that location to extrapolate where the function is going. But in reality, the function's slope may change along the way. The higher-order terms of the Taylor series would address this problem, but they contain awkward squares and cubes. Instead of dealing with such terms, we simply repeat the linear process: we use the new location, $x + \Delta x$, as our starting point for another iteration. Since $x + \Delta x$ is presumably much closer to our target than the original x—which we can verify by checking that $f(x + \Delta x)$ is closer to zero than $f(x)$—the next time it should be easier to get even closer, with a smaller Δx. Depending on the precision we desire in getting f to zero, we can repeat the process again with more iterations after that, or call it a day. This, in essence, is Newton's method for finding the zero-crossing of a function.

In any case, we will probably have many x's and Δx's around and ought to keep track of which iteration they belong to. One way to label them is with a superscript like x^v, where v (Greek lowercase nu) stands for the number of iterations ($x^1, x^2, \ldots$ and $\Delta x^1, \Delta x^2, \ldots$) and is not to be mistaken for an exponent. The process of approaching a value of x for which $f(x) \approx 0$ is illustrated in Figure 7.6 for two iteration steps. Clearly, the more the slope of the curve changes between x^0 and the solution (i.e., the more the function is curved), the more steps will be required to get close. Based on the diagram, the reader can also visualize how this approximation process can still succeed even if the slope of the line drawn to choose the next x^v is not precisely equal (but bears a reasonable resemblance) to the actual slope of the curve; this property is used in shortcut methods such as the *decoupled power flow* discussed later in this chapter that avoid some of the tedious computation of the exact derivatives.

The Jacobian matrix is essentially a very large version of the derivative $f'(x)$ used in Newton's method, with multiple f's and x's tidily summarized into the single, bulky object labeled $\mathbf{J}$. The function $f(x)$ itself appears essentially as the power flow equations $P(\theta,V)$ and $Q(\theta,V)$. We write $\mathbf{f}(x)$ in boldface to indicate that it is a vector containing an organized set of several numbers (one P and one Q for each bus except the slack). However, in order to keep with the format of searching

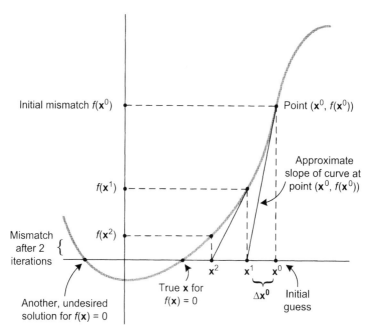

Figure 7.6 Iterative process of approximating $f(x) = 0$.

for $f(x) = 0$, we define $\mathbf{f}(x)$ as the difference between the $P(\theta,V)$, $Q(\theta,V)$ computed and the actual values of P and Q at each of the buses, which are known at the outset. In other words, $\mathbf{f}(x)$ represents the *mismatch*, which we want to get as close as is reasonable to zero.

In principle, we are now able to combine all the information at our disposal into improving our guess for the θ's and V's (the x's). Let us label our initial guess for θ and V as $\mathbf{x}^0$, where the boldface $\mathbf{x}$ indicates the vector composed of one θ and one V for each bus (except the slack) and the superscript 0 indicates the zeroth iteration. Now we might simply adapt the expression for Δx from Newton's method (or some variation that corresponds to other solution methods, though the basic idea is always the same),

$$\Delta x = -f(x)/f'(x)$$

and substitute our matrix and vector quantities

$$\Delta \mathbf{x} = -\mathbf{f}(\mathbf{x})/\mathbf{J}$$

But stop! We have just made every mathematician cringe, because dividing by a matrix is not something one does. Our two options are to use the proper *inverse* of the matrix, $\mathbf{J}^{-1}$ (which is obtained by a cumbersome but tractable procedure in linear algebra),

$$\Delta \mathbf{x} = -\mathbf{f}(\mathbf{x})\mathbf{J}^{-1}$$

7.4 POWER FLOW EQUATIONS AND SOLUTION METHODS

or to rearrange the equation to read

$$\mathbf{J}\Delta\mathbf{x} = -\mathbf{f}(\mathbf{x})$$

The latter choice does not get us around the algebraic awkwardness, because we must still multiply vectors and matrices rather than individual numbers. Moreover, the $\Delta\mathbf{x}$ no longer stands by itself, and extracting it amounts to solving a system of many equations (one less than the number of buses). As with inverting a matrix, this is mostly a tedious bookkeeping process. In the days of paper and pencil, half a dozen rows and columns would have easily defeated a diligent scribe. Computers, though, have turned repetitive procedures like linear algebra operations into a trivial task, and we can now cheerfully approach systems with hundreds if not thousands of buses (a capacity that is still best appreciated through the character-building experience of solving smaller systems by hand).

Having thus identified the $\Delta\mathbf{x}$, we simply add it to the old $\mathbf{x}$ and proceed to the next iteration. To be precise, then, let us rewrite the last equation as

$$\mathbf{J}^{\nu}\Delta\mathbf{x}^{\nu} = -\mathbf{f}(\mathbf{x}^{\nu})$$

to indicate that we are on the νth (nuth) iteration. Eventually, we will find that $\mathbf{f}(\mathbf{x}^{\nu})$ is close enough to zero to satisfy us.

When $\mathbf{f}(\mathbf{x}^{\nu}) = 0$, it means that there is no more mismatch, and the $\mathbf{x}^{\nu}$—that is, the θ's and V's from the νth iteration—give us the operating state of the power system that is consistent with the P's and Q's we specified initially. We may need to verify, though, that $\mathbf{x}^{\nu}$ is a realistic and true solution for our power system as opposed to some mathematical fluke, which can occur when a function has more than one zero-crossing.

We have now found θ and V for each P,Q bus. For any P,V buses, we would have found θ and Q as part of our computational process. There are several steps left to produce the complete output of the power flow analysis. First, by writing the power flow equation for the slack bus, we determine the amount of real power generated there. This tells us how many MW of losses there are in the system, as we can now compare the total MW generated to the total MW of load demand. Also, by using Ohm's law for every transmission link, we solve explicitly for each line flow, in amperes or MVA. Through the θ's and V's at each bus, we have information about the real and reactive power both going into and coming out of each link, and by subtracting we can specify the real and reactive losses on each link. Finally, we format the output and compare it to external constraints such as line flow limits so as to flag any violations. We have now completely described the system's operating state based on a given generation dispatch and combination of loads.

Although it is instructive to work through a numerical example, solving even a three-bus system entails several pages of algebraic manipulation and is therefore omitted here. With this conceptual introduction, though, any reader familiar with

7.4.3 Decoupled Power Flow

The general rule that relates voltage angle mainly to real power and voltage magnitude mainly to reactive power flow derives mathematically from two reasonable assumptions: first, that the reactive properties of transmission lines tend to outweigh the effect of their resistance, and second, that the voltage angle differences between buses are small (usually less than 10°). Specifically, we can pose the question, Which variable does the real (or reactive) power coming out of a bus depend on most—voltage angle or magnitude? In mathematical terms, we are asking, What is the partial derivative of P or Q with respect to θ or with respect to V? These are the partial derivatives which, evaluated at each bus in relation to the θ or V from each of the other buses, constitute the four partitions of the Jacobian matrix.

Without inserting any numerical values, we can examine the mathematical form of these partial derivatives and conclude, based on the two assumptions just cited, which terms ought to be large and which ought to be small. What we find is that the dependence of real power on voltage angle, $\partial P/\partial \theta$, ought to be substantial, while the dependence of real power on voltage magnitude, $\partial P/\partial V$, ought to be much smaller by comparison. For reactive power, we find the opposite: $\partial Q/\partial \theta$ ought to be small, but $\partial Q/\partial V$ should be substantial.

Readers familiar with calculus can follow the process of taking each of the four types of partial derivatives of the power flow equations. In fact, we must also distinguish whether the index k of the independent variable (θ_k or V_k) is the same or different from the index i of P_i or Q_i—in other words, whether we mean the dependence of real or reactive power on voltage angle or magnitude at the same bus, or at a neighboring bus. We are especially interested, of course, in the voltage relationships between neighboring buses (because we will want to draw conclusions about power flow from one bus to another as a result), so we will consider only the derivatives with unequal indices (such as $\partial P_2/\partial \theta_3$, as opposed to $\partial P_2/\partial \theta_2$) for now.

Let us examine the derivatives for a sample bus pair 2 and 3. First, we write out the derivatives $\partial P_2/\partial V_3$ and $\partial Q_2/\partial \theta_3$, which we will show to be small:

$$\frac{\partial P_2}{\partial V_3} = |V_2|[g_{23}\cos(\theta_2 - \theta_3) + b_{23}\sin(\theta_2 - \theta_3)]$$

$$\frac{\partial Q_2}{\partial \theta_3} = |V_2||V_3|[g_{23}\cos(\theta_2 - \theta_3) + b_{23}\sin(\theta_2 - \theta_3)]$$

We now observe the implications of our two assumptions. If a transmission link's reactive effects substantially outweigh its resistive effects, this means its conductance g_{23} (to be general, we would write g_{ik}) is a much smaller number than its susceptance b_{23}. This makes the cosine terms small, as they are multiplied by the

g's. The sine terms are multiplied by the b's, so they could be substantial based on that consideration. However, the sine terms are also small, for a different reason: if the voltage angle difference $\theta_2 - \theta_3$ between buses is small, then the sine of $(\theta_2 - \theta_3)$ is also small. Thus, each of the preceding derivatives consists of the sum of two small terms, and we might deem them small enough to be negligible.

By contrast, consider the derivatives $\partial P_2/\partial \theta_3$ and $\partial Q_2/\partial V_3$. Here, the g's multiply the sine terms, so these terms vanish on both accounts. But this leaves us with the cosine terms multiplied by b's, neither of which are negligible (since the cosine of a small angle is nearly 1).

$$\frac{\partial P_2}{\partial \theta_3} = |V_2||V_3|[g_{23}\sin(\theta_2 - \theta_3) - b_{23}\cos(\theta_2 - \theta_3)]$$

$$\frac{\partial Q_2}{\partial V_3} = |V_2|[g_{23}\sin(\theta_2 - \theta_3) - b_{23}\cos(\theta_2 - \theta_3)]$$

Indeed, if we consider the sine terms negligible and the cosine roughly equal to 1, we obtain the following approximations:

$$\frac{\partial P}{\partial \theta_3} \approx -|V_2||V_3|b_{23}$$

$$\frac{\partial Q}{\partial V_3} \approx -|V_2|b_{23}$$

Thus, the partial derivatives $\partial P_2/\partial \theta_3$ and $\partial Q_2/\partial V_3$ make up the "meat" of the Jacobian matrix, and they inform us about the variables to which real and reactive power are most sensitive.

We can take advantage of the these observations in what is called *decoupled power flow analysis*, where the term "decoupling" refers to the separation of the two pairs of variables (P and θ, on one hand; Q and V, on the other hand). By assuming the small derivatives to be zero, we greatly simplify the Jacobian matrix, which is used in the iterative procedure to find a better $x + \Delta x$, and thereby considerably reduce the volume of computation. Of course, now that we are not using quite the correct derivatives (i.e., we are only approximating the slope), we may not be pointed toward the best Δx in each iteration. Nevertheless, if the decoupling was a reasonably good assumption, then we will still be headed in the right general direction (toward the x that makes $f(x) = 0$), and with enough iterations we will still arrive at the same solution.

An even more radical simplification of the Jacobian matrix is possible and makes for *fast-decoupled power flow*. Here we make a third assumption: that the voltage magnitude profile throughout the system is flat, meaning that all buses are very near the same voltage magnitude (i.e., the nominal system voltage). We then observe the effect of this assumption, combined with the previous two assumptions about transmission lines and voltage angles, on the Jacobian matrix. By a process of approximation and cancellation of terms, the assumption of a flat voltage profile leads to a much handier version of the Jacobian, including a portion that stays the same during each iteration and therefore saves even more computational effort.[17] Again, this should affect only the process of converging on the correct solution, not the solution itself. If the simplifying assumptions were reasonable—in other words, if the simplified derivatives did not lead us in a grossly wrong direction—the computation is vastly sped up.

Note that the power flow solution obtained by the decoupled or fast-decoupled algorithm will expressly produce a certain profile of voltage angle and magnitude throughout the system, which contradicts the literal assumption that these profiles should be flat. Thus, we should think of the flat profiles as merely a procedural crutch along the way to discovering what the true profiles are. The reason we can get away with this apparent conflict is that the iteration process is self-correcting in nature. We can thus incorporate a statement that we know to be false when taken literally (i.e., the voltage profiles are flat) into the directional guidance for where to go with our next iteration (the derivatives in the Jacobian matrix), without contradicting the solution at which we ultimately arrive. This is a bit like giving driving directions along a detour route with fewer stop signs, which leads to the same destination—only faster.

Likewise, note the apparent contradiction between the existence of line losses, which can result only from line resistance, and the approximation that the conductances are negligible.[18] Again, the simplifying assumption of ignoring the g's is only a crutch for the process of approaching the correct power flow solution, and the solution itself will be consistent with the actual, nonzero values of conductance and resistance. This solution combined with the explicit resistance values—which are now are anything but negligible—then yields the losses for each transmission link.

7.5 APPLICATIONS AND OPTIMAL POWER FLOW

Power flow analysis is a fundamental and essential tool for operating a power system, as it answers the basic question, What happens to the state of the system if we do such-and-such? This question may be posed in the context of either day-to-day operations or longer-term planning.

---

[17]In the preceding discussion of partial derivatives we have only considered pairs of variables from neighboring buses, that is, the rate of change of real or reactive power at bus i (in our example, 2) with respect to voltage angle or magnitude at one neighboring bus k (in our example, 3). Having chosen that neighboring bus $k = 3$ and taking the derivative, we were able to cheerfully drop the summation sign with all its various k's, since our rate of change is independent of what happens at all these other buses. However, the Jacobian matrix also contains partial derivatives of power with respect to voltage at the same bus; for example, $\partial P_2/\partial \theta_2$. These terms, which appear along the diagonal of the matrix, are the ones affected by the assumption of a flat voltage profile. These diagonal terms look different in that they retain the summation over all the other k's. They succumb, however, to approximation and cancellation of various b's, leading to vastly simplified expressions. A thorough discussion appears in Arthur Bergen, *Power Systems Analysis* (Englewood Cliffs, NJ: Prentice Hall, 1986).

[18]Recall from Section 3.2.4 that $G = R/Z^2$, so that when R approaches zero and there is only reactance ($Z \approx X$), $G \approx 0$ as well.

In the short run, a key part of a system operator's responsibility is to approve generation schedules that have been prepared on the basis of some economic considerations—whether by central corporate planning or by competitive bidding—and scrutinize them for technical feasibility. This assessment hinges on power flow studies to predict the system's operating state under a proposed dispatch scenario; if the analysis shows that important constraints such as line loading limits would be violated, the schedule is deemed infeasible and must be changed.

Even with feasible schedules in hand, reality does not always conform to plans, requiring operators to monitor any changes and, if necessary, make adjustments to the system in real time. Power flow analysis is the only comprehensive way to predict the consequences of changes such as increasing or decreasing generation levels, increasing or decreasing loads, or switching transmission links and assessing whether they are safe or desirable for the system. Specifically, operators need to know impacts of any actions on voltage levels (are they within proper range?), line flows (are any thermal or stability limits violated?), line losses (are they excessive?) and security (is the operating state too vulnerable to individual equipment failures?). Similarly, power flow analysis is a fundamental tool in the planning context to evaluate changes to generation capacity or the transmission and distribution infrastructure.

Sometimes it is necessary to compare several hypothetical operating scenarios for the power system to guide operating and planning decisions. Specifically, one often wishes to compare and evaluate different hypothetical dispatches of generation units that could meet a given loading condition. Such an evaluation is performed by an optimal power flow (OPF) program, whose objective is to identify the operating configuration or "solution" that best meets a particular set of evaluation criteria. These criteria may include the cost of generation, transmission line losses, and various requirements concerning the system's security, or resilience with respect to disturbances.

An OPF algorithm consists of numerous power flow analysis runs, one for each hypothetical dispatch scenario that could meet the specified load demand without violating any constraints. (Clearly, this makes OPF more computation-intensive than basic power flow analysis.) The output of each individual power flow run, which is a power flow solution in terms of bus voltage magnitudes and angles, is evaluated according to one or more criteria that can be wrapped into a single quantitative metric or *objective function*, for example, the sum of all line losses in megawatts, or the sum of all generating costs in dollars when line losses are included. The OPF program then devises another scenario with different real and reactive power contributions from the various generators and performs the power flow routine on it, then another, and so on until the scenarios do not get any better and one is identified as optimal with respect to the chosen metric. This winning configuration with real and reactive power dispatches constitutes the output of the OPF run. OPF solutions may then provide guidance for on-line operations as well as generation and transmission planning.

Especially for applications in a market environment, where planning and operating decisions may have sensitive economic or political implications for various parties, it is crucial to recognize the inherently subjective nature of OPF. Power flow analysis by itself basically answers a question of physics. By contrast, OPF answers a question about human preferences, coded in terms of quantitative measures. Thus, what is found to constitute an "optimal" operating configuration for the system depends on how the objective function is defined, which may include the assignment of prices, values, or trade-offs among different individual criteria. In short, "optimality" does not arise from a power system's intrinsic technical properties, but derives from external considerations.

It is also important to understand that the translation of an OPF solution into actual planning and operating decisions is not clear-cut and has always involved some level of human judgment. For example, the computer program may be too simplistic in its treatment of security constraints to allow for sensible trade-offs under dynamically changing conditions, which then calls for some engineering judgment in adapting the OPF recommendation in practice. At the same time, the computational process is already complex enough that different OPF program packages may not offer identical solutions to the same problem. Therefore, the output of power flow analysis including OPF constitutes advisory information rather than deterministic prescriptions. Indeed, the complexity of the power flow problem underscores the difficulty of managing power systems through static formulas and procedures that might some day lend themselves to automation, especially if a system is expected to perform near its physical limits.

CHAPTER 8

System Performance

8.1 RELIABILITY

Reliability generally describes the continuity of electric service to customers, which depends both on the availability of sufficient generation resources to meet demand and on the ability of the transmission and distribution system to deliver the power. Historically, though, the analysis of reliability has emphasized the generation aspect, especially at the system level. In part, this is because transmission systems have been designed with sufficient excess capacity to merit the assumption that generated power could always be delivered, anywhere. However, as transmission systems are being more fully utilized due to a combination of economic pressures, demand growth, interconnections between territories, and difficulties in siting new lines, transmission is playing an increasingly important role in system reliability. The integrity of the transmission system is specifically analyzed in terms of *security* (Section 8.2).

8.1.1 Measures of Reliability

The simplest way in which utilities have traditionally described their system's reliability is in terms of a *reserve margin* of generation resources that are in excess of the highest anticipated load. Before the economic pressures that began to appear during the 1970s, reserve margins of 20% were standard, and some as high as 25%. One weakness of this approach is that it does not take into consideration the characteristics of specific generation units, notably their various failure rates, which may differ significantly.

A more refined measure that has come into use since then is the *loss-of-load probability* (*LOLP*), which states the probability that during any given time interval, the systemwide generation resources will fall short of demand. This probability is derived from the failure probabilities of the individual generators (i.e., the chance of that generator being unavailable) by summing up the probabilities of all the possible combinations in which total capacity is less than the anticipated load. The LOLP may be considered on a daily basis (looking at the peak load for that day) or for each individual hour.

Electric Power Systems: A Conceptual Introduction, by Alexandra von Meier
Copyright © 2006 John Wiley & Sons, Inc.

A closely related measure is the *loss-of-load expectation* (*LOLE*), in which the probability of loss-of-load for each day is summed up over a time period and expressed as an inverse, to state that we should expect *one* loss-of-load event during this period. The smaller the LOLP, the longer on average we will go until an outage happens. For example, if the LOLP is 0.00274 (1/3650) every day, this corresponds to a LOLE of one day in ten years. In other words, the systemwide generation capacity is expected to fall short of demand, presumably at the peak demand hour of that day, once every ten years. This latter figure has traditionally served as a benchmark value for reliability—the "one-day-in-ten-years criterion"—throughout the utility industry in the United States. Note that the loss-of-load probability or expectation say nothing about the duration of an outage; in particular, one day in ten years does not mean the load will be interrupted for all 24 hours of that day.

Finally, the *expected unserved energy* (*EUE*) can be calculated by combining the probability of loss-of-load with the actual megawatt (MW) amount of load that would be in excess of total generating capacity. This process assumes that the excess load would be *shed*, or involuntarily disconnected so as to retain system integrity and continue to serve the remaining load.

As measures of systemwide properties, the terms just cited describe the entire grid (as defined traditionally by an individual utility's service territory[1]) and consider only outages due to generation shortfall, not local disturbances in the transmission and distribution system. In truth, however, transmission and distribution failures are by far the most common cause of service interruptions. For this reason, service reliability varies regionally, depending in large part on topography and climate as well as population density. In the mountains, for example, power distribution lines are much more prone to storm damage, and it will take service crews longer to reach and repair them. Moreover, where only a small number of customers are affected by a damaged piece of equipment, its repair will tend to be lower on the utility's list of priorities, especially after a major event when line crews are working around the clock to restore service. In downtown areas, by contrast, many loads are considered so sensitive[2] that special features are incorporated into the design of distribution systems to minimize the LOLP, and the additional costs are justified by the high load density. The actual service reliability for specific customers within a power system is therefore a highly variable quantity that depends on a large set of heterogeneous factors.

This actual service reliability can be quantified in terms of how often service to certain loads is interrupted (an *outage* occurs), and how long the interruption lasts: *outage frequency* and *outage duration*. The product of outage frequency and average

---

[1] The territory can be defined at any scale for the purpose of this analysis, since generation resources outside the territory are simply considered as imports. The imports, like native generators, have some probability of being available at any given time.

[2] Such sensitive loads might include street lights in busy intersections, elevators in high-rise buildings, commercial customers where loss of power may have a big economic impact, and government or corporate offices with perceived political significance. In hospitals, where service interruptions have immediate life-threatening consequences, back-up generators or "uninterruptible power supplies" (UPS) are usually in place, although these loads remain a top priority to restore.

duration gives the total outage time. Since the most typical service interruptions are those associated with events in the distribution system, many of which are very brief (for example, the operation of a reclosing circuit breaker that remains open for a half-second to clear a fault), outage frequency may be computed so as to include only interruptions lasting longer than a specified time. However, given the increasing number of sensitive appliances, such as computers or digital clocks that reset themselves after even a momentary fluctuation, the nuisance of frequent small outages has become a growing concern in the area of customer satisfaction.

8.1.2 Valuation of Reliability

Generally speaking, the sudden loss of electric service entails some combination of nuisance and/or loss of economic productivity for different customers. In some cases, there is an actual threat to human health: classic situations where electricity may be absolutely vital include winter heating, air conditioning on extremely hot days, and traffic signals. Medical life support services are another case of vital electricity, although hospitals are typically equipped with UPS fed by their own backup generators.

Although the situations in which human lives depend on electric service constitute only a small fraction of electricity uses, they characterize the urgent importance generally ascribed to reliability. Considering all electric demand to be vitally important makes sense in that an outage does not usually discriminate among more and less important loads. For example, plugged in somewhere among the lights, televisions and refrigerators on a city block might be a dialysis machine. Unable to isolate the most vital loads to serve exclusively, power system operators often find themselves responsible to maintain or restore service to all customers with similar urgency.

In turn, this customary high standard of reliability has led industrialized societies to become increasingly dependent on uninterrupted electric service without considering this a risk or vulnerability. For example, appliances such as clocks, VCRs, and computers tend not to be equipped with batteries to safeguard information in the event of an outage, and gas-fired furnaces and water heaters with electronic ignition make their owners dependent on electricity to stay warm even though it is not needed as the primary energy source.

As mentioned earlier, the one-day-in-ten-years criterion has served as a benchmark for service reliability in the U.S. electric utility industry for many years. From a market perspective, though, the concept has been criticized for its arbitrariness and overgeneralization. An influential study in 1972 charged that much more was being spent on overdesigning equipment than could rationally be justified through the value of that increased reliability to consumers, and that, in this sense, utilities were "gold-plating" their assets.[3]

---

[3]Michael Telson, "The Economies of Alternative Levels of Reliability for Electric Power Generating Systems," *Bell Journal of Economics* **6**(2), 1975.

Historically, utilities' pursuit of very high levels of service reliability has had several reasons. One reason is their legal obligation to serve, as demanded by the contractual regulatory contract that grants them a territorial monopoly in return for the promise to serve all customers indiscriminately and to the best of their ability. Associated with this obligation has been a ratemaking process that allowed utilities to recover a wide range of reliability-related investments and expenses through the rates they charge customers, in which persuading the Public Utility Commissions of the "prudency" of these investments has not traditionally been very difficult.

The commitment of utilities and regulators alike to investments in system upgrades must also be viewed in the context of electric demand growth, which in the United States was very high during the period following World War II until the energy crises of the 1970s, and which subsequently tended to be overestimated by analysts who projected continued exponential growth at similar rates. The historical experience of continuous growth in combination with the fear of energy shortages explains the readiness to invest large sums of money in added generation capacity, as well as the focus on reliability measures that emphasize generation shortfalls over other causes of service interruptions. Finally, commitment to service reliability can also be understood in terms of a culture of workers who see themselves in the role of providing a vital public service, and who have long cultivated a sense of ownership of their vertically integrated system in which they take considerable personal pride. The implications of changing this cultural variable in the restructured market environment are far from clear.

From the economic perspective, it becomes necessary to explicitly consider customers' willingness to pay, which implies disaggregating various aspects of service quality and distinguishing among customer groups with different preferences. Analytically, the problem is to determine what level of reliability is "optimal" for a given type of customer, in that the amount of money spent on providing this level of service is commensurate with the amount this customer would be willing to pay for it if given the option. Such a determination of course requires a mechanism by which customers can express their preferences, and restructured electricity markets aim to achieve this goal by providing customers with more and increasingly differentiated choices. The most common approach is to offer rate discounts in exchange for an agreement to disconnect loads whenever the utility deems necessary, up to some maximum number of instances or duration per year. To actually provide different levels of service reliability for various sets of customers then requires a technical mechanism to discriminate among them, or selectively interrupt their service, in order to verify or enforce compliance with the interruptible load agreement (it is not unheard of for customers to gladly receive the discount, but fail to switch off their load when the call comes at an inconvenient time). To date, interruptible service contracts are relatively common for large commercial and industrial customers, but not at the residential level.

There exists a literature on the valuation of electric-service reliability that attempts to identify and distinguish how much service reliability is worth to different types of customers, or to specifically estimate the costs these customers incur as a result of outages. The simplest approach assumes a linear relationship between outage cost and duration. Here, outage cost is expressed in terms of dollars per kilowatt-hour (kWh) lost, where the lost kilowatt-hours are those that would have been demanded over the course of the outage period. Such a cost might be derived, for example, from the lost revenues of a business during that time. A more refined approach estimates cost components of both outage frequency and outage duration. In the absence of real choices, though, these estimates suffer the same uncertainties as any contingent valuation data that are based on people's responses in surveys, which may differ from the preferences they would reveal in an actual market. In any case, the nuisance and economic cost associated with outages must be assumed to vary widely among different types of customers.

The application of value-of-service data in actual policies and markets is still limited. The pricing system in the restructured electricity market of the United Kingdom actually incorporates a figure for the value of service, which is used in a calculation of payments to generators for providing capacity to enhance system reliability, but it is a relatively simplistic and arguably subjective measure: it uses a single, systemwide figure for the cost per kWh lost that is adjusted on an hourly basis.

8.2 SECURITY

Security is a measure of the width of the operating envelope, or set of immediately available operating configurations that will result in a successful outcome, that is, no load is interrupted and no equipment is damaged. In other words, security describes how many things can go wrong before service is actually compromised. A system in a secure operating state can sustain one or several *contingencies*, such as a transmission line going down or a generator unexpectedly going off-line, and continue to function without interruption, by transitioning into a new configuration in which the burden is shifted to other equipment (the load on other lines and/or generators is suddenly increased). Such a transition also requires *transient stability* (see Section 8.3), which, in the most general analysis of security, is assumed as given: the focus here is not on whether the system is capable of making a smooth transition to an alternative configuration, but on whether such alternatives exist at all.

Obviously, as a power system serves an increasing load, the number of alternative operating configurations diminishes, and the system becomes increasingly vulnerable to disturbances. In the extreme case, with all generators fully loaded (and all options to purchase power from outside the system already exhausted), if one generator fails, some service will inevitably be interrupted. To avoid this type of situation, utilities have traditionally retained a reserve margin of generation. Increasing interconnections among service territories over the past decades have enabled the confident operation with lower reserve margins than the traditional 20%, since reserves are in effect "pooled" among utilities. At the same time, this approach to providing reliability through scale implies an increased dependence on transmission links, as well as an increasing vulnerability to disturbances far away.

Analogous to generation reserve, system security relies on a "reserve" of transmission capacity, or alternate routes for power to flow in case one line suddenly goes out of service. The analysis of such scenarios is called *contingency analysis*. A standard criterion in contingency analysis is the *N-1 criterion* (for "normal minus one"), which holds that the system must remain functional after one contingency, such as the loss of a major line. (For even greater security, an *N-2* criterion may be applied, in which case the system must be able to withstand two such contingencies.) Note that the networked character of the transmission system makes such criteria possible, which cannot be achieved in a strictly radial distribution system where the failure of one line interrupts the only service link to all "downstream" sections.

Security criteria find expression in the form of *line flow limits*, which state the amount of current or power transfer permissible on each transmission link. The implication is that, as long as the currents on all the lines are within their limits, then even if one line is lost, the resulting operating state does not violate any constraints. This means that loading on the other lines and transformers will not exceed their ratings, and all voltages can be held within the permissible range. Line ratings, in turn, are based on either thermal or stability limits (see Section 6.5 on ratings and Section 8.3 on stability).

The computational part of contingency analysis is to run load flow scenarios (see Chapter 7) for a set of load conditions, including peak loads, each time with a different contingency (or combination of contingencies, if desired), and check that all constraints are still met. Usually, the contingencies chosen for this analysis are from a list of "credible contingencies" prepared by operators based on experience. The results may then be used both to set limits for secure operation and to suggest necessary reinforcements in transmission planning.

In the general form just described, this is a *steady-state* analysis, meaning that it considers the system operating state before and after the contingency, but not during the event and the transition into the new state. However, that transition itself may pose potential problems; this is assessed in a *dynamic* analysis. Here, contingencies are selected from a shorter list of more serious "dynamic contingencies," and the system is analyzed for transient and voltage stability (see the next section).

8.3 STABILITY

8.3.1 The Concept of Stability

In general, *stability* describes the tendency of an alternating current (a.c.) power system to maintain a synchronous and balanced operating state. Most often, the term stability refers to *angle stability*, which means that all the system's components remain locked "in step" at a given frequency (as opposed to *voltage stability*, which is discussed separately in Section 8.3.4). While the voltage and current throughout the interconnected a.c. system are oscillating at exactly the same frequency, there are differences in *phase* or *angle* (the relative timing of the maximum) at various locations that are associated with the flow of power between them (see Chapter 7). Stability analysis is concerned with these differences in phase and their implications for keeping the system locked in step. This "locking" phenomenon is based on the electrical interaction among generators.

We distinguish *steady-state* and *transient* or *dynamic stability*. In the steady state, we evaluate a system's stability under some fixed set of operating conditions, including constant generator outputs and loads. A crucial factor here is the length of transmission lines in relation to the amount of power they transmit. Transient stability has to do with the system's ability to accommodate sudden changes, such as faults (short circuits), loss of a transmission link, or failure of a large generating unit, and return quickly through this transient condition to a sustainable operating state.

"Stability" here is a rigorous application of the term from physics, where one distinguishes different types of equilibria, stable or unstable, that describe the tendency of a system to depart from or return to a certain resting condition in response to a disturbance. A simple mechanical analog is useful. Consider a round bowl and a marble, as shown in Figure 8.1a. Inside the bowl, the marble rests at the bottom in the middle. If we displace the marble from this resting point, by moving it up toward the rim of the bowl, its tendency is to return to the equilibrium location: it will roll around or back and forth in the bowl until it settles again at the bottom. This is a *stable equilibrium*. Gravity acts here as a *restoring force* that pushes the system back toward its equilibrium state (i.e., pulls the marble to the bottom).

Now imagine the bowl turned upside down, with the marble precariously balanced on top. As long as the marble is situated precisely at the highest point, without so much as a breath disturbing it, it will stay. But the slightest displacement will make the marble roll off the side of the bowl, away from equilibrium. (Gravity no longer acts as a restoring force, but rather pulls the marble away from equilibrium.) This type of equilibrium is *unstable*.

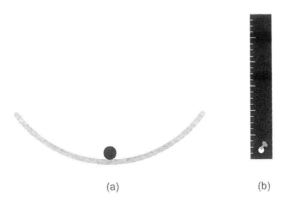

Figure 8.1 Bowl and marble, ruler in equilibrium. (a) Marble in a bowl: stable equilibrium; (b) balanced ruler: unstable equilibrium.

Another simple example is a ruler with a hole at one end, supported by a nail on the wall. In its stable equilibrium, the ruler hangs down from the nail. Displace it by lifting the other end, and the ruler will swing back and forth, eventually coming to rest again in the vertical position. How do we make an unstable equilibrium with this system? Turn the ruler upside down so that it stands vertically above the nail (Figure 8.1b). Again, if positioned carefully enough and left alone, the ruler will stay, but any disturbance will cause it to swing back around.

In an a.c. power system, the thing that is "moving" is the relative timing of the voltage maximum, expressed as the *voltage angle* or *power angle* δ. As we saw in Section 4.3 on controlling the real power output of synchronous generators, a generator's power angle can move ahead or behind in time, and this movement is associated with an exchange of power between this and other generators, mediated by a circulating current. Further on, in Section 7.3, we came to think of the power angle as a variable that, at any point in the system, indicates how much real power is being injected into or drawn out of the system at that point. Our conceptual challenge now is to combine these ideas and understand the power angle as a dynamic variable—that is, one whose changes over time interest us—that relates different and remote parts of a power system to each other.

8.3.2 Steady-State Stability

When considering steady-state stability, we are basically asking whether a particular operating configuration of an a.c. power system represents a stable equilibrium. In other words, is it possible to supply a given set of loads with a given set of power contributions from generators through a given network of transmission lines, and maintain synchronism among all components? *Synchronism* means that both the *frequency* and the *phase* of two or more oscillating components match. When synchronous generators are connected together, they must be spinning at the same rate,[4] and they must also be in step with each other, or their voltage output peak at about the same time. Only in this way can all generators simultaneously contribute to feed power into a network: if the timing of voltage and current did not match up for each generator, it would have to alternately inject and absorb power over different portions of the cycle (in practice, this would cause overloading of parts of the generator windings, if the circuit breakers had not opened first).

Synchronism also requires a stable equilibrium condition where there is a restoring force that tends to slow down a generator that has sped up and to speed up a generator that has slowed down; otherwise, synchronism could not be maintained, because the slightest disturbance would throw off individual generators and have them go at different speeds. Such a restoring force indeed exists: it was explained in terms of the power exchange between generators in Section 4.3.3. The force results from the fact that a generator whose relative timing or *power angle* is ahead of others must supply additional power (thus tending to restrain the turbine more), whereas one whose power angle is behind supplies less power

---

[4]The frequency that matters here is that of the electrical a.c. output, not the frequency of mechanical rotation.

(thus relieving the restraint on the turbine). This interaction provides for a negative-feedback effect that serves to control and "hold steady" each generator.[5]

When we say that a power system is operating within a regime of steady-state stability, we mean that it is in a regime where these stabilizing, restoring forces exist. In the marble-in-a-bowl analogy, the marble being displaced represents the individual generator rotor, and the marble's position is the generator power angle, δ. The generator system is harder to conceptualize, though, because the equilibrium position here is not one where things are simply at rest; rather, it is where the generator is spinning at synchronous frequency and at a certain power angle, δ_0, that corresponds to the set power output. The point at the bottom of the bowl represents this equilibrium power angle, δ_0. The generator may now speed up and slow down relative to its synchronous speed, thus pulling the power angle ahead or behind its equilibrium value. This relative movement of δ is represented by the movement of the marble. In terms of this analogy, then, steady-state stability addresses whether the shape of the bowl is indeed concave, with the marble inside it.

What happens now is that depending on the differences in phase or power angles at various locations in the system, the shape of the bowl changes. This is because the effective exchange of power between generators depends on the relative timing of their voltages and currents. As we illustrate in more detail later, the most effective negative feedback or stabilizing interaction between generators occurs when their phases are very close together. As the difference between their phase angles grows, which corresponds to a greater difference in power generation, and thus a greater transmission of power between them, the interaction, and thus the stabilizing effect, weakens. Thus, the question for steady-state stability analysis is, How much power can we transmit—say, along a given transmission line—and still maintain a sufficiently concave shape of the bowl?

It is helpful here to consider the mathematical expression for the transmission of power along a line, with generators (and loads) at either end. Each end is characterized by a power angle, δ, that indicates the relative amount of power being injected or consumed at that location and that also describes the relative timing of the generator there (the voltage angle θ in Chapter 7). We are interested now in the difference between power angles at the two ends, which we will call δ_{12}. This difference is directly related to the amount of power transmitted on the line: the bigger the difference, the more power is transmitted. Using a reasonable approximation,[6] it is possible to write a simple expression for the (real) power transmitted as a function of the power angle (which is further discussed in Chapter 7). This general expression is

$$P = \frac{V_1 V_2}{X} \sin \delta_{12}$$

---

[5] This is called a "negative feedback" because the force acts opposite (or in the negative direction) to the displacement: when the generator speeds up, slow it down; when it slows down, speed it up.
[6] The approximation requires that the resistance of the transmission line is negligible compared to its reactance, which physically implies that there are no losses on the line. Mathematically, it means that real and reactive power can be considered separately. This approximation is the same one used in power flow analysis for "decoupled load flow" (Chapter 7).

where V_1 and V_2 are the voltage magnitudes on either end, X is the reactance of the line (see Section 3.2), and $\sin \delta_{12}$ is the sine of the power angle difference. Note that when the power angles on either side are the same, δ_{12} and its sine become zero and no real power is transmitted.

In order to transmit the maximum amount of power along the line, the question is, How big can we make δ_{12}? In the preceding equation, it looks as if we could make δ_{12} equal to $90°$, where $\sin \delta_{12}$ would take its maximum value of 1. However, this is not generally safe to do. Rather, there is a *stability limit* on δ_{12}, which might be somewhere in the neighborhood of $45°$. This stability limit arises because, as δ_{12} increases, the negative feedback between the generators at nodes 1 and 2 diminishes. The object is to have δ_{12} large enough to transmit plenty of power, but not too large so as to risk losing the negative feedback between generators. This reasoning extends to an entire network, where each transmission link is examined for the δ_{12} between its two ends.

Let us now discuss in more depth why the negative feedback between generators should depend on their difference in power angle. There are two approaches to illustrating this: first, referring to the mathematical expression just given, and second, referring back explicitly to the circulating currents responsible for exchanging power between generators, as introduced in Section 4.3.3. Consider the graph of the preceding equation for the power P transmitted, which, if V_1, V_2, and X are constant, is just a simple sine curve, $\sin \delta_{12}$, as shown in Figure 8.2. Our question now is, for a given change in δ_{12}, how does P change? We can immediately see that if $\delta_{12} = 90°$, this entails a problem, because if, for any reason, δ_{12} were to fluctuate slightly and increase beyond $90°$, the power would actually *decrease*. This would represent an unstable equilibrium, since decreasing power would speed up the generator and lead to a further increase in δ_{12}. Therefore, it is unrealistic to operate a transmission line with a δ_{12} of $90°$.

Furthermore, we note the slope of the sine curve, which is steep for small δ_{12} and gets flatter as δ_{12} increases. A flatter slope means that, for a given increment in δ_{12}, there is only a small increment in P. However, for a good, stabilizing feedback effect on the generator, we want the increment in P to be large. This is analogous

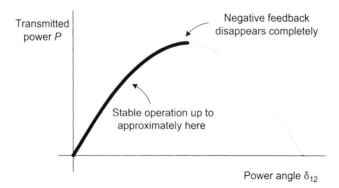

Figure 8.2 Power transmitted versus power angle.

Figure 8.3 (a) Deep and (b) Shallow equilibria.

to the slope of the sides of the bowl with the marble: a weak restoring force would correspond to a shallow soup dish, which still tends to return the marble to the center, but not as quickly and reliably as a deep salad bowl (Figure 8.3). Therefore, from the stability standpoint, it is preferable to operate with a small δ_{12} where the slope of sin δ_{12} and thus the incremental change in P is large.

In practice, this stability consideration is weighed against the incentive to transmit larger amounts of power on a given line, which forces an increase of δ_{12} toward the shallower region, unless it is decided to upgrade the transmission line by decreasing X, in which case, one can achieve higher P for the same δ_{12}. The stability limit comes at the point where the slope of sin δ_{12} is still just steep enough for comfort. Based on experience, power engineers generally consider 40° to 50° a reasonable limit on δ_{12}. Note that this stability limit tends to be relevant only for long transmission lines that have a substantial reactance, whereas the ability to transmit power on shorter lines tends to be limited by heating (see Section 6.5).

The second explanation of negative feedback between generators as a function of power angle refers directly to the exchange of power between (and among) generators, which lies at the heart of stability. In Section 4.3.3, we introduced the notion of a "difference voltage" that arises due to a difference in the timing of the voltage maximum because one generator has pulled ahead (greater power angle) or fallen behind (smaller power angle). The difference voltage results in a current that circulates between the armature windings of the respective generators.

The key requirement is that this circulating current have a timing such that it increases the load of the generator that is ahead of the other (Unit 1) and reduces the load of the one that is behind (Unit 2). In Section 4.3.3, we argued this was the case, based on the reasoning that follows: we assumed the difference in timing between the two voltages maxima was a small fraction of a cycle. Accordingly, the difference voltage is approximately 90° ahead of the main voltage wave, as seen from the perspective of the ahead generator. Next, because the impedance between the two sets of generator windings is almost purely inductive reactance, the circulating current resulting from the difference voltage lags about 90° behind this voltage. This brings it just about *into* phase with the main voltage output of the ahead generator, and also with the previously existing current (which is in phase with the voltage if the power factor is unity). Thus, in Unit 1, the circulating current is associated with an *additional* positive power output. By contrast, from the perspective of Unit 2, the circulating current appears to be negative or 180° *out of* phase with its regular voltage and current, and therefore acts to *diminish* its power output; in other words, power is being supplied to the behind generator, or its load is reduced.

But what if the difference in the timing between the voltage maxima—that is, the difference δ_{12} in the power angle—grows larger? The effect is illustrated in Figure 8.4. When the difference in power angle is small, the difference voltage (i.e., the difference between the two voltage curves at any given instant) is also small and almost exactly 90° ahead of the two main voltage waves, meaning that its associated circulating current is almost exactly in phase with the generator voltage that is ahead (V_1). As δ_{12} increases, the magnitude of the circulating current increases dramatically, meaning that more power is transferred from Unit 1 to Unit 2. At the same time, the circulating current is shifted slightly out of phase with V_1. As δ_{12} increases further, the current magnitude still increases somewhat, but the phase shift becomes more pronounced. With the power angles of the two generators 45° (one-eighth of a cycle) apart, the circulating current instead peaks at 22.5° behind V_1. As a result, the product of these two curves, which describes the power transferred, is sometimes positive and sometimes negative, which means that the generator is supplying additional power during some portions of the cycle and absorbing power (from Unit 2) during other portions of the cycle. Because there is now an oscillation of power back and forth, rather than a continuous transfer of power from one generator to the other, the stabilizing effect is gradually lost as the difference in power angles increases. The greater magnitude of the circulating current at large δ_{12} also becomes problematic in terms of overloading the generators, especially to the extent that it is out of phase with the load current and may be associated with eddy currents and uneven heating of the windings. For these reasons, the effective control of interconnected synchronous generators becomes more difficult with increasing difference in power angle, and it is not desirable to let δ_{12} become too large.

8.3.3 Dynamic Stability

In keeping with our marble-in-the-bowl analogy, while steady-state stability is concerned with the shape of the bowl, dynamic stability is concerned with how far we can displace the marble (i.e., the power angle) and how quickly it comes to rest again. Dynamic stability is also called *transient stability*, because the displacement of the power angle is generally due to a temporary, transient disturbance. The question then is, How large a disturbance can the generator sustain and still return to equilibrium? And will it return to equilibrium in a reasonable amount of time?

To explain dynamic stability, most engineering texts use a different mechanical analog to the generator, namely, a spring: as the spring is stretched or compressed, it tends to return to its original shape, with some bouncing back and forth until it comes to rest in its equilibrium position. If one translates all the variables, the

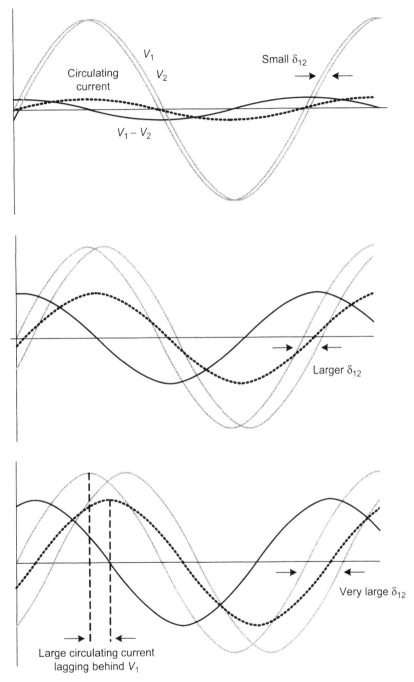

Figure 8.4 Effect of increasing δ_{12} on circulating current.

mathematical representation of such a spring system quite accurately describes the situation inside a generator, where the displacement of the spring is analogous to the displacement of the power angle. For the purposes of intuitive understanding, and especially for readers who are not already familiar with the analysis of the spring system, let us stick with the marble in the bowl even though it is less accurate mathematically.

As the marble moves around in the bowl—say, if we were to release it near the rim and let it roll down, up again on the other side, and so on—we can describe its behavior in terms of energy: it possesses a certain amount of gravitational potential energy, depending on its height at any given moment, and it possesses a certain amount of kinetic energy, depending on its speed. The sum of kinetic and potential energy remains constant (owing to energy conservation), except that gradually, as the marble keeps rolling, this energy will be consumed by friction (turning it into heat). Friction thus acts as a *damping* force that slows the oscillation and allows the marble to finally come to rest at the bottom.

Similarly, the electric generator has a certain amount of energy, a restoring force, and a damping force. Mathematically, this situation is described by a power balance equation, which is a statement of energy conservation: at any instant, the power going into the spinning rotor must equal the power going out. Thus, in equilibrium, the mechanical torque from the turbine shaft equals the electrical power that is pushed out the armature windings through the magnetic field, plus a certain amount of damping power that acts to slow down the rotor. This damping power includes friction as well as an effect from the control system (excitation). When there is excess mechanical power supplied from the turbine, the rotor speeds up; when the power supplied from the turbine is less than the electrical and damping power drawn out, the rotor slows down.

This power balance can be written as the generator *swing equation*, which is a differential equation[7] that implicitly describes the behavior of the power angle δ. This swing equation is conventionally written as

$$M\ddot{\delta} + D\dot{\delta} + P_G(\delta) = P_M^0$$

Let us examine each term in this equation, from right to left: P_M^0 is the mechanical power input from the turbine, where the superscript 0 indicates the value at equilibrium (i.e., constant power generation at the equilibrium power angle δ_0). We

---

[7] A differential equation is one that relates a variable (say, the position of an object) to its rate of change (say, the velocity and/or acceleration of the object). Such an equation can be written down by considering basic laws of physics and the various forces acting on the object. A "solution" to a differential equation is a function that states explicitly how the variable behaves (say, the object's position as a function of time), where this function must satisfy all the criteria stipulated by the differential equation. Usually more than one function meets these criteria, but the solution can be further narrowed down by identifying *boundary values* that the function must take on at certain points in order to fit a specific situation. For the differential equation describing an oscillating object, the general solution is some sort of sinusoidal function (see Section 3.1).

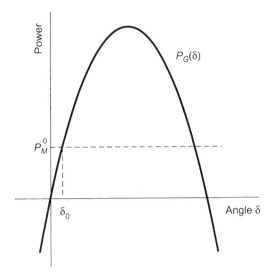

Figure 8.5 Power generated as a function of power angle.

assume that P_M^0 does not change over the time period of interest (i.e., the turbine is pushing with a constant force). This mechanical input power on the right must equal the total on the left-hand side because of energy conservation.

The electrical power output is represented by $P_G(\delta)$. The notation with parentheses indicates that the power generated P_G itself varies as a function of the power angle δ; this function is shown in Figure 8.5. Indeed, to a reasonable approximation, P_G varies as the sine of δ. (This is essentially the same as the equation for power transmitted and Figure 8.2 in Section 8.3.2.) Therefore, over a reasonable range of δ, as δ increases, so does P_G.

However—and this is crucial—P_G will not continue to increase indefinitely for increasing δ. In fact, for extremely large δ, P_G eventually becomes negative. It makes sense that P_G should eventually become negative because the power angle δ describes the position or timing of a generator *relative to others*, and because we are dealing with a cyclical motion. At some point, it will lead so far ahead of others that it appears to lag behind them. Mathematically, the fact that P_G does not just steadily increase with δ is referred to as a *nonlinear effect*. In terms of our previous analogy, the bowl with the marble flattens out near the rim and then begins to curve downward. Thus, if the generator pulls ahead of others by too great a phase angle, at some point it will be producing less instead of more electrical power. Physically, we can think of P_G as the restoring force that pushes back on the rotor through the magnetic field: the farther we displace δ, the harder the magnetic force pushes back—up to some point. At the equilibrium point δ_0, where the forces are balanced and δ holds steady, P_G equals P_M^0, which is seen graphically as the intersection of the P_G curve with the horizontal line that marks the value P_M^0.

The other two terms in the swing equation involve rates of change of the power angle. The single overdot indicates the rate of change (the first derivative, for readers familiar with calculus) of δ with respect to time, which is analogous to velocity and which in this case corresponds to the relative rotational *speed*: when $\dot{\delta}$ is positive, the rotor's frequency is greater than the system frequency of 60 cycles (i.e., in the process of moving δ ahead of other generators); when $\dot{\delta}$ is negative, the rotor frequency is less than 60 cycles. The constant D is a measure of the damping force, whose tendency is to resist any changes in δ, and the term $D\dot{\delta}$ represents the power absorbed by friction.

The double overdot indicates the rate of change of the rate of change of δ (the second derivative in calculus), which represents the *acceleration*, or change of speed, of the rotor. The constant M is a measure of the generator's inertia, whose effect is to resist changes in rotational speed. The power that goes into speeding up or that comes out of slowing down the generator rotor is represented by $M\ddot{\delta}$.

Note how the swing equation accounts for the equilibrium condition: when δ just sits at the equilibrium value δ_0 that makes P_G equal to P_M^0, there is no change of δ with respect to time, meaning that the damping and acceleration terms in the equation are zero. The generator keeps spinning and supplying electrical power at a constant rate, which in this context means that the situation is basically at rest.

But when analyzing dynamic stability, we want to know how the generator will respond to a disturbance of some sort, and whether it will be able to settle again in its equilibrium. In other words, what will happen if δ is somehow displaced from equilibrium? The essence of this analysis is to determine just how far δ can be displaced before there is trouble.

The classic case for study is where a transmission link is momentarily interrupted and then quickly reconnected. This type of event occurs, for instance, when a reclosing circuit breaker operates. Suppose the transmission link is a generator's only connection to the grid. Thus, during the time period where the link is interrupted, the generator cannot send out electric power. But because the time interval in question is very short—perhaps half a second—the steam turbine output cannot be adjusted. The turbine therefore continues to push the rotor with constant mechanical power input, P_M^0. Because no power goes out in the form of electricity, all of P_M^0 goes into accelerating the rotor, except for a small amount to overcome friction.

During this interval where the generator is disconnected from the grid, the rotor gains momentum, which means that the power angle δ increases, as does its rate of increase $\dot{\delta}$ (because it continues to *accelerate* under the turbine's torque for the duration of the disconnect). The generator thus acquires a certain amount of excess energy, which is just the accumulation of turbine power over that time interval. Aside from the part that is dissipated by friction, this excess energy manifests as kinetic energy of the rotor, which is now spinning at a higher than normal frequency.

The amount of this accumulated excess energy is crucial for determining stability. Up to a certain amount, it can be dissipated; more than that, the generator cannot return to equilibrium. Transient stability analysis is concerned with determining that critical amount of excess energy. Naturally, the longer the generator is accelerating in its disconnected state, the more kinetic energy it will build up.

Therefore, the problem is often stated in terms of the length of the time interval of disconnection. In other words, how long can the generator be disconnected before it gains so much speed that it will not return to normal when reconnected?

This question can be answered by turning to the solution of the swing equation. This equation was derived from the general principle of energy conservation, but it also dictates very specifically how δ may evolve as a function of time. Its mathematical solution says that, if δ is displaced a bit and then let go, it will oscillate back and forth, going alternately ahead and behind of δ_0. Owing to the damping force, the oscillation will diminish over time and δ will eventually settle down into its equilibrium, δ_0. This is analogous to the marble in the bowl, a pendulum, or a spring, which will oscillate back and forth until they finally settle at the bottom (or in the middle).

To see how this oscillation comes about, consider what happens at the end of the transient disturbance, when the generator is reconnected. As the transmission link is reestablished, the speeding generator can relieve itself of its excess energy into the grid. Indeed, it suddenly encounters a $P(\delta)$ that is very large, corresponding to the now very large δ, and that exceeds the turbine input power P_M^0. The rotor decelerates. However, even after the rotor has begun to decelerate, there is a period during which δ still increases, because the rotor is still spinning faster than the other generators in the system. This is analogous to the marble that has been given a good push and rolls up the side of the bowl even though it is already under the influence of gravity pulling it back. The rotor continues to decelerate until δ is less than δ_0, at which point $P(\delta)$ is less than P_M^0 and the rotor begins to accelerate again. As before, though, δ continues to decrease despite the fact that it already has positive acceleration, which is why it overshoots δ_0 until it reaches a minimum where it turns around again. This movement would continue back and forth indefinitely, save for the damping force that slows the motion of δ and causes the excursions to gradually diminish until δ settles at δ_0.

If the initial displacement of δ was too far, however, it will not return. Essentially, this is the point where the additional power that the generator supplies to the grid when δ is ahead is not sufficient to slow the generator back down because it has already built up too much momentum, or acquired too much energy. Such a point exists because of the nonlinear characteristic of $P(\delta)$, which does not continue to increase once δ gets too large, and indeed eventually becomes negative. The analogy is that the marble has been given too big a push so that the force of gravity can no longer confine it below the rim of the bowl.

This situation can be more specifically analyzed in terms of the exchange of potential and kinetic energy as the object (marble or rotor) oscillates. For the marble in the bowl, we can easily see that a certain amount of energy corresponds to the marble being confined to the bowl, while a greater amount of energy would imply that the marble jumps over the rim. Notably, it does not matter whether we are talking about kinetic or potential energy, since either can propel the marble out of the bowl: if the marble has too much potential energy, this means it will be located too high, that is, past the rim of the bowl; if it has too much kinetic energy, it will move too fast and overshoot the rim on its next upward roll. We can thus state that the marble will stay inside the bowl as long it has no more than a certain maximum total energy, where this energy is the sum of potential and kinetic.

For the generator whose power angle is oscillating or swinging, there is an analogous limit of total energy. In engineering texts, this is also described as the sum of a potential and kinetic energy. However, while this nomenclature establishes an easy mathematical analogy with other physical systems, it can also be confusing. Therefore, while keeping in mind the rest of the grid, let us first consider what physically happens to the energy during the generator's oscillation.

Because the turbine power inputs and loads throughout the network do not change, and because energy is conserved, the only place where energy can increase or decrease is in the rotational kinetic energy of the generator rotors. During the interval that the generator in question is disconnected and speeding up, other generators—for simplicity, let us say just one other generator—elsewhere in the system is supplying the extra load and is therefore slowing down. After the connection is reestablished, the two generators now have a symmetric power imbalance: one is too fast, the other too slow. The ensuing oscillation is nothing other than the exchange of energy back and forth between these two generators (by means of the circulating currents we discussed in Section 4.3.3), in which the generators alternately speed up and slow down until they again share the load according to their set points.

We now focus only on the individual generator that is immediately affected by the transient disturbance. This individual generator alternately gains and loses rotational kinetic energy as its rotational speed increases and decreases. It has a maximum amount of kinetic energy when the rotor is spinning fastest, meaning that δ is increasing most rapidly. Perhaps counterintuitively, this point does not coincide with the point at which δ itself is maximum (where δ actually holds steady momentarily and the rotor spins at its nominal 60 cycles), but rather when the rotational frequency or $\dot{\delta}$ is maximum, while δ increases and passes δ_0. This point corresponds to the bottom of the bowl where the marble has maximum speed.

On the way back, as δ decreases and passes δ_0, the rotor is physically moving the slowest. If we were describing the rotational kinetic energy of the generator in strict physical terms, we would say that it is at a minimum at this instant. However, for the purpose of analyzing the oscillation, one construes a mathematical quantity called "kinetic energy" that does not care whether the speed *relative* to the nominal 60 cycles is positive or negative, as if the power angle itself (which is really only a label marking the position of the rotor relative to that of other generators) were an actual physical object in motion. Given by $\frac{1}{2}M\dot{\delta}^2$, this quantity is mathematically analogous to the kinetic energy of $\frac{1}{2}mv^2$ of a rolling marble (where m is mass and v is velocity), and it is the same regardless of the direction in which the marble is rolling. Therefore, the so-called kinetic energy is again at a maximum as δ passes δ_0 in the opposite direction, when $\dot{\delta}$ is greatest in the negative direction.

Similarly, we can define a "potential energy" for the rotor, which is analogous to the gravitational potential energy of the marble. When δ is equal to δ_0, this "potential energy" is zero (the marble is at the bottom of the bowl). As δ is displaced, in either direction, the generator acquires "potential energy." We can think of this potential energy as the accumulation of restoring power, or work the generator has performed by squeezing excess power into the grid. As δ is displaced farther, we can imagine more and more capacity built up to push δ back in the opposite direction, with a maximum potential energy at the point of maximum δ. The restoring power thus accumulated is the difference between the electrical power output and the mechanical power input. This difference is represented in Figure 8.6 as the portion of the $P_G(\delta)$ curve that extends above the line representing P_M^0. Mathematically, the potential energy is the *integral* of the restoring power over δ, or the area under the curve between δ_0 and the given δ. Note that the units on the horizontal axis are degrees, which represent time, so that the area (power · time) has dimensions of energy. The maximum potential energy occurs at the point of farthest displacement of δ, just like the marble has maximum potential energy at its highest point. The curve of cumulative potential energy as a function of δ, conventionally labeled $W(\delta)$, is shown in Figure 8.6. This curve is analogous to the bowl.

It may be counterintuitive that the generator ought to have a maximum of potential energy at both maximum and minimum δ (just as it has maximum kinetic energy at both maximum and minimum physical speed). At maximum δ,

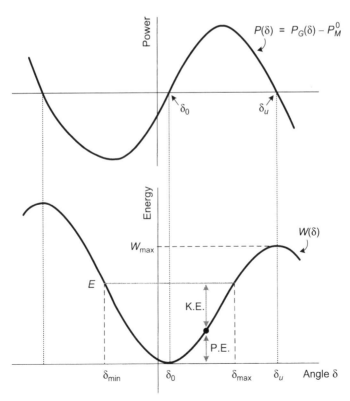

Figure 8.6 Restoring power and "potential energy" $W(\delta)$. (K.E. = kinetic energy; P.E. = potential energy.)

it has the greatest capability of doing work on other generators, that is, to relieve their load by carrying extra power and thereby slowing down, sacrificing its own "lead" in δ. At minimum δ, the situation is reversed, where the generator now has the greatest capability to absorb the extra work of others. For the purposes of stability analysis, these two situations are symmetric and are given the same label.

The peak of the potential energy curve in Figure 8.6—specifically, the one on the right if we are concerned about a forward displacement of δ—indicates the maximum amount of restoring work that can be done on the generator to bring it "back in line." It is labeled W_{max}. Physically, this means the maximum energy that the generator can dump into the grid by running ahead of others. (Conversely, the peak on the left shows the maximum amount of energy that the generator can absorb from others if it is behind.) This corresponds to the accumulated restoring power from δ_0 to the upper limit, labeled δ_u, or the area between the $P_G(\delta)$ curve and P_M^0.

Finally, these definitions provide us with a concise way to state the transient stability criterion. During the transient condition, the generator acquires both potential energy (because δ is displaced) and kinetic energy (because δ is in the process of changing). The generator is transient stable if the total energy acquired during the transient period is no more than the maximum amount of energy that can be gotten rid of, W_{max}. The beauty of this articulation is that it is very general and assumes nothing about the peculiarities of the transient disturbance.

There are two ways of illustrating this condition graphically. The first is on the $W(\delta)$ curve, where the excess energy acquired by the generator is represented in terms of height. It gains height as it moves out along the curve with increasing δ. During the oscillation, the total energy remains constant (save for damping), which is indicated by the horizontal line E. This total energy is composed of a potential energy component (the height of the W curve at a given δ, labeled P.E.) and a kinetic energy component (the difference between total and potential energy at the same δ, or the distance between the W curve and the horizontal line E, labeled K.E.). Stability requires that the horizontal line E must not go above W_{max}. Note that this does not allow us to displace δ as far as δ_u by the end of the transient period, because the displacement will also entail kinetic energy (i.e., still be in the process of increasing) at the moment that it is "let go." As a result, given a certain total energy E, δ will increase up to δ_{max} and then swing back to δ_{min} before hopefully settling at δ_0. The stability limit is where δ_{max} will never exceed δ_u, which is to say that the combination of potential and kinetic energy always remains less than W_{max}.

This representation establishes a perfect analogy to the condition that the marble must not overshoot the rim of the bowl. In particular, we must not let go of the marble near the rim if it has too much upward velocity! It also illustrates the fact that instability can result from excessive slowing down of the generator, that is, moving it out on the left side of the W curve, which would cause it to then oscillate and overshoot the peak on the right. In practice, though, this situation is more difficult to construe as it requires a sudden, enormous load holding back the generator.

The other representation refers to the graph of electrical power $P_G(\delta)$ and turbine power P_M^0, shown in Figure 8.7. This representation provides us with a specific δ as the stability limit. Here, the stability criterion is articulated as the *equal-area*

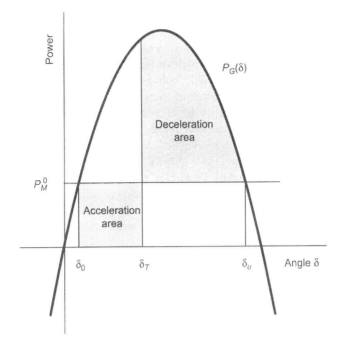

Figure 8.7 Equal area criterion.

criterion. It states that the generator is transient stable if the area below P_M^0, between δ_0 and the displaced δ_T at the end of the transient, is no greater than the area between $P_G(\delta)$ and P_M^0 up to their intersection point, which occurs at δ_u and represents the maximum δ for which there is still any restoring force at all. The first area represents the amount of excess energy acquired by the generator during the transient period (where the full turbine power is being absorbed); this is why it is called the "acceleration area." The second area, called the "deceleration area," represents the cumulative deceleration power, or the total amount of energy that the generator can dump into the grid, minus the amount to which it is already committed due to its displacement of δ. In other words, there must be enough deceleration power left to absorb the kinetic energy that is left once the potential energy has been accounted for.

For a given generator, the manufacturer will specify a curve $P_G(\delta)$ as a basis for this type of analysis.

8.3.4 Voltage Stability

While until now we have taken "stability" to refer to stability of the voltage angle or power angle (*angle stability*), the stability of the voltage magnitude, referred to as *voltage stability*, is also an issue. While the two are in fact interrelated, it is pedagogically helpful to consider them separately. Justification for doing so can be found in the "decoupling" approximation often used in power flow analysis (Chapter 7), where voltage angle and magnitude are taken as unrelated variables, the former being associated only with real power and the latter only with reactive power. In any case, the interactions between voltage and angle instability are so complex that they are well beyond the scope of this text. Empirically, it has been found that a system can become unstable with respect to only voltage magnitude but not angle, or vice versa, and voltage instability is therefore often treated as a separate subject.

As in the previous discussion, the basic concept of stable versus unstable equilibria applies. Recall that angle stability hinges on the condition that power output increases with power angle in a generator. Voltage stability means that as load increases (i.e., load impedance goes down), power consumption also increases (i.e., voltage does not go down). Generally, this is the case up to some limit of load, beyond which the voltage cannot be maintained as the load attempts to draw more power. Voltage stability is enhanced, and thus the limit on power that can be sent to the load is increased, by placing reactive compensation or voltage support near the load.

Like the power angle, voltage exhibits an oscillatory behavior when displaced from equilibrium. Usually, a departure from equilibrium (a voltage spike) is expected to dampen out very quickly, though sometimes continuing oscillations are observed, especially in the event of major disturbances. Voltage oscillations can propagate far, and power systems appear to be more vulnerable to voltage oscillations the larger they become. These effects on a large scale are not entirely well understood.

When voltage and power cannot be controlled, this condition is termed *voltage instability*. It may be possible to continue to operate a system at a reduced voltage. In practice, though, other interactive effects may come into play. For example, generator field excitations may increase, leading generator fields to be overloaded and then deliberately reduced or the generator tripped; transmission lines may trip, transformer tap changers and voltage regulators may attempt to restore load voltage to normal and thereby further increase the load power. Angle instability may also result, and finally, a smaller or larger part of the system may be lost completely. This is the condition known as *voltage collapse*. Voltage instability played a role, for example, in the August 14, 2003, blackout in the Northeastern U.S.[8] Analogous to the term *security* as it refers to the ability to generate and deliver real power (Section 8.2), there is the concept of *voltage security*, or width of the operating envelope with respect to voltage control.

8.4 POWER QUALITY

Power quality encompasses voltage, frequency, and waveform. From a theoretical point of view, good power quality can be taken to mean that the voltage supplied by the utility at the customer's service entrance is steady and within the prescribed

---

[8]The very readable Final Report by the U.S.–Canada Power System Outage Task Force can be found at http://www.nerc.com/~filez/blackout.html.

range; that the a.c. frequency is steady and very close to its nominal value (within a fraction of a percent); and that the *waveform* or shape of the voltage curve versus time very much resembles the smooth sine wave from mathematics textbooks (a condition also described as the absence of *harmonic distortion*). In practice, however, it makes more sense to consider power quality as the *compatibility* between what comes out of an electric outlet and the load that is plugged into it. The development and proliferation of electronic technologies over the past several decades has turned "electric load" into a wide spectrum of devices that vary tremendously in their response to changes in voltage, frequency, and waveform: what is perfect power quality for one appliance may be devastating to the next, or cause it to do goofy things.[9] Not surprisingly, how much power quality is needed and by whom—and how much money it is worth—is the subject of some controversy.

8.4.1 Voltage

The voltage received by a utility customer varies along with power flows in the transmission and especially the distribution system. Initially, generators inject their power at a fixed voltage magnitude, which would translate through several transformers into a fixed supply voltage for customers. But as consumption and thus line current increases, there is an increasing *voltage drop* along the power lines according to Ohm's law (Section 1.2). This means that the difference between the voltage supplied at the generation end and that received by a given load varies continuously with demand, both systemwide and local. The utility can take diverse steps to correct for this variance, primarily at the distribution level (see Section 6.6 on voltage control), but never perfectly. The traditional norm in the United States is to allow for a tolerance of $\pm 5\%$ for voltage magnitude, which translates into a range of 114–126 V for a nominal 120-V service.

Low voltage may result if a power system's resources are overtaxed by exceedingly high demand, a condition known as "brownout," because lights become dim at lower voltage. Aside from the nuisance of dimmer lights, operation at low voltage can damage electric motors (see Section 5.2). Excessively high voltage, on the other hand, can also damage appliances simply by overloading their circuits. Incandescent light bulbs, for example, have a shorter life if exposed to higher voltages because of strain on the filament.

For utilities whose revenues depend on kilowatt-hour sales, there exists a financial incentive to maintain a higher voltage profile, as power consumption by loads

---

[9]With regard to the variation of power quality tolerance among electronic equipment, it is interesting to note that electronics, and especially the power supplies that represent the crucial interface with the grid, are usually designed and tested by engineers in laboratories where power quality tends to be better than at the consumer's location out on a typical distribution circuit. Thus, absent a specific testing protocol for ensuring a new gadget's resilience with respect to the voltage, frequency, and waveform variations that might be expected, the design engineers may have no idea they are building a fragile piece of equipment that will behave annoyingly, or worse, in real-life conditions. I am indebted to Alex McEachern for this and other observations concerning power quality.

252 SYSTEM PERFORMANCE

generally increases with voltage. Conversely, it is possible in principle to reduce electric energy use by reducing service voltage. Just how well this works has been the subject of some debate throughout the industry, ever since conservation voltage reduction (CVR) was proposed as a policy for energy conservation in the late 1970s and early 1980s. Some of the difficulties in predicting the effects of voltage reduction have to do with motor loads that respond differently to voltage changes, and with the prevalence of loads such as thermostat-controlled heaters or refrigerators that deliver a fixed amount of energy as demanded by the user.[10] In practice, given the vintage of most distribution system hardware currently in place, there tends to be relatively little room for discretion in choosing a preferred operating voltage; rather, most operators are probably glad simply to keep voltage within tolerance everywhere.[11] This situation could change with the installation of new voltage-control technology, including equipment at customer locations.

Beyond the average operating voltage, of concern in power quality are voltage swells and sags, or sudden and temporary departures from normal voltage levels that result from events in the distribution system. Abrupt voltage changes can be caused by lightning strike, or by large inductive loads connecting and disconnecting (like the familiar dimming of lights when the refrigerator turns on), but are most often related to faults on nearby distribution circuits. Specifically, in the time interval between the appearance of a fault and its isolation by a fuse or circuit breaker, the fault current (which is much greater than the usual load current) results in a significantly greater voltage drop along the entire feeder, as well as other feeders connected to it in a radial distribution system (see Section 6.1.4 on distribution system design and Section 6.7 on protection). This time interval before the circuit protection actuates may be anywhere from a fraction of a second to several seconds, depending on the fault's impedance, and thus the magnitude of the fault current; this relationship explains why small voltage sags are usually observed to last longer than large ones.

To describe temporary voltage increases, the term *swell* is generally used to denote a longer event, whereas an *impulse* would last on the order of microseconds; the word *spike*, though used colloquially and found in the common "spike protector," is frowned upon by experts because of its ambiguity, having been used historically to describe rather diverse electrical occurrences. It is easy to see how a voltage swell or impulse can actually damage loads because the proportionally increased current may quickly overheat a small component inside an appliance, or an arc may form between components that are insufficiently insulated, and the equipment is "fried." To protect against this obvious risk, power strips with spike protectors have become very popular, especially for computers and other expensive

---
[10]For example, a resistive heating unit that draws less power as a result of lower voltage would simply stay on for a longer time period until the desired temperature was reached.

[11]For example, a conservation voltage reduction (CVR) program in California in the 1980s, intending for utilities to narrow their voltage tolerance range from $\pm 5\%$ to $+0\%$—5% so as to reduce energy consumption, could not be implemented as planned because of limitations in utilities' voltage-control hardware, in addition to the logistical challenge and high cost of measurement and compliance verification.

circuitry. While they are probably a wise precaution, the actual incidence of consumer equipment damaged by utility voltage spikes appears to be quite low.

Indeed, the more important job of a spike protector may be to mitigate temporary decreases in voltage—*sags* to Americans and *dips* to the British—that can cause electronic loads to shut off or otherwise behave strangely. One might expect voltage sags to be essentially a nuisance, noticeable as a brief dimming of lights and the occasional rebooting of a computer, but they surprisingly constitute the most common power quality problem by an order of magnitude. Owing to the large number of sensitive commercial and industrial loads, economic losses from voltage sags in the United States are estimated at a staggering $5 billion per year.[12]

It is difficult to obtain empirical data on how often voltage sags and other power quality problems occur because utilities, which are in a unique position to measure and record them, are typically not keen on publishing this information. More than just a matter of image, there is an issue of accountability and liability if a particular performance level is quoted in writing but turns out not to be met at some future time. It is important to recognize that many of the factors that substantially affect power quality—from squirrels and winter storms to drunk drivers and digging backhoes—lie beyond the utility's control, putting them in the awkward position of being responsible for system performance, yet unable to make firm guarantees such as zero voltage sags. This intrinsic problem also limits the feasibility of contractual agreements with customers willing to pay more for power quality, because an electric distribution system in the real world, though it can be made relatively more robust, simply cannot be guaranteed to operate without any disturbances whatsoever.[13] As a result, customers with important sensitive loads generally have to look to power conditioning equipment on their own side of the utility service drop.

8.4.2 Frequency

Frequency departs from its nominal value if generation and demand are not balanced. If the demand exceeds available real power generation, energy will be drawn from the rotational kinetic energy of the generators, which will slow down (see Section 4.3). At lower frequency, the amount of real power transmitted will

---

[12] Alex McEachern, personal communication. An example that vividly illustrates this point is a carpet manufacturing plant, where the tension on each of the myriad threads is a function of the voltage across the control equipment: we can easily imagine the mess, as well as the cost, when one thread tears or bunches up momentarily, and the entire roll of carpet has to be thrown out....

[13] Alex McEachern offers the following amusing and instructive analogy: Imagine a secretary who does a fantastic job in every respect, except for one very strange habit. Every so often, and completely unexpectedly, he comes up from behind and pokes you in the back with a knitting needle. Suppose this happens, on average, six times per year. Since the secretary is so competent otherwise and you wish to retain him, you ask whether you could perhaps raise his salary so that he would not poke you any more. The answer comes back that with a raise, he could reduce the number of knitting needle incidents from six to three per year. This deal does not seem very attractive! Similarly, a utility customer plagued by occasional voltage sags might be willing to pay more money for better power quality, but what they want is *zero* incidents—which the utility cannot offer.

be less, and so the loads are prevented from consuming more power than can be generated. Conversely, in the case of overgeneration, frequency will increase.

Drifting frequency presents a risk mainly for synchronous machines, including generators and synchronous motors, as some of their windings may experience irregular current flows and become overloaded. For their own protection, synchronous generators are equipped with relays to disconnect them from the grid in the event of over- or underfrequency conditions. The sensitivity of these relays is a matter of some discretion, but would typically be on the order of 1%.

Similarly, sections of the transmission and distribution systems may be separated by over- and underfrequency relays. For example, a transmission link in a nominal 60-Hz system may have an underfrequency relay set between 58 and 59 Hz. Such a significant departure from the nominal frequency would indicate a very serious problem in the system, at which point it becomes preferable to deliberately interrupt service to some area and isolate functioning equipment, rather than risking unknown and possibly more prolonged trouble. A key objective is to prevent cascading blackouts, in which one portion of the grid that has lost its ability to maintain frequency control pulls other sections down with it as generators become unable to stabilize the frequency and eventually trip off-line. The situation is somewhat analogous to a group of mountain climbers who are roped up to support each other in case of a fall—but under some dire circumstances, it is necessary to cut the rope.

Unlike the large frequency excursions associated with crisis events, smaller deviations can be treated by utilities or system operators with some degree of discretion. In highly industrialized areas, the choice of tolerance is driven more by cultural and regulatory norms than by the technical requirements of the grid itself. In areas facing serious supply shortages, the frequency tolerance may be much wider as the risk of damaging equipment is weighed against the need to provide any service at all. Accordingly, there are international differences in the precision with which nominal a.c. frequency is maintained.[14] In U.S. power systems, frequency can be expected to fall between 59.9 and 60.1 Hz, barring any major disturbances, and usually within a much tighter range of 59.99 to 60.01 Hz.

One practical and intuitive reason for maintaining a very exact frequency is that electric clocks (and any technology that depends on timekeeping based on the a.c. frequency) will in fact go slower if the frequency is low and faster if it is high. Grid operators in highly industrialized countries, where people and their equipment might actually care about a fraction of a second lost or gained, keep track of cycles lost during periods of underfrequency over the course of a week, and

---

[14]For example, the tolerance for frequency deviations from the nominal 50 Hz in East and West Germany was so different that their grids could not be synchronized after the 1990 reunification, and the transmission of electricity across the former border first required conversion to direct current. In another example, a team that installed wind generators in a North Korean village in 1998 describes a failed attempt to synchronize the output of an inverter with the nominal 60-Hz grid; they ended up using an inverter designed for 50-Hz systems (Jim Williams and Chris Greacen, personal communication). The remarkable thing is that the North Korean grid can physically continue to operate at all, presumably in an islanded condition, in the face of extreme generation shortages.

make up those cycles on a certain evening or weekend, outside regular business hours.[15]

Example

Suppose the nominal 60-Hz frequency remains low at 59.9 Hz for one day. How much time is "lost" on a.c. clocks?

Normally, there are 60 complete a.c. cycles in each second. The time lost corresponds to the number of cycles lost, where each cycle represents 1/60 of a second. At 0.1 Hz below nominal, 1/10 of a cycle is lost every second, or one full cycle lost every 10 s. This is equivalent to 60 cycles lost every 600 s, or one second every 10 min. Over the course of an hour, 6 s are lost, which comes to 144 s or about two and a half minutes per day. Repeating the exercise for a more realistic frequency of 59.99 Hz, the error comes to one-tenth that, or 14.4 s per day.

A few seconds per day is a level of inaccuracy in timekeeping that most people would hardly notice—indeed, it might compare to the error in a mechanical clock of average quality. It would be a gross error, however, in any advanced technological application that requires synchronization of components like communication instruments, global positioning systems, cellular phones, and such. Any technologies today that depend on a precise time reference rely not on the a.c. grid but on subatomic oscillators, like the quartz crystal in a wristwatch, powered by uninterruptible power supplies. Thus, for practical purposes, the implications of timekeeping errors due to frequency variations in the electric grid are relatively minor.[16] This statement does not apply to backup generators and uninterruptible power supplies, whose output not uncommonly departs from the nominal frequency by several hertz.

8.4.3 Waveform

A clean waveform means that the oscillation of voltage and current follow the mathematical form of a sine or cosine function. This conformance arises naturally from the geometry of the generator windings that produce the electromotive force or voltage. Aside from transient disturbances, this sinusoidal waveform can be altered by the imperfect behavior of either generators or loads. Any a.c. machine, whether producing or consuming power, can "inject" into the grid time variations of current and voltage, which can be observable some distance away from the offending machine. Voltage waveform distortions are typically created by generators, while current distortion results from loads. These distortions of voltage or current occur in the form of oscillations that are more rapid than 60 Hz and are called *harmonics*, as in music, where a harmonic note represents a multiple of a

---

[15]Proving that time does, in fact, go by faster on the weekends.
[16]An insightful commentary on the preoccupation with precise timekeeping in our culture can be found in James Gleick, *Faster: The Acceleration of Just About Everything* (New York: Pantheon, 1999).

256 SYSTEM PERFORMANCE

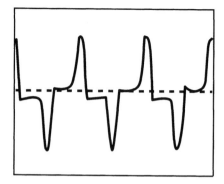

Figure 8.8 Poor waveform.

given frequency. While there are infinitely many possible higher-frequency components that can distort a wave, we care here about those that are *periodic*, that is, recurring with every cycle, and are therefore steadily observable. For this to be true, they must be exact multiples of the basic a.c. frequency, thus entitling them to the name "harmonic."

When superimposed onto the basic 60-cycle wave, harmonics manifest as a jagged or squiggly appearance instead of a smooth curve. Mathematically, such a jagged periodic curve is equivalent to the sum of sinusoidal curves of different frequencies and magnitudes.[17] The relative contribution of these higher-frequency harmonics compared to the base frequency can be quantified as *harmonic content* or *total harmonic distortion* (*THD*). THD is expressed as a percentage indicating the amount of power carried by the harmonic frequencies, as opposed to the power contained in the basic 60-Hz wave. A common standard for power generation equipment such as inverters is to produce voltage THD below 5%, although the actual utility waveform may include much higher distortion levels (see Figure 8.8) due to loads that cannot easily be identified or regulated.

A desirable waveform is one with low harmonic content, which is synonymous with a smooth, round sine wave. While resistive loads are unaffected by waveform, harmonics can cause vibration, buzzing, or other distortions in motors and electronic equipment, as well as losses and overheating in transformers.

The key physical fact to recognize is that alternating currents of different frequencies behave differently as they pass through electrical equipment because the magnitude of reactance (inductive or capacitive) is frequency dependent. Transformers, which are essentially large inductive coils, present a higher impedance to higher frequency current and therefore do not transmit power as readily. When a transformer is supplied by power with high harmonic content, some fraction of that power becomes "stuck" and converted to heat in the conductor coils, as well as in the magnetic core, which resists rapid reversals. The overall result is that the

---

[17]The process of converting between the two representations is known as *Fourier analysis*.

transformer not only operates at lower efficiency but also faces a shortened life span due to chronic overheating.

In addition, certain harmonics can exhibit special behavior owing to symmetries. The most important such case involves any transformer with a delta connection on its primary side (see Section 6.3.3). The voltage on each of its three primary windings is the voltage *difference* between one pair of phases, $A-B$, $B-C$, or $C-A$, where the 120° phase (or time) difference between the same sine wave on A, B, and C creates the *relative* voltage between them. But now consider the third harmonic of the a.c. base frequency: in a 60-Hz system, this means a small oscillation at 180 Hz. As illustrated in Figure 8.9, this third harmonic of Phase A is indistinguishable from that of Phase B or C; the waves are superimposed on each other. The same is true for all multiples of the third harmonic (6th, 9th, 12th, 15th, etc.). But this means that, to the extent that a voltage component alternates at a multiple of 180 Hz on all three phases, there is no phase-to-phase difference to be had, and no power to be transferred by the transformer. How much does this matter? One-third of all the integer multiples of the base frequency are also multiples of three. Therefore, one-third of all the power contained in harmonic components of a wave is blocked by any delta connection, doomed to keep cycling through conductors until it meets its fate as waste heat. If THD represents, say, 3% of the power carried by a wave, this implies an immediate 1% loss, which is not insignificant.

Even beyond the practical and economic implications, a clean waveform entails a certain degree of engineering pride; after all, the a.c. wave is the final product delivered to the customer. Of course, everyone in the industry discovers at some point how the reality of a.c. power systems differs from its textbook abstractions, be it the waveform, balanced phases, or the behavior of generators. Because so many of the factors and events that bear on power quality and performance are beyond the control of the utility or system operator, the only realistic goal is a working compatibility between the system and the job it is expected to do, not mathematical perfection; this goes for waveform, frequency, and voltage as well as for reliability. Yet there is also a human desire to at least approximate the ideal condition as much as possible. Power quality has an additional poignancy in that a problem may not be readily solvable by spending more money.

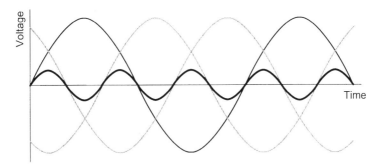

Figure 8.9 Third harmonic.

This tension may explain some of the past reticence among utility engineers to welcome distributed generation into the grid—specifically, inverters whose so-called "sine wave" output is in fact a stepwise assemblage of rapidly switched signals and therefore has a certain intrinsic harmonic content (see Section 4.6). The amount of this distortion may or may not be significant or even observable compared to that already present in the circuit, but the stepwise waveform might nevertheless look like an abomination to the power engineer. Conversely, some discriminating utility customers might be outraged if they saw their waveform as an oscilloscope trace, whether or not they actually experience a problem with their load. The value of power quality thus includes an intangible aspect rooted in the undeniable aesthetic appeal of a perfect, smooth sinusoid.

CHAPTER 9

System Operation, Management, and New Technology

The electric grid is often referred to as a *complex* system. This makes perfect sense colloquially, as the visual image of a vast array of metal hardware and moving machinery resonates with the standard definition of complex as "consisting of interconnected or interwoven parts" or "involved or intricate, as in structure; complicated."[1] Indeed, a system's complexity can be defined academically as the product of the number of its components, their diversity, and the "tightness of coupling" among them, meaning the strength, immediacy, and importance of their interactions. Electric power systems certainly score very high on all of these criteria.

But asking what this complexity might mean in practice brings us to a more poignant definition that captures a key property of the electric grid: a complex system is one where no individual can, at any one time, understand the entire thing.[2] What might seem "complicated" or hard to understand to the lay person is in fact not fully understandable even to the expert, because there is just so much of it that by necessity every power system professional has a limited scope of expertise and responsibility, and keeping track of the interactions of all these domains—the performance of the system as a whole—is by necessity a team process. This remains true regardless of whether the terms "individual," "expert," and "professional" are taken to represent human beings or advanced computers. No single entity can simultaneously monitor, control, and troubleshoot every generator, every load, every piece of conductor in between, and every possible external disturbance, any one of which, nevertheless, has the intrinsic ability to affect every other system component almost instantaneously and sometimes severely. Such complexity makes it possible for a system to surprise both its operators and its designers. This essential property underlies and motivates this chapter.

---

[1]*American Heritage Dictionary*, Tenth Edition (Boston: Houghton Mifflin Company, 1981).
[2]This definition follows Barbara Czarniawska, who writes that "an organization becomes complex when no one can sensibly and comprehensibly account for all of it." [B. Czarniawska-Joerges, *Exploring Complex Organizations* (Newbury Park, CA: Sage Publications 1992).]

Electric Power Systems: A Conceptual Introduction, by Alexandra von Meier
Copyright © 2006 John Wiley & Sons, Inc.

259

9.1 OPERATION AND CONTROL ON DIFFERENT TIME SCALES

The central challenge in the operation of electric power systems is often cited: electricity must be generated in the exact moment that it is consumed. To move electric power through a grid is to obey the law of energy conservation; what goes in must come out. Unlike a natural gas pipeline that can accommodate a variation in gas pressure and thus serve simultaneously as conduit and storage reservoir, a transmission line cannot store electricity. Some grids include storage facilities like pumped hydroelectric plants to smooth out diurnal or seasonal variations in demand and supply. But while a power plant that alternately absorbs and releases power can help with saving up capacity for when it is needed, it does not allow us to circumvent the fundamental problem of coordinating generation and load in real time.

If we look carefully, we find that some physical energy storage is in fact provided by the standard components of a power system; there just is not very much. In reality, if there were exactly zero energy storage, it would be literally impossible to operate a grid: every control action would need to be perfectly precise and instantaneous, or the alternating voltage and current would collapse. What we really mean when we say that generation and load have to be exactly balanced is that they have to be equalized *within the time scale* permitted by the system's capacity to buffer the discrepancies, that is, to store and release energy, and to adapt to new operating conditions. This time scale happens to be very short—short enough that in most types of analysis it seems instantaneous.

The intrinsic energy storage capacity in a conventional power system resides within large rotating machinery. As discussed in Section 8.3, generators provide stability to the system by means of their rotational inertia, in effect absorbing and releasing kinetic energy in response to changes in the electric load. This process takes place in a fraction of a second, meaning that the amount of energy involved is comparatively small, but it represents the first line of defense against power imbalances in the grid.

On the load side there is also a certain degree of flexibility, because power consumption is not precisely fixed. Although a load's power demand is generally modeled as an externally given, wholly independent variable that determines everything else, it does in fact afford a modicum of stabilizing response to power imbalances.

Specifically, if real power into the grid is less than real power out, the alternating-current (a.c.) frequency will drop (as stored kinetic energy from the generators is being used). But at a lower frequency, some loads (motors, in particular) will also consume somewhat less power. The system is not designed to be operated in such a state of underfrequency, as it implies poor power quality and ultimately the risk of physical equipment damage. The point, however, is that on a very fine scale load is not perfectly rigid.

Even more so than frequency, voltage level affects power consumption by loads. A power deficit—particularly of reactive power, but ultimately real power as well—causes voltage levels to sag and make lights dimmer. In the extreme, this produces the oft-cited *brownout* condition. The fortunate tendency of the load is to respond to low voltage by drawing less power, and vice versa in case of excess generation. While power system operators do not take advantage of this feedback by design, it represents a buffer of last resort that prevents immediate system collapse during moderate, inadvertent mismatches of generation and load.

These small degrees of flexibility notwithstanding, the law of energy conservation is strictly enforced by nature, and what little wiggle room there is can be quickly used up. It remains true, then, that the prime directive for power system designers and operators is to balance generation and load at every instant. This balancing act occurs on multiple levels, with control methods appropriate to each time scale.

9.1.1 The Scale of a Cycle

The most sensitive aspects of maintaining equilibrium in a power system happen on the time scale of a fraction of a second, where time is measured in cycles, from a fraction of a cycle to several cycles (one cycle at 60 Hz measures 1/60 of a second or about 16 ms). To the human observer, these events appear instantaneous, and their speed certainly demands automatic responses of the technical system components.

As described earlier and in Section 8.3 on stability, frequency regulation occurs on this time scale. The first level of frequency regulation is the passive negative feedback effect built into the physics of the synchronous generator: if the generator speeds up, the torque holding it back increases as the result of an increasing magnetic field; if the generator slows down, the restraining torque decreases. Here, the term "passive" means that this effect requires no intervention on the part of any human or machine to take place; it is intrinsic to the device and guaranteed by natural law.

By contrast, the second level of frequency regulation involves an active intervention, though also automatic. This is the negative feedback between generator revolutions per minute (rpms), which are directly proportional to the a.c. frequency produced, and the rate at which power is delivered to the turbine—the flow rate of steam in a thermal plant, water in a hydroelectric plant, or exhaust gas in a gas turbine. The device connecting the generator speed measurement with the steam or fluid valve is called the *governor* and is discussed further in Section 4.3.1. If the generator is spinning too fast, the governor closes the valve to reduce the flow rate; if the generator is too slow, the governor opens the valve.

By definition, measuring the frequency requires a time interval of several cycles. The actual physical response of the fluid flow and thus the mechanical torque will lag somewhat behind the governor's frequency measurement, but still take place on the order of seconds or fractions of a second. While some generators are set to produce a fixed amount of power, operating at least some units in a power system *on the governor* allows the system to follow loads—that is, respond to changes in demand—just about instantaneously and without the need for immediate supervision.

Another important control function that occurs on the time scale of a cycle is circuit protection, discussed further in Section 6.7. Protection means that in the event of a fault, or an accidental contact or short circuit on any system component, the current flow is interrupted automatically and as soon as possible or practical in order to prevent harm to people or equipment. The challenge lies in distinguishing a fault current from an unusually high but tolerable current, and maximizing safety on the one hand while avoiding nuisance interruptions on the other hand. While some devices, such as fuses, are triggered by heat and may require several seconds of high current to melt, more sensitive solid-state relays can interrupt a current within one or several cycles. Like load following, it is essential that circuit protection occurs instantaneously and automatically without the need for supervision or intelligent intervention.

9.1.2 The Scale of Real-Time Operation

Real-time operation refers to the time scale on which humans perceive and analyze information, make decisions, and take action. Generally, this means on the order of minutes, though some actions may be executed within seconds. Human intervention in real-time may be called for at individual generation units, at the system operator level where systemwide generation and load are balanced, and in transmission and distribution switching.

While power plants are generally designed to provide constant power output without human intervention, real-time operations are required during start-up or shutdown and sometimes to implement changes in output. Starting up a steam generation unit is a demanding procedure that involves coordinating numerous pumps, valves, flow rates, pressures, and temperatures throughout the plant. From a cold start, it takes hours to bring a large steam plant into its hot operating equilibrium, at which point it can be electrically connected or *paralleled* with the grid. Once steady output is reached, operators focus on the general monitoring of automated processes and stand by until major output changes are required.

It is possible but not part of standard procedure for human operators to manually match a generation unit's output with load when the load variations exceed the normal range of the governor system. An experienced operator can do this by watching frequency and voltage levels and adjusting valve settings accordingly. Such skill might be called upon in emergency situations where a plant is supplying a power island or part of a severely disrupted grid.[3]

At the level of system operator or dispatcher, the object is to arrange for the correct amount of real and reactive power actually demanded by the system, as opposed to the amount previously estimated and contracted for. While power is scheduled administratively on an hourly basis, these schedules cannot be physically

---

[3]For example, operators at Pittsburg Power Plant in California recalled keeping their units online "by the seat of their pants" after the 1989 Loma Prieta earthquake. In a nearly complete communications blackout, their only guide as to how much load might still be connected out there were their local frequency and voltage measurements.

accurate for several reasons: first, load depends on consumer behavior and cannot be known with certainty ahead of time; second, even if the forecast is generally correct, the load may still vary throughout the hour; and third, generators may not actually produce what they claimed they would. Another minor contribution to the uncertainty comes from line losses. As a result, neither supply nor demand, nor any discrepancies between them, can really be known until the moment they are measured empirically.

To determine the actual relationship between generation and load, it is necessary to define the boundary of the "system" to be balanced. While countries, states, or regions within states may be separate politically and administratively, electric power flows along any physical link connecting them. Ultimately, then, the control of a synchronous grid like that in the eastern or western United States is a team effort among a number of system operators with neighboring jurisdictions.

Historically, each regulated utility had its own system operator or control center to coordinate its service territory. In the restructured market environment, this function is being shifted to a separate administrative entity, an independent system operator (ISO) responsible for a given region that typically spans several utilities' territories. In either case, the system operator's jurisdiction or control area has a clearly defined geographic boundary with a finite number of transmission links crossing into neighboring areas. Whether contractually scheduled or not, these transmission links carry a certain amount of power between jurisdictions according to Kirchhoff's laws. To determine the actual real-time balance of supply and demand within their own territory, the system operator keeps track of these flows to and from each adjacent one. The real-time difference between actual and scheduled imports or exports can be expressed in a single number, the *area control error* (*ACE*). A positive ACE means that there is more generation than load within the territory, or that the exports resulting from actual power flow are greater than scheduled, so that local generation can be reduced; a negative ACE conversely means that local generation should be increased.

Based on the ACE, information about actual system needs must then be sent to selected generators to respond. This information transfer occurs both automatically and through human negotiation. Units equipped with automatic generation control (AGC) can receive a signal directly from the system operator to their governor, requesting an increase or decrease in output. This signal overrides or preempts the response of the governor to its own measurement of generator rpms. When the departure of actual system conditions from expected conditions exceeds the range of automatic controls, it becomes the system operator's task to call upon generation operators to take action. Such calls are often made by telephone, and in critical situations their success may depend on personal rapport.

If generation remains insufficient, the system operator's options also include *shedding load*. This may mean selectively disconnecting large customers with specific contracts and remunerating them for being *interruptible loads*, or local groups of customers in *rotating outage blocks* that are assigned to spread the burden of power shortages evenly, with outage blocks taking turns of an hour or so of blackout. The disconnection is physically carried out at the distribution level.

In addition to load shedding performed at the system operator's request, transmission and distribution switching in real-time includes reconfiguring the system for maintenance and restoration purposes, or to preempt local problems such as overloading a particular circuit. These processes essentially involve opening and closing specific switches or circuit breakers according to a carefully mapped sequence, so as to isolate and connect parts of the grid without overloading any component and while maintaining appropriate circuit protection. Examples of switching operations are given in Chapter 6 on transmission and distribution.

9.1.3 The Scale of Scheduling

Whereas real-time operation is dominated by technical criteria such as security and stability, the scheduling context emphasizes optimization around economic criteria, specifically, which generation units to operate when and at what power level so as to minimize overall cost. While these economic and technical objectives reside under the same administrative roof in the regulated utility setting, they are explicitly separated in the restructured environment so as to prevent the system operator from making economically motivated decisions (such as preferentially calling upon certain generating units) under the guise of technical necessity; thus the ISO appellation. Instead, responsibility for the key economic decisions is assigned to a separate organizational entity known as a *scheduling coordinator*.

Scheduling generation units to match the forecast load, also known as *unit commitment*, is done on a daily and hourly basis. In the "Old World" of vertically integrated utilities that own and operate all the generation and transmission assets within their service territory, the decision of which generator contributes how much and when is traditionally made in a central scheduling process by means of an *economic dispatch* algorithm.

This process uses the *load duration curve* (*LDC*) (see Section 5.5) as a reference, "filling in" the area under the curve with various types of generation so as to minimize overall cost while meeting all operating constraints. The cost minimization is done by an optimization algorithm that takes into account the marginal cost of each unit's output in dollars of fuel and operational expense per additional megawatt-hour, as well as the approximate line losses associated with supplying power from each location (expressed as a mathematical *penalty factor* to adjust that generation unit's cost).

In practice, there are three general categories of generation: *baseload* generation units, which produce the cheapest energy and are best operated on a continuous basis (for example, coal or nuclear plants); *load-following* units that respond to changes in demand (typically, hydroelectric and selected steam generation units); and *peaking* units that are expensive to operate and are used to meet demand peaks (for example, gas turbines).[4] These are scheduled with baseload as a priority, load-following

---

[4]There may also be nondispatchable, intermittent generation such as solar or wind power, but it is generally treated as negative load in the sense of combining its statistical variability with that of demand.

9.1 OPERATION AND CONTROL ON DIFFERENT TIME SCALES

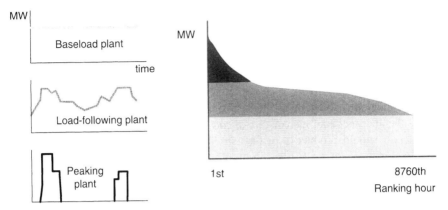

Figure 9.1 Generation scheduling with the load duration curve.

plants added in as required, and peakers kept in reserve for the extreme days or hours.

The allocation process is shown graphically in Figure 9.1. Note that as the vertical axis of the LDC measures units of power (MW) and the horizontal axis measures units of time (hours, where it does not matter that the hours do not occur sequentially), the area under the curve corresponds to units of energy (MWh). The energy output of baseload units throughout the year is represented by the rectangular area filling the lower portion of the curve, with the combined power output of baseload units remaining constant. Load following and other units operated at variable power levels are shown together as filling in the central portion of the curve. Their power output actually varies from hour to hour and day to day, as can be seen on the left-hand portion of the diagram where time is shown sequentially; the LDC simplifies this temporal profile while emphasizing the overall energy contribution. Finally, the contribution from peaking units is immediately recognizable as the area that fills the top of the curve.

The scheduling process as shown here is obviously somewhat idealized. In reality, generation units have specific constraints on their operation that must be taken into account. These constraints include scheduled outages for maintenance (and refueling, in the case of nuclear units), unscheduled outages, and limitations on the *ramp rates* at which particular units can safely increase and decrease their output power. For large thermal plants, the ramp rates may significantly affect scheduling. The combination of constraints on unit availability produces a continually changing menu of generation capacity throughout the system, from which the optimal contribution levels are to be determined.

The underlying assumption in the competitive "New World" of electricity is that an optimal allocation of generation resources is better achieved through market transactions as opposed to some organization's centralized planning. In restructured markets, unit commitment can be determined through some combination of *bilateral contracts* between individual generators and consumers and a *power pool* that serves as a clearinghouse for power bought and sold. The detailed institutional arrangements vary among systems and are beyond the scope of this text.

From the technical perspective, what is important is that some entity serves as a scheduling coordinator that keeps track of megawatts to be bought and sold (there may be more than one scheduling coordinator in the same geographical area). The scheduling coordinator may solicit and schedule generation through some form of auction, calling upon the lowest bidders to generate during each hour, where the auction could include *day-ahead* and *hour-ahead* markets. Ideally, such an auction should produce a picture similar to the one depicted in Figure 9.1, with each part of the LDC filled in by the least expensive generation available at that time.

While day-ahead and longer-term planning or contracting are thus intended to create an economically optimal schedule, some modifications and adjustments are always necessary on short notice to accommodate factors like changing unit availability or maximum ramp rates. In an auction, these factors can in principle be accounted for by each generator bidding its available power specifically for each hour. In addition, though, the grid relies on generators to be responsive to real-time changes in demand, or any other factors (say, an outage somewhere) that affect the overall power balance. Such responsiveness may be specifically contracted for as *spinning reserve* or automatic generation control, in which generating units are remunerated not for the megawatt-hours of energy they provide but for being "on call" to respond instantaneously to the grid's needs. (The "spinning" property of the generation reserve means that the turbine and rotor are literally spinning at synchronous speed and able to pick up megawatt load without having to warm up or even accelerate.)

Spinning reserve falls in the category of *ancillary services* provided by generators to the grid. Another component of ancillary services is the provision of reactive power or MVAR (see Section 3.3). It usually does not cost a power plant very much to provide reactive power (since, to a first-order approximation, it uses no fuel energy) and one might therefore expect MVAR allocation to be a somewhat casual, ad hoc process. On the contrary, though, it is necessary to schedule reactive contributions from specific generators ahead of time in order to achieve an optimization similar to the way real power is allocated—in this case, maintaining a certain voltage profile across the grid while minimizing overall line losses. In a competitive market, these services make for additional business opportunities.

Given a proposed generation schedule, the system operator must ascertain that it does not violate any technical operating constraints, such as transmission line loading. The system operator requires schedules to be revised if a violation can be identified in advance, or makes adjustments in real time as needed. With the large number of generation and load variables, their intrinsic uncertainty, and the difficulty of controlling them directly, it is not uncommon for much of what would ideally be considered a scheduling task to become effectively a real-time operations task, presenting a challenge to system operators that is not to be underestimated.

Any crisis situation in a power system means that the time horizon for decisions shortens, and the focus in operation shifts accordingly from economic optimization to meeting the physical requirements of the grid at any cost. This shift is implicit in the shared assumption by all participants that the electric grid as a whole is to be kept functional at all times and under any circumstances if humanly possible. As long as that assumption holds, the ideal of cost minimization in scheduling always remains subject to being trumped by events in real-time, regardless of the institutional framework.

9.1.4 The Planning Scale

Hourly and daily generation scheduling as well as the real-time operation of power systems take place within a set of boundary conditions that include extant generation, transmission, and distribution capacity and loads. These boundary conditions are addressed in the realm of planning, on a time scale of years.

Historically, the planning process has been driven almost entirely by load forecasts.[5] The premise of the traditional regulated monopoly framework is that all demand within a given territory is to be served by means of prudent investments sufficient to satisfy this demand. Planning then comes down to estimating load growth in megawatts, locally and systemwide, and accommodating this growth with appropriate upgrades in transmission and distribution hardware, new construction of generation units, or securing of electricity imports.

In view of consistently increasing electric demand and guaranteed future revenues under the regulatory framework, it made sense for utilities to adopt a long time horizon in their planning. This meant getting ahead of load growth with oversized transmission and distribution (T&D) capacity that would come to be utilized over the years (the analogy of children's clothing comes to mind). Such investment was justified by the cost structure of T&D upgrades, whose price is high but not necessarily dominated by the physical size of installed equipment (in other words, you are better off building it large the first time than to have to go back and replace it), and by the operational benefit of a robust grid, recognizing that power delivery is essential for serving loads and any constraints can present an insoluble problem in the short term.

The case for anticipating load growth in generation planning is somewhat different, but leads to a similar conclusion. In generation, the cost of excess capacity tends to be more significant on a per-megawatt basis. Generation investments have also received much more public scrutiny than those for transmission and distribution, despite their comparable overall magnitude. For both technical and institutional reasons, however, the lead time for conventional generation projects is considerable,

---

[5]The term "load forecast" is generally viewed within the industry as a purely technical parameter determined by population growth and consumption levels as independent variables. Of course, it can also be understood as the expression of a more complicated social dynamic in which the industry itself plays a role—say, through advertising campaigns or incentive programs. The latter sort of "strategic" planning tends to be organizationally separate and conceptually remote from the "tactical" planning by power system engineers.

with at least several years and sometimes a decade passing between a unit's inception and its operation. It was also considered part of prudent practice in the regulated U.S. industry to maintain a substantial generation reserve margin of 20% above peak demand. Therefore, unit construction plans were typically based on conservative demand projections for ten or twenty years out. This approach made sense until the 1980s, when electric demand growth fell behind projections and many U.S. utilities suddenly found themselves with excess capacity.

The problem of anticipating demand and implementing appropriate levels of generation capacity remains difficult and controversial in the restructured environment. For example, the design of the deregulated California market in the 1990s assumed plenty of excess generation capacity to be in place and therefore failed to carefully consider the system's behavior under a hypothetical generation shortage. When a dramatic shortage appeared in 2001 that resulted in extreme wholesale price spikes and rotating outages, it was not immediately obvious to what extent this crisis resulted from an actual physical generation scarcity, or to what extent it was manufactured through profit-maximizing behavior on the part of power generators. In theory, a competitive market ought to provide incentives not only for short-term production but long-term investment, including generation and power delivery. How such investment signals will occur in practice and how closely the results will match society's expectations is anything but clear; some of the associated difficulties are outlined in Section 9.3.

9.2 NEW TECHNOLOGY

9.2.1 Storage

Energy storage alleviates the need to generate power at the same time as it is demanded. This may be desirable both for economic reasons and to guarantee sufficient supply during times of peak demand or when resources are unavailable.

On the subutility scale, electricity storage is practically synonymous with batteries. Stand-alone power systems, whether residential or commercial, typically use banks of lead–acid batteries to store intermittently generated energy from renewable sources or to provide a reliable backup in case of a generator failure. They are similar to car batteries, but designed to be more tolerant of repeated deep discharge. Batteries of different chemical makeup exist, but are very expensive and rarely used on the scale of building energy supply. Even the cost of standard lead–acid batteries to supply loads on the order of kilowatts for any appreciable duration is significant compared to the cost of generating the energy itself. Besides being expensive, batteries are toxic, corrosive, potentially explosive, and bulky; also, their performance is sensitive to proper treatment and maintenance. If a convenient and affordable alternative existed, it would no doubt revolutionize the field.

Batteries intrinsically work with direct current (d.c.), so that their use for a.c. systems always requires an inverter. The basic principle of energy storage in a battery is that an exothermic (energy-releasing) chemical reaction proceeds subject to the flow of ions through the battery and electrons through the external circuit. (Even though the electrons are tiny, they are unable to travel independently through the battery fluid.) When an electric load provides a path between the positive and negative terminals, the current can flow and the reaction proceed until the stock of chemical reactants is depleted. The chemical energy released in the reaction is equivalent to the work done by electrons in moving through the load, plus any heat losses. The battery is recharged by forcing the chemical reaction to reverse itself, which means forcing ions to flow backwards through the battery by pushing a reverse current into the terminals with an electric power source. This recharging process consumes an amount of electric energy commensurate with the chemical energy to restore the original reactants, plus losses. The round-trip efficiency for electric energy into and out of a well-functioning battery bank is typically in the neighborhood of 75%.

Example

A car battery has a storage capacity of 80 amp-hours (Ah) at 12 V. Neglecting losses, how many of these batteries would be needed to supply a residential load of 5 kW for 24 hours?

The energy demanded is 5 kW × 24 h = 120 kWh. The energy stored in one battery is readily computed by combining volts with amp-hours, recalling that 1 volt × 1 ampere = 1 watt. Thus, 80 Ah × 12 V = 960 Wh. The number of batteries required is 120,000 Wh/960 Wh = 125.

On the scale of utility power systems, the amount of energy storage required to have any operational impact is so huge by comparison that batteries are simply not a realistic option. The most common and practical form of storage here is in the form of pumped hydroelectric energy. A "pumped hydro" storage unit requires a reversible turbine–generator and reservoirs uphill and downhill for water to be stored. The idea is to draw electric power from the grid when it is readily available and cheap, operating the turbine–generator as a motor to pump water uphill. This water is then available to flow downhill and generate power, by pushing the turbine as in a standard hydroelectric power plant, at a later time when electricity is scarce and expensive.

The size of the reservoir determines the total amount of energy and thus the time period for which power can be stored. Although water reservoirs also provide seasonal storage, as in saving spring runoff for peak summer loads, the typical application of pumped storage is for a diurnal cycle, pumping at night and generating during the day. The difference in the value of electricity between night and daytime peak hours easily compensates for the conversion losses: with the efficiencies of pumping and generating each in the neighborhood of 80–90%, a round-trip efficiency of 75% is readily attained. In addition to the energy capacity, another relevant measure of a storage unit is its power capacity, or the rate at which it can absorb and redeliver the energy; this is a function of flow rates and machine ratings.

While the equipment is straightforward and the application convenient, the construction of pumped hydro storage units is constrained by topography, because large volumes of water and significant elevation gains are required to reach the scale of megawatt-hours. One prominent example of using pumped hydropower to great advantage is Switzerland, which serves as a storage bank of electric energy for the western European grid. Alpine elevations with plenty of water, centered geographically among dense loads, obviously make this an ideal situation, not to mention a great business opportunity. In the United States there are only few pumped hydro units, though their contributions are operationally significant.

If storage capabilities in U.S. grids were to be expanded, some newer technologies in addition to pumped hydro storage might be drawn upon. Absent water reservoirs, it is possible to use compressed air as the storage medium. In compressed-air energy storage (CAES), electric energy is used to operate pump motors that fill a confined space such as an underground cavern with air at high pressure. To retrieve the energy, the pumps are operated in reverse as generators. As with water, the storage location is the key constraint on siting large CAES facilities.

Different approaches to electricity storage currently under research and development include flywheels and superconducting magnetic energy storage (SMES). The principle of a flywheel is that energy is stored in the form of rotational kinetic energy of a spinning disk or wheel, where the energy can readily be supplied and extracted by standard motor-generators. The fundamental design challenge is to produce a wheel with high rotational inertia—which intrinsically requires it to be large and massive—capable of withstanding extremely high rotational speeds while minimizing the risk and hazards due to fracture, along with the ability to scale up the system at a reasonable cost.

SMES hinges on the ability to sustain extremely high currents within superconducting material, which are associated with a strong magnetic field. The so-called "high-temperature" superconducting materials known at present still require cooling by liquid nitrogen, which has a boiling point of $-195°C$. Interestingly, SMES provides the ability to inject and withdraw electric power between the grid and the storage unit with extreme sensitivity, on the time scale of a cycle. This affords the option to correct power quality in addition to providing bulk-energy storage.

Another way of storing electricity directly, without conversion into another energy form, is by capacitor. A capacitor is a simple device designed to store electric charge. Its opposite plates effectively coax like charges (positive or negative) onto the same piece of conductor (see Section 3.2.2); their mutual repulsion, or the electric field between the plates, represents stored potential energy. Capacitors are widely used in power systems for reactive power compensation. Physically, this represents an extremely short-term energy storage, where power is alternately absorbed and released within the duration of each a.c. cycle. Capacitors are also used in uninterruptible power supply systems to bridge the very brief gap during switching from one power source to another. However, expanding the storage capacity from the scale of a single cycle to time spans of hours or even minutes would involve a tremendous increase in scale and cost (one minute is 3600 times as long as one cycle!). Capacitors therefore do not seem practical for bulk-storage applications where the sheer quantity of energy is what matters, as opposed to the ability to rapidly absorb and release power.

Electrical energy can also be stored by electrolyzing water into hydrogen and oxygen, where the hydrogen gas becomes a convenient and clean chemical fuel that can be stored in quantity, transported, and converted back to electricity by means of fuel cells. While the technology for each conversion step is commercially available, the efficiencies range in the sixties and seventies of percent, making for a low round-trip efficiency. The key advantage of hydrogen is its suitability for mobile as well as stationary applications, combined with the ease of extending the storage time without local constraints.

With the exception of pumped hydroelectric storage, none of these technologies are currently inexpensive enough to be implemented on a large scale. Conversely, we might say that the value of energy storage to electric power systems is not high enough to warrant large investments in such facilities at the present time. It is important to note, however, that electric energy storage is fundamentally possible, with technologies ranging from experimental to very well proven, and that the valuation of storage capability in electric grids might change in the future. Possible reasons for an increased value of storage include both generation and transmission constraints leading to a high cost of accommodating demand peaks, as well as a growing contribution of intermittent, nondispatchable generation such as solar and wind power.

9.2.2 Distributed Generation

Distributed generation (DG) describes electric power generation that is geographically distributed or spread out across the grid, generally smaller in scale than traditional power plants and located closer to the load, often on customers' property. As an emerging and increasingly popular set of technologies, DG is associated with interesting questions about the overall design of the grid, operating strategies, economics, environmental impact of electricity production, and, ultimately, energy politics. This section is intended as a brief and general introduction to the main characteristics of DG and key related issues.

Since the beginning of the electric power industry in the late 1800s, there have been essentially two types of power plants: hydroelectric turbines, which convert the downhill movement of water into rotation of an electric generator, and steam generation plants, which burn fossil fuels (coal, oil or natural gas) to boil water and push a turbine–generator with the force of hot, expanding steam. Nuclear reactors added uranium to the industry's repertoire of fuels, but while the technology for controlling the fission reaction and safely extracting its heat is highly sophisticated, it is, from the power system point of view, just another way to boil water; the turbine–generator components of a nuclear facility are essentially identical to those of conventional fossil-fuel plants.[6] The same presumably applies to any nuclear fusion reactors built in the future. Likewise, solar thermal power plants with mirrors that focus sunlight onto a collector, whether in the configuration of a central tower, parabolic troughs, or parabolic dishes, are just another way to harness a heat source and, again, boil water to make steam. Other resources for steam generation include biomass (essentially, anything organic that burns) and geothermal power (where steam or hot water is extracted directly from the ground). Another renewable energy alternative, tidal power, is technically just a variation on conventional hydroelectric power, with the turbines submerged in a tidal basin instead of a river. Thus, a wide range of energy resources, while very diverse in their economic, environmental, and political characteristics, are compatible with the mainstay technology—the synchronous generator—to interface with the electric grid.

By contrast, there exists another set of electric generation technologies quite different in character from steam or hydro turbines; the most important of these today are wind turbines, photovoltaics (PV), microturbines, and fuel cells. They differ from conventional resources in the electrical properties of the generator component, the patterns of resource availability (for the case of wind and solar), their scale and the range of suitable locations for their deployment, making them classic distributed generation technologies.

The generation of grid-connected wind generators that saw increasing deployment in the 1980s relied primarily on inexpensive induction generators. As discussed in Section 4.5, the induction generator is not capable of controlling bus voltage or reactive power output; it always "consumes" VARs in the course of injecting watts to the grid. It also is not capable of controlling a.c. frequency or starting up without an a.c. signal already present at the bus. Although this poses no particular problem as long as the *system penetration* or percentage contribution to the grid of induction generators is small,[7] it does mean that they cannot be considered a fully controllable generation resource on a par with synchronous machines. More recently, however, wind generators have been built with an a.c.-to-d.c.-and-back-to-a.c. inversion step, which affords rather complete control over reactive power, output voltage, and frequency. This change was motivated by the increase in mechanical efficiency attained by permitting the wind rotor to operate at variable rotational speeds as a function of wind speed (thus varying the a.c. frequency),

---

[6]This sentence calls for several qualifications. A minor technical one is that for safety reasons, the steam temperatures and pressures used in nuclear plants are characteristically lower than in fossil-fuel plants, implying some changes to turbine design and operating efficiency. More significantly, nuclear facilities are unique in that the continuity of their operation has general public safety implications. These warrant special efforts on the part of the entire power system to guarantee a continuous receptivity for their output, avoiding rapid shutdowns or frequent changes of power level at nuclear plants. These subtleties notwithstanding, the power system sees a nuclear generator essentially as a synchronous machine moved by steam.

[7]Just how small is "small" in this context remains an interesting technical research subject. The reactive power problem is generally addressed by adding capacitance on location; the main constraint here is economic, not technical.

which easily outweighs the electrical losses from rectifying and then inverting the output.

Similar conversion processes are used with microturbines, which tend to operate at very high rotational speeds and whose a.c. output therefore must be adapted to the grid frequency. Microturbines are powered by natural gas [essentially methane (CH_4)], which may be derived from renewable sources such as wastewater, landfill, or manure digester gas. They are considerably smaller than steam turbines or even gas turbines,[8] with units in the range of tens to a few hundred kilowatts that can readily fit into a basement. Although atmospheric emissions from microturbines tend to be higher, they can provide systemic efficiency gains by way of a *cogeneration* or *combined heat and power* (*CHP*) option, allowing customers to simultaneously make use of electricity and waste heat, on a smaller scale than conventional cogeneration facilities that use steam in industrial settings.

Unlike wind and microturbines, solar PV and fuel cells contain no rotating parts. They naturally produce d.c., not a.c., and accordingly require inverters as an interface with the grid (see Section 4.6). PV cells consist of a specially treated semiconductor material (usually silicon) that produces an electric potential when exposed to light;[9] the working PV cell appears to the electric circuit somewhat like a battery with a positive and a negative terminal on either side. A fuel cell resembles a battery even more in that a chemical reaction forces electrons to one side, except that the reactants (hydrogen and oxygen) are continually supplied from an external source.[10] The d.c. voltage supplied by both PV and fuel cells is determined by the electrochemical properties of the materials and is on the order of one volt per cell. In both cases, multiple individual cells are connected together in series to form *PV modules* or *fuel cell stacks* with a convenient operating voltage on the order of tens of volts. The inverter, then, is the crucial step that converts this d.c. input to a.c. output of the desired voltage to connect to the power system.

PVs epitomize the notion of a *modular* technology in that a PV system can be built at just about any scale, from pocket calculator to megawatt utility array.

---

[8]*Gas turbines*, fueled by natural gas, resemble jet engines in that the turbine blades are propelled directly by the combustion exhaust gases. Being smaller but more expensive to operate than steam generation units, the key advantage of gas turbines is being able to ramp power up and down very quickly; they are characteristically used for peak generation. *Combined cycle* plants use a combination of steam and gas turbines that affords greater overall efficiency.

[9]Incoming bits of light or photons interact with electrons inside the semiconductor, making them free to travel. The special treatment of the material creates an asymmetry called a p-n junction (for positive–negative) which compels free electrons to move in one direction, toward the already negative terminal. This asymmetry is never used up; the work done by the moving electrons comes entirely from the light energy. A wide variety of PV technologies is both on the market and under development, using different materials and manufacturing processes.

[10]The electrical energy is released from the chemical reaction where hydrogen and oxygen combine to form water. Although hydrogen as an energy storage medium can be produced by the reverse process—electrolysis—from water and any electrical energy source, commercial hydrogen today is obtained from natural gas in a process called *reforming*. Some types of fuel cells incorporate the reforming step and use methane directly as a feedstock. The electrochemical process is more efficient in principle than burning the gas to drive a turbine because it avoids the thermodynamic limitations of converting heat into motion. A variety of fuel cell technologies are under development and commercially available.

Large arrays are made by simply combining more modules in parallel to obtain arbitrarily large amounts of current and power. Aside from volume purchasing discounts, there are no intrinsic *economies of scale* associated with PV technology that would make one large plant cheaper than, say, 10 small ones. This is a very important contrast to steam generation, where the optimal unit size in terms of cost per output is typically in the hundreds of megawatts. It means that the technology can be used in location- and load-specific applications that tend to be of smaller size, without an intrinsic cost penalty compared to centralized, bulk power.

Although some economies of scale do exist for wind turbines and fuel cells, these are still at a much smaller level than for steam generation. Wind turbines of various designs are built on scales ranging from one or more kilowatts to megawatts, with a somewhat lower cost per unit output for the larger machines. Fuel cells, depending on type, have similar optimal size ranges between kilowatts and a few megawatts.

The absence of pronounced economies of scale is one key characteristic of distributed generation. To be suitable for applications very near loads, however, generation must also be environmentally compatible. The easiest technology to site is PV, which entails minimal nuisance: zero emissions, no noise, minimal aesthetic impact, and options to integrate power installations with buildings or other structures (for example, shading for parking lots). Without any moving parts whatsoever, the maintenance requirements for PV systems are also minimal, allowing the technology to be installed in remote locations. Fuel cells, whose benign waste products are heat and water, require somewhat more supervision, but are not too hazardous to operate in occupied buildings. The higher power density (owing to the fuel's energy content) makes for compact units compared to solar generation, and installations in spaces such as office basements are becoming standard. Siting considerations for microturbines are similar to fuel cells, except that the emissions from combustion include undesirable components. Wind power is more constrained with respect to siting because of resource availability—it is generally easier to find a sunny than a windy spot—and because of the intrinsic hazard of moving rotor blades, but nevertheless lends itself to many distributed locations.

Siting generation close to loads rather than in a centralized manner has significant technical implications for the grid, most of them probably positive. First, we should expect thermal line losses throughout the grid to be reduced. Although a quantitative estimate of loss reduction due to distributed generation requires an explicit power flow analysis comparing specific scenarios, we can say qualitatively that when generation occurs next to a load of comparable size, effectively negating this load, it would tend to have the effect of lowering current flow in the transmission and distribution lines that connect this load to major generation sources in the grid. Thermal energy losses are then reduced in proportion to the square of the current and the resistance of the all the affected lines.

Second, distributed generation can offer voltage or reactive power (VAR) support, offsetting the need for other devices such as capacitors and voltage regulators (see Section 6.6). Note that while much of the grid's overall reactive power demand is met inexpensively through centrally located synchronous generators, some portion of it must be injected locally, that is, near loads, in order to maintain an acceptable voltage profile throughout the distribution system. If distributed generation can provide voltage support in a reliable manner such that other hardware installations or upgrades can be avoided—for example, it might obviate the need for a new capacitor bank, or it might reduce wear on a load tap changer—there are systemwide economic savings. Realizing these savings requires that the DG reliably coincides with demand, which tends to be true especially for solar PV.[11]

Third, to the extent that distributed generation coincides with load, it tends to reduce the demand on transmission and distribution capacity such as conductors and transformers. Because this equipment has to be sized to accommodate the peak, not average load, offsetting a small amount of load with distributed generation during key hours can provide a significant relief for the T&D infrastructure. In areas with substantial load growth, local generation can serve as an alternative to T&D capacity, making it possible to avoid or at least defer capital intensive upgrades (for example, the replacement of a substation transformer with a larger model) while maintaining reliability.

Fourth, generation in the distribution system impacts protection needs and coordination. This is generally regarded as a negative property, especially by utility engineers faced with the challenge of adapting or redesigning circuit protection to guarantee that any fault will be properly interrupted and cleared (see Section 6.7). The reason this can be problematic is that in the traditional radial system, power flows in only one direction at the distribution level, from substation to load, and the extant protection is designed and coordinated accordingly. Distributed generation introduces the radically new possibility of power flowing in the other direction, and any section of line or piece of equipment therefore being energized from the load side, on which there is no fuse or circuit breaker to protect it. Beyond feeding power into a fault, this implies an electrocution risk for utility line workers.[12] This protection issue remains controversial in the industry today. While manufacturers, owners, and advocates of DG tend to feel confident that built-in protection features of the generation equipment guarantee that it will separate from the grid in the event of any disturbance, utilities typically require additional safety disconnect switches accessible only to their crews.

---

[11] A second-order effect that is small but mentioned here for the sake of completeness is a reduction in systemwide reactive losses due to distributed generation. As described in Section 7.2.5, reactive losses represent the difference between the total reactive power Q injected to the grid at generator buses and that withdrawn at load buses. Physically, this difference does not represent any actual energy like the analogous I^2R real power losses that are radiated into the air as heat. Nevertheless, reactive losses are associated with some economic cost.

[12] Note that the electrocution hazard is associated with voltage level, not necessarily the amount of power provided by DG. For example, as long as a residential PV system remains connected to the grid by a 12-kV-to-120-V transformer, it energizes the primary side of that transformer with 12 kV regardless of the fact that it could never by itself sustain the loads in the neighborhood. Thus, even a very small DG installation could be deadly to line crews if not electrically isolated.

Another technical and controversial aspect of distributed generation is the problem of control and availability. On the one hand, DG is generally *nondispatchable*, meaning that power system operators cannot call on it to provide power on demand or at specified times. The resource may be intermittently available (solar or wind) or controlled by the owner of the generation equipment; in either case, the utility or system operator has no control over operating schedules. Indeed, active centralized control over a large number of small DG units may be unrealistic due to the sheer information and communications volume, unless it were completely automated.

On the other hand, the lack of control is not necessarily problematic, because DG can often be expected to coincide fairly well with local load. This is true in the case of solar power in areas of summer-peaking demand, where the diurnal and seasonal pattern of sunshine can match the utility demand profile almost perfectly. It also tends to be true wherever customers use DG to meet their own demand or to receive credit on net metering arrangements during peak demand hours when utility rates are higher. To the extent that DG behaves in a predictable manner and tends to level out rather than amplify load peaks, there is little reason for active intervention on the part of system operators. Instead, they would tend to consider DG as "negative load," combining its statistical uncertainty with that of the demand, which as a whole is also beyond operators' direct control. Obviously, this approach is sensible when the relative amount of DG in the system is small, but may become problematic at some greater level of system penetration in the future.

Last but not least among the systemic technical properties of distributed generation is that it would seem to reduce the grid's vulnerability to sabotage. In part, this is because any one smaller individual facility has less impact on the stability of the power system as a whole, and thus would be less effective at causing broad disruption (besides making a less attractive target in the first place). In part, too, DG implies less reliance overall on long-distance transmission links, and therefore would tend to reduce the scope of disruption caused by sabotage of individual transmission lines or substations. In any case, DG introduces the possibility of local self-sufficiency in power generation, which may dramatically reduce the social impact of grid failures by making electricity available locally for vital applications during a crisis.

DG also has institutional implications for power systems at large. As generation is distributed geographically, it enters the jurisdiction of power distribution, as opposed to transmission, where power plants traditionally interface with the grid. For distribution engineers and operators, dealing with generation is a fundamentally new responsibility—one, it might be noted, they cannot be expected to embrace with unqualified enthusiasm, as it entails new demands, complexities, and failure possibilities. Finally, there are important social and political dimensions of ownership of resources and generation assets.

In sum, an electric power system with significant amounts of distributed generation represents a radical departure from the centralized and strictly hierarchical power system of the 20th century. Such a transition, which many analysts consider all but inevitable, entails some important unanswered questions: while it seems clear that the grid's character is changing, it is not obvious what, exactly, it is changing into.[13]

While the social and political aspects are beyond the scope of the present discussion, we can identify some specific open questions about the grid's evolution in the face of distributed generation. One has to do with the problem of *islanding*, or operating parts of the system while disconnected from others (see Section 6.1.4). On the one hand, the option of islanding allows extracting the full benefit from distributed generation resources, that is, supplying power locally while the rest of the grid is unavailable. On the other hand, an interconnected system that could routinely operate with local or regional power islands—beyond individual customers with self-generation who are carefully isolated from the distribution system—represents yet another leap of technical and institutional transformation, with a host of controversial aspects including safety, liability, accounting, and control.

Another issue of great practical consequence concerns the economic cost–benefit analysis of distributed generation in relation to transmission and distribution infrastructure. While it was suggested earlier that DG offers systemic savings (for example, on line losses or T&D investments), it is not obvious how this information can be used for strategic investment decisions when different corporate entities pay out the capital and realize the savings, respectively. Even if markets or regulation offer incentive mechanisms for DG, there remains the analytic problem of comparing and trading off what are qualitatively different investments. Limited experience exists in this area because, in vertically integrated and restructured market environments alike, generation and transmission planning have almost always been carried out by separate organizational entities, with separate accounting and separate responsibilities for justifying investments. This problem also relates to the economic evaluation of electric storage capacity within the context of the grid. The kind of systematic and integrated three-way comparison of generation, storage, and transmission capacity that one might imagine guiding the strategic development of power systems is simply not yet part of the analytic arsenal within the electricity sector.

Yet such an analytic framework will become crucial if future power systems are to include increasing contributions from intermittent, renewable resources. For example, we might want to compare the cost of solar thermal electricity produced at two locations with different weather and different transmission distances to major loads, or that of a PV plant adjacent to a load with a fossil-fuel resource some distance away. Alternatively, consider a region where wind is the cheapest energy source available and supplies a substantial portion of local demand. Here we might have to assess the maximum percentage of load that can be met by wind power before it becomes too unpredictable, and then determine to what extent the unpredictability ought to be compensated for by either

---

[13]This transformation is discussed in A. von Meier, "Integrating Supple Technologies into Utility Power Systems: Possibilities for Reconfiguration" in Jane Summerton (ed.), *Changing Large Technical Systems* (Boulder, CO: Westview Press, 1994).

introducing storage capacity, adding more expensive generation capacity from other sources, strengthening transmission ties to other regions, or simply oversizing the wind power plant. The analytic challenge for sensible decision making, whether by market mechanism or an integrated planning entity, thus increases with the introduction of distributed and intermittent resources because of a sensitivity to time and geographic variables that 20th century technology simply did not have.

How distributed generation technologies fare in view of these systemic challenges, parallel to their own intrinsic developmental hurdles, remains to be seen. The results will have societal implications far beyond electric power, as the use of distributed and renewable generation relates to diverse issues of resource scarcity, environment, and the politics of ownership. With respect to technical control issues, the role of DG will necessarily be intertwined with that of automation technology, the subject of the following section.

9.2.3 Automation

Electric power systems today harbor a fascinating juxtaposition of old and new technology. Finely tuned gas turbines and state-of-the-art inverters inject power into the grid along with hydroelectric units dating back to the early 20th century. The hardware of electric power systems represents some of the oldest industrial machinery still in general use today. The fundamental task of transmission and distribution—that is, to connect pieces of conducting metal so as to form electric circuits—remains unchanged, as does a transformer's job in stepping voltage up and down. While advances have been made in the design and materials used in transformers, capacitors, and circuit breakers, the essential functions of these power system components are stunningly simple. Synchronous generators also have experienced only refinement, not fundamental changes of design. Many of today's power system components could be replaced with well-preserved models from the early 1900s without compromising their role in the grid.

The simplicity of the hardware, especially for T&D, stands in remarkable contrast to the complexity of the system it constitutes. When the electric grid is considered as a whole that requires operation and coordination, it becomes a subtle and sophisticated entity whose behavior and sensitivities are the subject of careful and elaborate study. As the grid's functioning has come under increased scrutiny since the energy crises of the 1970s, the restructuring efforts of the new millennium, and the occasional blackout, there is ever more incentive to tune its operation and maximize its performance.

One key step toward streamlining or automating the operation of T&D systems involves increased monitoring of circuit data by remote sensing along with remote operation of equipment such as switches and circuit breakers to reconfigure the system topology. This technology is known as *supervisory control and data acquisition* (*SCADA*). The sensing and control nodes in the field may be connected to a staffed control room by one of several types of communication including dedicated telephone lines, microwave radio, or power line carrier signals.[14] SCADA has been implemented to various extents by U.S. utilities over the past decades, first in transmission and subsequently at the distribution level. In its absence, operators rely on field personnel to physically travel to each location in order to report equipment status or undertake switching actions. Many areas, especially rural distribution systems, are still operated in this manner.

It is interesting to note that without added communications systems, a transmission or distribution operator has no way of determining whether customers are connected or whether a problem has occurred somewhere in the field. Unless the problem has proliferated to the level of a manned substation (say, by tripping a circuit breaker), nobody but the customer might notice a local disturbance, as the grid offers no intrinsic way of measuring from one location what is happening in another. To reduce both the hazards of undetected faults and the duration of service interruptions, many utilities have installed outage report systems that help identify trouble spots by correlating customers' telephone numbers with their locations on the grid.

Beyond information gathering and remote control, a more radical or comprehensive approach to automating power systems involves operation through *expert systems* that either recommend actions to the operator (*open-loop*) or execute them as well (*closed-loop*). Systems of various scope have been implemented or proposed in power generation, ranging from straightforward automated mechanisms to perform limited modifications (such as load following) to the highly controversial substitution of computer systems for human decision making in hazardous places like nuclear power plants.

In transmission and distribution, expert systems mean augmenting SCADA with "intelligence" that allows for rapid computational analysis and thereby makes possible a more interventionist operating approach. For example, *load balancing* involves reconfiguring distribution circuits in real time with the goal of increased efficiency, as measured by reduced electric losses or greater asset utilization. Another example is automated service restoration, where rapid data analysis and execution of switching procedures can make for much reduced interruption times. Despite the potential savings and performance improvements by means of such expert systems, considerations of safety and human factors present significant challenges to this type of innovation; these challenges are discussed further in Section 9.3.

What seems certain at this time is that automation of various components of electric power systems will continue to see both lively debate and increased implementation. Aside from human factors, both the major justifications and the major constraints on automation have been economic, with the bases for cost and benefit calculations shifting rapidly. In general, as the cost of information decreases in relation to the cost of energy, it makes ever more economic sense to carefully

---

[14] In communication by power line carrier, the information signal is simply superimposed onto the 60-Hz power signal along the conductor. Having a smaller amplitude but much higher frequency, this signal is readily isolated by equipment on the receiving end; nevertheless, the equipment must be designed to safely interface with high voltage.

survey the grid and intervene in its operation through an increasing number of data points and control nodes, deferring to computers where necessary to handle the information volume or achieve the required reaction speed. In the context of DG, the sheer volume of information and control opportunities associated with a hypothetically large number of generators of diverse behavior would seem to require some form of automation if resources are to be actively monitored, coordinated, and their contributions optimized. It remains to be seen which areas best lend themselves to the integration of information technology with extant and new hardware, and how the overall character of power systems will change as a consequence.

9.2.4 FACTS

Another dimension of automation in the case of transmission systems is the direct modification of the grid's properties with the aid of solid-state technology—essentially, various types of transistors[15] scaled up and combined to handle large power applications—in a new category of hardware called *flexible a.c. transmission systems* (*FACTS*). Transmission lines generally have physically fixed parameters such as length and impedance that become firm constraints in modeling and analysis. Other components such as transformers and capacitors may have variable states or settings, but conventionally these settings are discrete and require mechanical switching. FACTS technology offers ways to modify the electrical characteristics of transmission components much more rapidly, even in real time, so as to increase operating efficiency and relieve constraints without the need for adding major hardware. FACTS devices include various types of reactive compensation, phase shifting, and power flow control. The idea is to effectively change the impedance of a given transmission link as seen by the system on an instantaneous basis by means of an appropriately designed solid-state electronic circuit.

For example, a long transmission line may have a stability limit less than its thermal limit, meaning that no more than a certain amount of real power can be transmitted along it without risking a loss of synchronicity between the generators at either end (see Section 6.5.2). Conventionally, this limit poses a firm constraint on permissible power flow scenarios. For instance, it could mean that generators in location A must not inject more than a certain amount of power based on generation at B, lest the flow from A to B (specifically, the voltage phase angle difference between A and B) become too great. The only way around the constraint would be to upgrade the link with a larger or additional conductor that provides a lower impedance. FACTS technology, by contrast, allows system operators to intervene

---

[15] A transistor is essentially a piece of semiconductor material that changes from a conducting to a non-conducting state in response to a small voltage signal, making it act like an electrical switch. Because the transition happens on the atomic level and involves no mechanical movement, it is not only very compact but extremely fast, allowing circuits to be switched back and forth many thousands of times per second. This process lies at the heart of information technology that represents data in terms of open and closed circuits (0s and 1s).

directly by modifying the behavior of the transmission link in question: in this case, a solid-state device would shift the voltage phase angle on the line between A and B in such a way as to reduce the difference, despite the generators' difference in production. In essence, we could say that FACTS amounts to cheating the power flow calculation by actively controlling flow along a particular link.

This should seem to contradict the basic principles emphasized throughout this text, especially the notion that current and power flows are determined solely by boundary conditions such as generation, load, and impedance, and are therefore beyond our immediate control. But the key to achieving control lies in the time dimension. In a conventional electric power circuit, we assume the properties of the hardware components, including the connections among them, to be unchanging. Solid-state technology, however, affords the ability to switch circuits around many times within a single a.c. cycle. This means being able to strategically open and close connections between elements such as transformers, inductors, or capacitors that add or subtract a.c. waveforms at specific points during the cycle. As a result, we can change the shape of a voltage wave (for example, to eliminate unwanted harmonics or resonance), its timing or phase angle (as in the phase-shifting example given earlier), its magnitude, or its relationship to current (in effect controlling reactive power flow).

Since a growing variety of such devices is under development and initial deployment, FACTS technology suggests the possibility of a profound change in how power systems operate. As with automation, the emphasis is on increased information and control in lieu of larger hardware, with robustness giving way to a more refined and interventionist approach to managing the grid. Also like automation, the implementation of FACTS is constrained by the relationship of capital cost to operational gains. It will be interesting to observe its course over the coming years.

9.3 HUMAN FACTORS

9.3.1 Operators and Engineers

This section characterizes some of the important differences between an academic and a practical view of power systems.[16] Specifically, it distinguishes between operations and engineering as two types of human expertise and activity that are essential to the grid's performance. The following is a summary of findings from research conducted by the author that set out to explain differences of opinion about the introduction of automation technology in distribution systems by examining how different people conceptualize the power system as a whole.

A majority of readers, it is assumed, will be more familiar with engineering as an activity and modeling framework. In the electric power industry, "engineering"

---

[16]Parts of this analysis have been previously published in A. von Meier, "Occupational Cultures as a Challenge to Technological Innovation," *IEEE Transactions on Engineering Management* **46**(1):101–114, February 1999.

encompasses a great variety of specific job tasks. Engineers make design drawings, calculate specifications, select components, evaluate performance, and analyze problems. Their work has an important idealistic aspect, finding innovative solutions and always striving to improve things. Some utility engineers are directly engaged with physical hardware (for example, overseeing its installation); others work with abstract models of the power system (for example, power flow analysis) or on its indirect aspects (for example, instrumentation or computer systems). Those engineers whose work is more remote from the field and of a more academic nature best match the archetype of the present description.

Operators of technical systems, be they power plants, airplanes, or air traffic control, must keep the system working in real time. In a thermal generation plant, for example, operators have to assure that steam flows at a multitude of temperatures and pressures between boiler, turbine, and condenser remain coordinated and in balance with electrical load on the generator unit throughout a range of conditions. In electric power transmission and distribution, operators monitor and direct ongoing reconfigurations of their system of interconnected power lines and components from switching stations and in the field. Unlike engineering, where the object is to optimize performance, the goal in operations is to maintain the system in a state of equilibrium or homeostasis in the face of external disturbances, steering clear of calamities. An operating success is to operate without incident. Depending on the particular system, maintaining such an equilibrium may be more or less difficult, and the consequences of failure more or less severe.

Three types of challenges are generally characteristic of the operations job: external influences, clustering of events, and uncertainties in real-time system status. Especially in the case of power distribution systems, a large part of the hardware is physically accessible and vulnerable to all kinds of disturbances, whether they are automobiles crashing into poles or foxes electrocuting themselves on substation circuit breakers. Events such as heavy storms or extreme loading conditions entail cascading effects in the system and require a large number of switching, diagnostic, and repair operations to be coordinated and carried out under time pressure. At the same time, system parameters such as loading status for certain areas or even hardware capabilities are often not exactly known in real time. Operators are quite accustomed to working in this sort of situation, and the way they view and imagine the electric grid, along with their values and criteria for system performance, can be seen as specific adaptations to these challenges.

9.3.2 Cognitive Representations of Power Systems

In the engineering framework, "the system" is considered as a composite of individual pieces, since these are the units that are readily described, understood, and manipulated. The functioning of the system as a whole is understood as the result of the functioning of these individual components: should the system not work, the obvious first step is to ask which component failed. Engineering is therefore analytic, not only in the colloquial sense of investigating a complex thing, but analytic in the very literal sense of "taking apart," or treating something in terms of its separate elements.

Like any analytic process, engineering requires modeling, or representing the actual physical system in abstracted and appropriately simplified terms that can be understood and manipulated. Abstraction and simplification also requires that the system elements be somehow idealized: each element is represented with its most important characteristics, and only those characteristics, intact. An engineering model will thus tend to consider system components in terms of their specified design parameters and functions. Each component is assumed to work as it should; components with identical specifications are assumed to be identical. Similarly, the relationships among components are idealized in that only the most important or obvious paths of interaction (generally the *intended* paths) are incorporated into the model. The parameters describing components and their interactions are thought of as essentially time-invariant, and invariant with respect to conditions not explicitly linked to these parameters.

The behavior of the system is thus abstracted and described in terms of formal rules, derived from the idealized component characteristics and interactions. These rules, combined with information about initial conditions, make the system predictable: from the engineering point of view, it should be possible in principle to know exactly what the system will do at any point in the future, as long as all rules and boundary conditions are known with sufficient accuracy. These rules also imply a well-understood causality: it is assumed that things happen if and only if there is a reason for them to happen. Of course, engineers know that there are random and unpredictable events, but in order to design and build a technical system, it is essential to be able to understand and interpret its behavior in terms of cause-and-effect relationships. Chains of causality are generally hierarchical, like if–then decision-making systems. Stochasticity is relegated to well-delimited problem areas that are approached with probabilistic analysis.

In summary, then, the classic engineering representation of a technical system can be characterized as *abstract*, *analytic*, *formal*, and *deterministic*. By contrast, the operator representation of a technical system can be typified as *physical*, *holistic*, *empirical*, and *fuzzy*. This representation is instrumental to operators in two important ways: it lends itself to maintaining an acute situational awareness, and it supports the use of intuitive reasoning.

Because operations involve much more immediate contact with the hardware, system components are imagined as the real, physical artifacts in the way that they are perceived through all the senses. For example, a particular overhead distribution switch has a certain dimension, offers a certain resistance to being moved, makes a certain noise, and shakes the pole in a certain way as it closes. Even when looking at abstract depictions of these artifacts on a drawing or a computer screen, operators "see" the real thing behind the picture. With all its physical properties considered, each artifact has much more of a unique individuality than its abstract representation would suggest: one transformer may overheat more than another of the same rating, or one relay may trip slightly faster than another at the same setting. Thus, components that look the same on a drawing are not necessarily identical to an operator.

To be sure, operators must also work with abstract representations. For distribution operators, for example, this means primarily circuit maps and schematic diagrams for switching. However, the abstractions they find useful and transparent may differ from those preferred by engineers. While good maps for engineers are those that do a thorough job of depicting selected objects and their formal relationships, the most useful maps for experienced operators are those that most effectively recall their physical image of the territory.

Another aspect of operators' cognitive representation is that they conceptualize the system more as a whole than in terms of individual pieces. Rather than considering the interactions among components as individual pathways that can be isolated, the classic operator model is of one entire network phenomenon. Every action taken somewhere must be assumed to have repercussions elsewhere in the system, even if no direct interaction mechanism is known or understood. This is consistent with operators' experience, where they are often confronted with unanticipated or unexplained interactions throughout the system.

Rather than using formal rules to predict system behavior, operators rely primarily on a phenomenological understanding of the system, based on empirical observation. The underlying notion is that no number of rules and amount of data can completely and reliably capture the actual complexity of the system. Therefore, though one can make some good guesses, one cannot really know what will happen until one has seen it happen. No component can be expected to function according to its specifications until it has been proven to do so, and the effect of any modification has to be demonstrated to be believed. While engineers would tend to assume that something will work according to the rules, even if it did not in the past, operators expect that it will work the way it did in the past, even if analysis suggests otherwise. Many arguments between engineers and operators can be traced to this fundamental difference in reasoning.

Finally, the operator representation is one that expects uncertainty rather than deterministic outcomes. Whether due to the physical characteristics of the system, insufficiency of available data, lack of a complete understanding of the system, or simply external influences, uncertainty or "fuzziness" is taken to be inevitable and, to some degree, omnipresent. Ambiguity, rather than being subject to confinement, is seen to pervade the entire system, and operators suspect the unsuspected at every turn. Thus, distribution operators have described their system as a "live, undulating organism" that must somehow be managed.

This physical, holistic, empirical, and fuzzy view of the system is adaptive to the challenge of operating the system in real time in that it allows one to quickly condense a vast spectrum of information, including gaps and data pieces with different degrees of uncertainty, into an overall impression or *gestalt* that can be consulted with relative confidence to guide immediate action.

Finally, operators tend to draw on intuitive reasoning, especially when data are insufficient but action is required nonetheless. Although there are manuals specifying operating procedures, many situations occur that could not have been foreseen in detail and courses of action recommended. To deal with the problem at hand, analytic tools may not be able to provide answers quickly enough. Worse yet, information on the books may be found to be untrustworthy under the circumstances; for example, if recent data appear to contradict what was thought to be known about the system. In order to come to a quick decision, the operator's main recourse then is to recall past experience with similar situations. How did the system behave then? Were people surprised? How did the particular equipment respond? Based on such experience, an operator will have an intuitive "feel" for the likelihood of success of a given procedure.

This experience-based approach is intuitive not because it is irrational, but because it is nonalgorithmic. An operator might have difficulty articulating all the factors taken into consideration for such a decision, and how, precisely, they were mentally weighed and combined. He or she might not be able to cite the reasons for feeling that something will work or not work. Nonetheless, the decision makes use of factual data and logical cause–effect relationships, as they have been empirically observed.

The use of intuitive processes is so deeply embodied in the culture of operations that they are often chosen over analytic approaches by preference rather than necessity. Obviously, both methods can fail; the question is about relative degrees of confidence. While engineers may frown on operator justifications that seem based on intractable, obscure logic or even superstition, operators delight in offering accounts of situations where their intuition turned out to be more accurate than an engineer's prediction. In fact, both approaches are adaptive to the work contexts of their proponents, and while both have a certain validity, either approach may turn out to yield better results in a given situation. The important point here is that substantive differences in cognitive representations and reasoning modes underlie what may appear to be trivial conflicts or petty competition between cultural groups. These differences will also have specific implications for the evaluation of system design, operating strategies, and technological innovations.

9.3.3 Operational Criteria

The most important general properties of technical systems, or goals and criteria for evaluating their performance, can be summarized as *efficiency*, *reliability*, and *safety*. These goals tend to be shared widely and across subcultures throughout an organization managing such a system. However, individuals or groups may hold different interpretations of what these general goals mean in practice and how they can best be realized. Accordingly, they will also have different expectations regarding the promise of particular innovations.

When there are trade-offs among safety, reliability, and efficiency, cultural groups may also emphasize different concerns, not only because they have different priorities, but because they have different perceptions of how well various criteria are currently being met. In the academic engineering context, it is often assumed that certain standards of safety and reliability have already been achieved, and the creative emphasis is placed on improving efficiency. In the case of power systems, safety and reliability are problems that were academically solved a long time ago, whereas new approaches to increase efficiency offer a continuing intellectual challenge.

The efficiency criterion thus takes a special place in engineering. Efficiency here can be taken in its specific energy-related sense as the ratio of energy or kilowatt-hour output to energy input, or in a more general sense as the relationship of output, production, or benefit to input, materials, effort, or cost. Efficiency is often a direct performance criterion in that its numerator and denominator are crucial variables of interest that appear on the company's "bottom line" (for example, electric generation and revenues). Even where efficiency measures something more limited or obscure (for example, how many man-hours are required for service restoration), a more efficient system will generally be able to deliver higher performance at less cost while meeting the applicable constraints. Conversely, low efficiency indicates waste or the presence of imperfections that motivate further engineering. A more efficient system will also be considered more elegant: beyond all its practical implications, efficiency is an aesthetic criterion.

In addition, there is a set of indirect or supporting criteria that, according to the cognitive framework of engineering, advance efficiency as well as safety and reliability. While these criteria can be taken as qualitative standards for the system as a whole, they also apply in evaluating technological innovations and judging their promise. One such criterion is *speed*. It is an indirect criterion because it does not represent an actual need or an immediate, measurable benefit. However, the speed of various processes offers some indication of how well the system is theoretically able or likely to succeed in being efficient. Generally, a system that operates faster will involve less waste. For example, restoring service more quickly means less waste of time, waste of man-hours, and waste of potential revenues. Responding and adapting to changes faster can also mean higher efficiency in terms of improved service quality or saved energy. Given the choice between a slow- and a fast-operating device, all else being equal, most engineers would tend to prefer the faster one.

Similarly, *precision* is generally considered desirable in engineering culture. Actually, the desired criterion is *accuracy*: not only should information be given with a high level of detail, but it should be known to be correct to that level. Accurate measurements of system variables allow for less waste and thus support efficiency; they also further safe and reliable operation. However, the accuracy of a given piece of data is not known a priori and is subject to external disturbances, while its degree of precision is obvious and inherent in design (e.g., the number of significant figures on a digital readout). Precision can be chosen; accuracy cannot. Although precision does not guarantee accuracy, it at least provides for the possibility of accuracy and is therefore often taken in its place (and sometimes confused). Given the choice between a less and a more precise indicator of system parameters or variables, most engineers would prefer the more precise one.

More fundamentally, *information* in and of itself is desirable. Generally, the more information is available, the better the system can be optimized, and information can in many ways advance safety and reliability as well. In the event that there are excess data that cannot be used for the purpose at hand, the cost to an engineer of discarding these data is typically very low: skipping a page, scrolling down a screen, or ignoring a number is no trouble in most engineering work. In selecting hardware or software applications, all else being equal, most engineers would prefer those offering more information.

Finally, the ability to control a system and its parts is another indication of how successfully the system can be engineered, managed, and optimized. This is because any variable that can be manipulated can also, in principle, be improved. As with information, in the engineering context, there is hardly such a thing as too much control. If the ability to control something is available but not needed, the engineer can simply ignore it. Most engineers would prefer to design systems and choose components that are controllable to a higher degree.

This set of criteria suggests a general direction for technological innovations that would be considered desirable and expected to perform well. Specifically, from the viewpoint of engineering, innovations that offer increased operational speed, precision, information, and control appear as likely candidates to further the overall system goals of efficiency, reliability, and safety. While such expectations are logical given the representational framework of engineering, the perspective of operations yields quite a different picture.

Of the three general system criteria—safety, reliability, and efficiency—safety takes a special priority in operations, while efficiency is less of a tangible concern. From the point of view of managing the system in real time, efficiency is an artifact of analysis and evaluation, a number tagged on after the fact, having little to do with reality as it presents itself here and now. Although it may indicate operating success, efficiency more directly measures the performance of engineers. Most operators would agree that having an efficient system is nice, as long as it does not interfere with their job.

Safety, on the other hand, takes on a profoundly tangible meaning for operators because the consequences of errors face them with such immediacy. In power systems, any single operation, performed at the wrong time, has the potential to cause customers to lose power. Immediately, telephones will ring, voices on the other end will shout and complain, and the control room may even fill with anxious supervisors. Because of the interdependence of power system components, the consequences may occur on a much larger scale than the initial error. Aside from causing power outages, incorrect switching operations can damage utility and customer equipment.

But even more serious is the risk of injury or death. Any steam generation plant, for example, harbors the intrinsic hazard of water vapor at up to $1000°F$ and several hundred atmospheres of pressure; it is difficult to stand next to a roaring primary steam valve without sensing that the smallest leak could be instantly fatal. In transmission and distribution, the risk of electrocution looms for utility crews as well as others who might be accidentally in contact with equipment (for example, people in a car under downed power lines). The one action T&D operators dread most in the course of switching operations is to energize a piece of equipment that is still touching a person. Like operators of other technical systems, they carry a personal burden of responsibility for injuries or fatalities during their shift that goes far beyond their legal or procedural accountability. The difference between an intellectual recognition and the direct experience of the hazards cannot be overemphasized: hearing an accident described is not the same as watching one's buddy die in a flash of sparks a few feet away, with the smell of burning flesh. This immediate awareness of the life-taking potential of system operation is omnipresent among operators of power systems and implicitly or explicitly enters any judgment call they make, whether about day-to-day procedures or about implementing new technology.

Their acute perception of safety colors operators' interpretation of other system goals and helps define their criteria for good system design and performance. The set of criteria—speed, precision, information, and control—that, from the engineering perspective, support not only efficiency but also safety and reliability may be seen by operators as less important or even counterproductive. Instead, operators value a different set of criteria that specifically support their ability to operate the system safely.

Speed, generally advantageous in engineering, is more problematic in operations because one is working in real time. Speed is desired by operators in the context of obtaining information. They may also wish for their actions to be executable quickly, so as to gain flexibility in coordinating operations. However, a system of fast-responding components and quickly-executed operating procedures, where effects of actions propagate faster and perhaps farther, also introduces problems: it will tend to be less tractable for the operator, provide less time to observe and evaluate events and think in between actions, and allow problems to become more severe before they can be corrected. Power systems are inherently fast in that electric effects and disturbances propagate at the speed of light, making cascades of trips and blackouts practically instantaneous. Any delays or buffering of such effects work toward the operator's benefit. Thus, from the perspective of operations, *stability* is generally more desirable than speed. Operators would prefer a system that predictably remains in its state, or moves from equilibrium only slowly, allowing for a greater chance to intervene and bring it back into balance.

Information can also be problematic in the context of operations. To be sure, there are many examples of information that operators say they wish they had, or had more of. But more is not always better. Because one is gathering information and acting upon it in real time, the cost of discarding irrelevant information is not negligible. Deciding which data are important and which are not costs time and mental effort; superfluous data may distract from what is critical. Specifically, too much data may interfere with operators' acute situational awareness. Operators often give examples of information overload: many computer screens that must be scanned for a few relevant messages, or many pages of printout reporting on a single outage event. Generally, instead of greater quantity of information, operators desire *transparency*, meaning that the available information is readily interpreted and placed into context. It is more important for them to maintain an overview of the behavior of the whole system than to have detailed knowledge about its components: in terms of maintaining situational awareness, it is preferable to lack a data point than to be confused about the big picture even for an instant. If more information has the potential to create confusion, then for operators it is bad.

Suppose, for example, that a computer screen is to display real-time measurements from throughout the system to operators in a control room. One option might maximize information delivery, say, providing constantly updated figures from 100 sensor nodes. In a situation where all 100 data points are equally likely to be relevant, where it is important that no detail be missed, and where the data need not be processed and acted upon with great time pressure, this may be most desirable. By contrast, suppose that much of this information is irrelevant to decisions that must be made very quickly. Here it may be appropriate to reduce the amount of information in the interest of transparency, for example, by limiting the number of points reported, or by displaying only those that changed recently. The idea is that data should quickly and correctly characterize the situation behind the numbers, even at the expense of breadth or depth.

Similarly, more precision is not always better for operators. While engineers can make use of numbers with many significant digits, the last decimal places are probably not useful for guiding operating decisions. In fact, operator culture fosters a certain skepticism of any information, especially quantitative. This skepticism is consistent with their keen awareness of the possibility of foul-ups like mistaking one number for another, misplacing a decimal point, or trusting a faulty instrument, and the grave potential consequences. Therefore, operators' primary and explicit concern about any given numerical datum is whether it basically tells the true story, not how well it tells it. Moreover, precision can be distracting or even misleading, suggesting greater accuracy than is in fact given. Thus, in operations, *veracity* of information is emphasized over precision. Rather than trusting a precise piece of information and running the chance of it being wrong, operators would generally prefer to base decisions on a reliable confidence interval, even if it is wide.

The difference between precision and veracity is that precision offers a narrow explicit margin of error, while veracity offers confidence that the value in question truly lies within that margin, and that the value really represents that which it is assumed to represent. Suppose one measurement is taken with a crude and foolproof device, while another is displayed by a sophisticated monitoring network. The latter offers more precision, and yet the skeptic may wonder: Is it possible that the instrument is connected to the wrong node? Could the display be off by an order of magnitude? Or might it be telling me yesterday's value instead of today's? And if so, would there be any discernible warning? If someone's life depended on our correct estimate of a quantity, we would prefer the former, more reliable measurement, even though it is less precise.

Finally, more control is not always better. Of course, there may be variables over which operators wish they had more control. But the crucial difference is that in engineering, control always represents an *option*, whereas in operations there may be an associated responsibility to exercise this control: the ability to control a variable can create the expectation that it *should* be controlled, and produce pressure to act. Operators tend to be wary of such pressure, primarily because it runs counter to a basic attitude of conservatism fostered by their culture: When in doubt, do not touch anything. Their reluctance to take action unless it is clearly necessary arises from the awareness that any operation represents a potential error, with potentially severe consequences. An interventionist approach that may allow greater optimization and fine-tuning thus inherently threatens what operators see as their mission, namely, to avoid calamities.

In pragmatic terms, more controlling options may mean that operators have more to do and keep in mind, and thereby increase stress levels. Alternatively, they may not have time to exercise the control at all, in which case their performance will be implicitly devalued by the increased expectation. Because time and attention are limited resources in operations, and because of the potential for error associated with any action, the option not to control can be more desirable than the ability to control. This option is provided by a system's *robustness*, or its tendency to stay in a viable equilibrium by itself.

Suppose that a technological innovation gives operators the ability to control some system parameter (voltage, for example) within a narrower band, closer to the desired norm. The potential downside of this innovation has to do with the following question: What happens if, for whatever reason, the control option is not exercised? For example, is the parameter liable to drift farther outside of the normal range if it is not actively controlled? In doing so, does it pose a safety hazard? Does the new technology raise expectations for system performance, leading to disappointment if control actions are not taken in the manner envisioned by system designers? Will pressure to exercise control options create extra work for operators? By contrast, robustness and stability characterize a situation where things will be fine without active intervention.

Another example of efficiency versus robustness in the power distribution context is the handling of load peaks in view of limited equipment capacity. An "efficient" approach might call for transferring loads in real time among various pieces of equipment so as to achieve the most even distribution and avoid overloading any one piece. This approach both maximizes asset utilization (and may even help avoid capacity upgrades to accommodate demand growth) and minimizes the inefficiency due to lines losses (since collective I^2R losses increase with uneven allocation of current among lines). Its success hinges, however, on constant vigilance and intervention. By contrast, a "robust" approach would emphasize strength and simplicity: the idea is simply to have enough extra capacity built into the equipment so that overloading is not an issue, and loads need not be tracked so carefully.

In summary, then, the system qualities that are most important for operators are stability, transparency, veracity, and robustness, which support them in their task of keeping the system in homeostasis. Not coincidentally, these criteria are generally associated with older technologies, designed and built in an era where operability by humans was more of a firm constraint than material resources. In power systems, stability and robustness were provided largely by oversized equipment and redundancy of components, while transparency and veracity were furnished through simple mechanical and analog instrumentation and controls. From the viewpoint of increasing the efficiency of such systems in today's world, process innovations guided by engineering criteria may be desirable indeed. From the operations perspective, however, such innovations may be expected to adversely affect performance reliability and especially safety. Thus, when steps are proposed toward more refined and sophisticated system operation, operators may identify potential backlash effects, in which opportunities for system improvement also introduce new vulnerabilities.

9.3.4 Implications for Technological Innovation

The cognitive representations used by engineers and operators, respectively, give rise to different ideas about what system modifications may be desirable, and divergent expectations for the performance of innovations. If one imagines a technical system in terms of an abstraction in which interactions among components are governed deterministically by formal and tractable rules, then (1) these formal relationships suggest ways of modifying individual system parameters so as to alter system performance in a predictable fashion according to desired criteria, and (2) it is credible that such modifications will succeed according to a priori analysis of their impacts on the system. From this point of view, innovative technologies hold much positive promise and little risk.

On the other hand, if one imagines the system as an animated entity with uncertainties that can never be completely isolated and whose behavior can be only approximately understood through close familiarity, then (1) modifications are inherently less attractive because they may compromise the tractability and predictability of the system, and (2) any innovation must be suspected of having unanticipated and possibly adverse consequences. From this point of view, efforts to modernize or automate may imply an attempt to squeeze the system into a conceptual mold it does not fit—treating an animal like a machine—and thus harbor the potential for disaster.

This is not to suggest that modernization efforts are ill-advised or necessarily conflict-ridden. Indeed, there are many examples of operators welcoming if not actively campaigning for improved and more efficient information and control technology. The preceding observations suggest that when concerns about technological innovation do arise, they may be based on legitimate professional considerations.

Consider the example of SCADA at the distribution level. Traditionally, distribution operators sitting in the control room at the distribution switching center have relied on their field crews as the main source—and sometimes the only source—of information about system status, whether switches (open or closed), loads (current through a given line or transformer), voltage levels, or the operating status of various other equipment (circuit breakers, capacitors, voltage regulators, etc.). The operators' lifeline to information has been the telephone or radio through which his "eyes and ears" in the field communicate. By the same method, operating orders (often written out in hardcopy beforehand) are verified, or modified orders communicated if necessary. The introduction of SCADA implied a transition from operating through field personnel to directly accessing the system via a computer terminal in the control room. The advantages are obvious: fewer man-hours are needed to execute a given procedure; things can be done much faster; the computer affords a clear, central overview—in short, the entire operation, still largely based on 19th- and early 20th-century technology, finally comes into the electronic age.

Nevertheless, the implementation of SCADA in the utility industry has not been entirely unproblematic. While many distribution operators are quick to point out the advantages of SCADA and indeed find it difficult to imagine how anyone ever worked without it, some have also offered critiques. The main areas of concern relate to safety (is the computer correct in reporting an open switch?), physical surveillance (sacrificed chances to discover any developing problems early on in site visits), time pressure (not having time to think while waiting for field crews to execute orders), loss of redundancy (not necessarily having a second person reviewing steps), and the loss of situational awareness (such as that afforded by audible communication among operators, as opposed to silent interaction with computer terminals).

As a result of such concerns, even when new information and control technology is successfully installed, operators may not always choose to make full use of the available capabilities. This must be expected especially if the technology pertains to a more automated operation that may threaten to make the system less stable, transparent, verifiable, and robust to operators. The key to successful innovations seems to be ample consultation and discussion across occupational and cultural groups, so as to make both engineering and operating concerns clear to all parties in the design process.

9.4 IMPLICATIONS FOR RESTRUCTURING

As of this writing, the most passionately debated subject in the field of power systems is the restructuring of the electric industry: the transition from a regulated, vertically integrated monopoly—the utilities of the 20th century—to some form of competitive marketplace for electricity. This can involve any aspect of production, delivery, and consumption, though the focus until now has been primarily on wholesale energy. While the tone of the debate often suggests a battle between two absolute alternatives—pure regulation versus pure market—every arrangement for the buying and selling of electric power to date has in fact been some hybrid of approaches, whether by economic incentives embedded within the regulated system or by governmental definition of market rules and boundary conditions. With efforts to deregulate electricity underway in the United States and abroad, there exists a growing literature on the merits and failings of various market designs, or ways of achieving a contemporary and effective hybridization. The one thing on which experts tend to agree is that the consequences to society, whether favorable or unfavorable, are important, profound, and potentially

9.4 IMPLICATIONS FOR RESTRUCTURING

far-reaching—so important that it is difficult to imagine publishing a text on electric power systems without mention of market restructuring.

Yet it also seems impossible to do the subject justice in the context of a technical reference book. Because electricity is such a central and indispensable part of our culture, the restructuring debate is about much more than kilowatt-hours and dollars. Our choice of means and institutions for dealing with electricity has to do with very fundamental and contentious questions about how society and business ought to be organized. A key set of questions concerns the extent to which competitive markets can produce socially optimal outcomes, what constitutes market failures, and when and how government intervention is required. At stake in the debate is an entire *weltanschauung* about appropriate roles of public and private sector, or degrees of faith in the "invisible hand" of the marketplace, issues obviously beyond the scope of this text.

This section aims for a compromise by outlining some of the intrinsic difficulties of electric power with respect to market function. Because of its physical nature and the attendant technical constraints in its production and distribution, electricity is not a commodity (like wheat, pork bellies, or petroleum) that one would find in the economics textbook. In an ideal competitive market, the intersection of demand and supply curves yields an equilibrium price and quantity at which the market clears and which, in theory, maximizes social utility or overall benefit to participants. The central requirement of such a market is that demand and supply can vary freely with respect to price. Yet for several reasons such a condition is very tricky to achieve in the case of electricity.

Demand for electricity tends not to respond very much to price signals—in technical terms, the *price elasticity of demand* is low—in part simply because electricity is so indispensable to our economy and our everyday lives. People and businesses are organized around being able to consume electricity at any time they choose, and many are willing to pay a considerable amount of money in order not to have their service interrupted.

Demand elasticity is further limited by institutional and technical factors. The entire planning and design philosophy of power systems and regulated utilities in the 20th century presumed that electricity should be available to anyone in essentially arbitrary amounts around the clock, at a known and fixed price deemed reasonable by public regulators. Electric load was defined as the independent variable whose satisfaction became the central objective of technical, organizational, and regulatory efforts throughout the power industry. Consumers only experienced rate changes on the time scale of years, from one utility rate case to another, rather than a "price signal" according to the much more rapid fluctuations of actual marginal generation cost. In many deregulated markets, consumers continue to be shielded from short-term price volatility and therefore receive no direct incentive to respond to power shortages by lowering their demand during critical hours, thereby making demand almost perfectly inelastic with respect to varying wholesale prices.

The ideal market solution would be to introduce *real-time pricing* or related incentive mechanisms to encourage *electric demand response*. Such programs may involve communication only or include direct control, either inviting a certain customer behavior or physically disconnecting designated loads under specific conditions. Demand response can be considered an extension of well-established mechanisms such as *time-of-use rates* that discriminate among predetermined on- and off-peak periods, or *interruptible tariffs* available primarily to larger customers willing to trade service continuity for reduced rates.[17] Nevertheless, generalizing such responsiveness with the expectation of routine implementation in real time for many small and diverse customers represents a considerable innovation. Fundamentally, demand response means turning upside down a century-old design philosophy and requires addressing a broad set of factors ranging from customer behavior, education, and economic preferences to control hardware, information management, and communications protocols—factors that had no reason to be considered in electric power systems until very recently. Thus, while it seems plausible in theory to imagine aggregate electric demand as a well-behaved function of price, such a transition in practice requires new ways of thinking as well as a new generation of technology.

The supply side harbors equally if not more serious challenges. A competitive market according to the economics textbook requires free entry and exit of firms in response to scarce or excessive supply. Yet in practice, market entry and exit are impeded by capital intensity, transaction costs, and construction schedules of generation projects.

The capital-intensive nature of electric generation along with economies of scale and a measure of risk make this industry intrinsically uninviting for smaller businesses. In theory, it should be possible for any entrepreneur to join the fray and supply kilowatt-hours when the market signals that they are needed. Yet in practice, the initial investment is too large and near-term profits too unpredictable for many small, independent companies to participate.

Interconnection arrangements, too, may be cumbersome if not prohibitive for small producers. Under *direct access*, an arrangement aimed at fostering competition in the generation sector while preserving a T&D monopoly, any generator can in principle avail themselves of the extant infrastructure by injecting electric power to be consumed by a purchaser elsewhere. However, the transaction costs of becoming an electricity seller through the grid are not negligible and may include fees, bureaucratic procedures, and special equipment to interface with the utility on its terms. At the same time, it remains illegal anywhere in the United States to run independent lines from a supplier to a consumer—say, a wire from my rooftop PV system to my neighbor across the street.[18]

---

[17]This is analogous to airline passengers accepting payment in return for yielding their seats on an overbooked flight, except that the electric customer faces only the possibility, not the certainty of interrupted service. Indeed, anecdotal evidence suggests that customers on interruptible tariffs are often surprised and dismayed when an interruption actually occurs.

[18]This restriction has been motivated by safety as well as the intention to avoid the inefficiency of duplicating infrastructure, though it is also arguably the greatest barrier to a truly competitive electricity market. One might imagine that the proliferation of distributed generation technologies could challenge this status quo; if so, it would amount to nothing short of a radical reinvention of "the grid" as we know it.

Finally, the time scale on which conventional power plants are planned, built, and licensed makes it all but impossible for new generation capacity to appear in response to acute shortages. And because existing generation facilities cannot arbitrarily increase output beyond their rated capacity,[19] the supply curve, too, becomes more and more inelastic as the number of megawatts approach the system capacity limit.

Not only is the quantity of supply constrained, so is the grid's ability to transmit it to the desired location. With increased utilization of transmission capacity and long-distance sharing of generation resources, transmission congestion has become as important as generation in limiting the available supply for many areas. At the same time, upgrading transmission capacity tends to be slow, expensive, and politically contentious, so that bottlenecks are not readily relieved by simply constructing more power lines. And congestion cannot be made to go away by administratively or economically "rerouting" power: as the previous chapters have emphasized, electric power flow obeys only physical law; it is difficult enough to predict, let alone control and direct. Therefore, the quantity of interest in electric markets is not just aggregate supply (as it would be in the market for a readily interchangeable and transportable commodity), but *available supply to any given region* that would need to respond to price signals in a truly competitive market.

As various designs for electricity markets have been discussed and tried, a central concern has been the potential exercise of *market power* by sellers. The question is whether market participants are *price takers* as in the textbook, or whether by their individual actions they have the power to influence the market clearing price and thereby manipulate outcomes in their favor—*gaming* the market. Anyone who has followed the news about the California electricity crisis and its political fallout is aware that gaming is possible; the interesting question is what factors enable or prevent it. It would stand to reason that as long as no individual supplier holds too large a market share, they cannot exercise market power. But because of the particular properties of the electric grid, "too large" could be a very small percentage. The extreme inelasticity of demand and supply as the system nears its limits makes it vulnerable to the withholding of even small amounts of generation capacity.

Moreover, because line flows are sensitive to the amount of power injected at specific locations, generation capacity can be used strategically not only to relieve but, in the case of unabashedly profit-maximizing firms, to deliberately create transmission congestion.[20] The unique problem here lies both in the utter dependence of the entire system on limited transmission assets and the subtlety of predicting power flow. The effect of generation at a given node on congestion at a certain

---

[19] It is quite possible for an electric generator to exceed its nameplate rating at the expense of equipment life span. Nevertheless, every generation unit ultimately encounters a firm, nonnegotiable physical limit on the amount of power it can deliver.

[20] Transmission congestion may raise regional wholesale prices, of which a company or financial interest group can take advantage if they also own generation facilities located in that region.

link is sometimes obvious from geography, in which case most experts agree that some form of regulatory intervention is warranted to prevent abuse of this information. At other times, it remains the subject of speculation, keen interest by calculating firms, and careful guarding by transmission system operators. The resultant paradox is that key information about the system must be withheld rather than openly shared in the interest of competitive market function, which is precisely the opposite of what standard economics would suggest.

Considering market equilibrium in the broader context of maximizing social utility, an additional problem pervades the energy sector: the *external costs* of many of its components, primarily the environmental impacts of various generation sources, that fail to be included in the apparent costs of production and consumption. This problem is not unique to the restructured industry, but has been grappled with in the context of regulated utilities since the 1970s and 1980s. Economists generally agree that the presence of such externalities warrants regulatory intervention in the form of financial incentives or explicit requirements to *internalize* these costs in the market, so as to allow participants to make decisions based on correct price signals. The "correct" price would then promote activities known to produce social benefit while discouraging those with hidden costs to society. Examples of such intervention range from portfolio standards for clean, renewable generation resources to publicly funded campaigns to promote energy efficiency. While analysts differ considerably on the type and extent of methods favored, it seems quite clear that externalities will need to be confronted in any future market design.

This list of problems does not rule out the successful application of market mechanisms to electricity in principle. Rather, it emphasizes that, owing to their technical characteristics and the reality of their present design, electric power systems pose serious, intrinsic challenges to the balancing of supply and demand at an equilibrium price and quantity. Without passing judgment on the ultimate possibility of overcoming them, it is fair to say that these challenges ought not to be underestimated. Moreover, appreciation of the inherent difficulties suggests particularly careful and explicit consideration of what is to be gained by restructuring the electric power industry.

Ultimately, a crucial problem in this regard concerns equity. Electric power systems have historically served as instruments for the reallocation of significant amounts of wealth across society—not in a revolutionary way but very much in line with cultural norms, like the graduated income tax. In the 1930s, federally subsidized rural electrification brought villages and farms onto the grid that never could have been connected economically, as revenues from these accounts clearly would not pay for the infrastructure investment. In fact, many of those rural areas might have been far more cost-effectively supplied with local sources such as wind power. But the electric grid was designed to reach out even to customers who were expensive to serve, while splitting the price tag among everyone. In this way, urban customers continue to subsidize rural customers, and large commercial and industrial customers, who are much less expensive to serve on a per-kWh basis, continue to subsidize smaller customers under regulated utility rates. A key impetus for deregulation has been the desire of such larger customers to procure less expensive electricity on their own through open market access, prompting considerable argument as to their obligation to share the cost burden of past utility investments. The debate about electric industry restructuring is thus very much about what is fair, who pays how much, and how society should decide.

The historical dimension brings to light just how much the complex technical system we call "the electric grid" is a social artifact as much as a fascinating incarnation of physics and engineering. We can view the electrification of industrial society not only as a logical result of technical and economic driving forces, but as an embodiment of ideas and values. The idea of connectedness in and of itself represented the sophistication and unity of an advanced, civilized society; being "on the grid" meant not only receiving electrons, but being part of progress in the modern world. By the same token, living "off the grid" today often implies independence, self-sufficiency, and sometimes environmentalism. Historically, the idea of equal access to the grid has represented a fundamental sense of social equality, with electricity considered not a privilege affordable to some but an entitlement of all citizens. As electric power systems evolve with new technologies and new organization, they will continue to embody social values, explicit or implicit. The better and the more widely electric power systems in all their complexity are understood, the greater the opportunity for people to guide this evolution with awareness and conscious choice.

APPENDIX

Symbols, Units, Abbreviations, and Acronyms

A	ampere	unit of current
A.C., a.c.		alternating current
ACE		area control error
AGC		automatic generation control
AM		amplitude modulation
amp	ampere	unit of current
ASD		adjustable speed drive
B		symbol for magnetic flux density
B, b		symbol for susceptance
BTU		British thermal unit, unit of energy
C, c		symbol for capacitance
C	coulomb	unit of charge
CAES		compressed-air energy storage
CAISO		California Independent System Operator
cal	calorie	unit of energy
CHP		combined heat and power
CVR		conservation voltage reduction
D.C., d.c.		direct current
DG		distributed generation
e		natural logarithm base, 2.71828...
E, e		symbol for voltage
E		symbol for electric field
ELF		extremely low-frequency fields
emf		electromotive force
EMF		electromagnetic fields
erg		unit of energy
EUE		expected unserved energy
F		symbol for force
F	farad	unit of capacitance
f		symbol for frequency

Electric Power Systems: A Conceptual Introduction, by Alexandra von Meier
Copyright © 2006 John Wiley & Sons, Inc.

FACTS		flexible a.c. transmission systems
FM		frequency modulation
G, g		symbol for conductance
G	gauss	unit of magnetic flux density
GFCI		ground-fault circuit interrupter
GMD		geometric mean diameter
GMR		geometric mean radius
H	henry	unit of inductance
H		symbol for magnetic field strength
hp		horse power
Hz	hertz	unit of frequency
i		imaginary number, $\sqrt{-1}$
I, i		symbol for current
ISO		independent system operator
j		imaginary number, $\sqrt{-1}$
J	joule	unit of energy
KCL		Kirchhoff's current law
KVL		Kirchhoff's voltage law
kV	kilovolt	unit of potential difference, 1000 volts
kVA	kilovolt-ampere	unit of apparent power
kW	kilowatt	unit of power, 1000 watts
kWh	kilowatt-hour	unit of energy
l		symbol for length
L, l		symbol for inductance
LCD		liquid crystal display
LDC		load duration curve
LED		light-emitting diode
LOLE		loss-of-load expectation
LOLP		loss-of-load probability
LTC		load tap changer
mmf		magnetomotive force
MRI		magnetic resonance imaging
MVA	megavolt ampere	unit of apparent power
MVAR	megavolt-ampere reactive	unit of reactive power
MW	megawatt	unit of power, one million watts
MWh	megawatt-hour	unit of energy, one million watt-hours
N	newton	unit of force in m–kg–s system
n		symbol for rotational rate
N, n		symbol for number of turns in a coil
N.C., n.c.		normally closed
NERC		North American Electric Reliability Council
N.O., n.o.		normally open
OCB		oil circuit breaker

OPF		optimal power flow
P		symbol for real power
PCB		polychlorinated biphenyls
p.f.		power factor
p.u.		per unit
PV		photovoltaic
PWM		pulse-width modulation
Q		symbol for reactive power
Q, q		symbol for charge
$\mathcal{R}$		symbol for reluctance
R, r		symbol for resistance
rad	radian	unit of angle
REC		recloser
rms		root mean squared
rpm		revolutions per minute
rps		revolutions per second
S		symbol for complex power
SCADA		supervisory control and data acquisition
SF_6		sulfur hexafluriode
SIL		surge impedance loading
SMES		superconducting magnetic energy storage
SVC		static VAR compensator
T	tesla	unit of magnetic flux density
T&D		transmission and distribution
THD		total harmonic distortion
UPS		uninterruptible power supply
V, v		symbol for voltage
V	volt	unit of voltage
VA	volt-ampere	unit of apparent power
VAR	volt-ampere reactive	unit of reactive power
VSD		variable speed drive
W	watt	unit of real (active) power
Wb	weber	unit of magnetic flux
X, x		symbol for reactance
Y	wye	symbol for set of three phase-to-ground connections
Y, y		symbol for admittance
Z, z		symbol for impedance
Δ	delta	symbol for set of three phase-to-phase connections
δ	delta	symbol for voltage or power angle
θ	theta	symbol for power or phase angle
η	eta	symbol for efficiency
λ	lambda	symbol for wavelength

μ	mu	symbol for magnetic permeability
μ	micro-	one-millionth
ν	nu	symbol for frequency of a wave
π	pi	ratio of a circle's circumference to diameter, 3.1415...
ρ	rho	symbol for resistivity
σ	sigma	symbol for conductivity
σ	sigma	symbol for surface-charge density
σ	sigma	symbol for Stefan–Boltzmann constant
Φ	phi	symbol for magnetic flux
φ	phi	symbol for magnetic flux or phase angle
Ω	ohm	unit of resistance
ω	omega	symbol for angular frequency

INDEX

Abstraction, 283, 291
Accuracy, 286
Adapter, 135
Adjustable speed drive (ASD), 134
Admittance, 40, 64, 215
Algebra, linear, 40, 222
Algorithm, 196, 264
Alternating current (a.c.), 7, 49, 127
  in generator, 89
Ambiguity, 284
Amp-hours, 269
Ampacity, 182
Ampere, 7
Amplitude, 50
Angular frequency, 51, 77
Antenna, 28
Apparent power, 115, 167, 208
Arc, 6, 190–191
Area control error (ACE), 263
Armature, 86
  current and load, 100
  reaction, 90, 96
  winding, 93
Automatic generation control (AGC), 263
Automation, 159, 278–281, 291–292
Availability, 266

Base-load, 100, 264–265
Batteries, 268–269
  and charge flow, 7, 12
Birds, 185
Black-start capability, 99
Blackout(s), 250, 263, 288
Brownout, 261

Bus(es), 100, 149, 197, 201, 207
  and load, 201
  P,V and P,Q, 204–206
  slack, 202, 207
  swing, 202
Busbar, 99, 197

California Independent System Operator (CAISO), 138
Capacitance, 55
  in loads, 127
  of transmission lines, 176
  shunt, 186
Capacitor(s), 58–60, 127, 270
  control of, 186
Capacity of equipment, 70. *See also* Ratings
  and distributed generation, 275
Characteristic impedance, 177
Charge
  electric, 1, 2
  static, 1, 11
Christmas lights, example, 40
Circuit, 12, 30
  behavior and analysis of, 30, 200
  breakers, 151, 157, 189
  closed, 13
  elements, 30–31
  lumped, 8
  magnetic, 46
  open, 13
  short, 16
Circulating current, 155, 239, 241
  and delta-wye connections, 175
  between generators, 110–111, 113

Electric Power Systems: A Conceptual Introduction, by Alexandra von Meier
Copyright © 2006 John Wiley & Sons, Inc.

Coaxial cable(s), 59, 167
Cogeneration, 273
Combined cycle units, 273
Combined heat and power (CHP), 273
Communication, 278–279. *See also* Telephone
Commutation, 133
Complex conjugate, 68, 84
Complexity, 228, 259, 278, 282, 284, 297
Complex numbers, 61–63
Complex plane, 62, 76
Complex power, 68–71, 214
Compressed air energy storage (CAES), 270
Computer(s), 223, 231, 280, 289
 as load, 135–136
Condensors, synchronous, 186
Conductance, 10, 33, 64, 66, 215, 224
Conductivity, 5, 6, 10
Conductor(s), 175
 diameter of, 10
 and heating, 98
Congestion, 154, 295
Conservation voltage reduction (CVR), 252
Contingencies, 233
Contingency analysis, 234
Contracts, bilateral, 265
Control, 289–290
Copper losses, 170–171
Corona, 178
Coulomb, 3
Cross product, 24
Culture(s), 281, 285
Current, 6–7
 alternating. *See* Alternating current
 circulating. *See* Circulating current
 direction of, 7
 induced. *See* Induced current
 in power flow analysis, 199
 inrush, 132
 propagation speed of, 8
 starting, 132
 time derivative of, 57
Cycle(s), 49, 95, 261

Damping force, 242, 244
Degrees of angle, 50
Delta and wye connections, 143, 164
 grounding of, 166
 in transformers, 171–174

Demand, 136, 268
 coincident and non-coincident, 137–138
 elasticity of, 293
Demand response, 293–294
Deregulation, 292, 296
Derivative, 57, 60, 80, 219
 partial, 218, 226
Device, 30
Dielectric material, 11, 59
Differential equation, 242
Dimension, 1, 82
Dimmer, 130
Diodes, 31
Direct current (d.c.), 49, 127, 195, 268
 circuit breaker, 191
 generator, 90
 motors, 133
 transmission, 160, 167
Direct access, 294
Dispatch, 206
Distributed generation, 271–278
 and radial distribution systems, 275
 availability of, 276
 siting of, 274
Distribution, primary and secondary, 148
Distribution system
 radial, 150, 194
 network, 150, 152

Earth, 5. *See also* Ground
Economic dispatch, 264
Economics, 137, 292, 296
Economies of scale, 144–145, 274, 294
Edison, Thomas, 49
Efficiency, 285–286
Electrification, 296–297
Electrocution, 287. *See also* Shock
Electromagnet, 44, 55, 92
Electromotive force (emf), 12, 85, 101
 in generator, 88, 102
 in transformer, 169
Electrons, 2, 5, 269
Energy, 3–4, 135–136
Energy conservation, 27, 56, 72, 90, 112, 243, 260, 261
Engineering, 281–282
Equal-area criterion, 248–249

304 INDEX

Equilibrium, 242, 261, 282
  deep and shallow, 239
  generator frequency, 108
  stable and unstable, 235
Equity, 296
Euler's equation, 79
Excitation, 92–93, 117, 132
Exciter, 93, 99
Expected unserved energy (EUE), 230
Expert systems, 279
Exponential base e, 78
Exponential function, complex, 79
Extension cord, 10, 41
External costs, 296

Farad, 59
Fault, 188
  current, 193
  transient, 191
Feeder, 149
  lateral, 149
  primary, 151
Field(s), 18–19
  electric, 19
  electromagnetic (EMF), 25
  extremely low-frequency (ELF), 25
  magnetic. See Magnetic field
Field lines, 20–21
Flat start, 217
Flexible a.c. transmission systems (FACTS), 280
Flicker, 50
Fluorescent lamps, 127, 130
Flux, magnetic, 23, 44, 87, 169
  density, 46
  in generator, 103
  leakage, 46, 169
  linkage, 47, 88, 120
Force on a charge, 20–21, 24
Frequency, 49–50, 204, 236, 253–254
  and power consumption, 260
  and stability, 244
  angular, 51
  choice of a.c., 50
  control, 261
  tolerance, 168, 254
Friction, 242, 244
Fuel cells, 273
Fuse(s), 189, 192–193, 262

Gas turbines, 273
Gauss, 23
Generator(s)
  capacity, 115, 295
  conversion of energy in, 85
  cooling, 98
  damage, 115
  induction. See Induction generator
  rating, 116, 295. See also Ratings
  synchronous. See Synchronous generator
  windings, 98
Geometric mean radius (GMR), 178–179
Governor, 101, 261
Gradient, potential, 20
Grid, design and evolution of, 144, 259, 277, 280, 296–297. See also Power system
Ground, 4
  in outlet, 140–141
Ground-fault circuit interrupter (GFCI), 190
Ground return, 160

Harmonic(s), 126, 255–256
Harmonic distortion, 53, 123, 251
Heat dissipation, 136, 171–172, 178, 182. See also power dissipation
Heating, resistive, 14–15. See also Losses, resistive
Helix, 76, 79
Henry, 57
Hertz, 49
High-pass filter, 59
Horsepower (hp), 131
Hydrogen
  energy storage, 271
  generator cooling, 98–99

Imaginary axis, 76
Imaginary number i or j, 61
Impedance, 64, 72, 127, 197
  characteristic. See Characteristic impedance
Impulse, 252
Independent system operator (ISO), 146, 263, 264
Induced current, 25, 56, 85, 87

INDEX 305

Inductance, 55
 as function of flux linkage, 48
 of transmission lines, 176
 mutual, 48
Induction, 24, 85
 generator, 86, 118, 272
Inductor, 55
Information, 279, 286, 288, 296
Innovation, 287, 291
Insulator(s), 10, 11
 on transmission lines, 180
Interconnection, 144–148
Intertie(s), 144, 168
Intuition, 284–285
Inverter(s), 123–126, 268
Ionization, 6, 191
Ions, 2, 6, 269
Iron losses, 170–171
Island(ing), 152–153, 277
Iteration, 195–196, 218, 223

Jacobian matrix, 218–219, 221, 225–226
Joule, 4

Kilowatt-hour, 66
Kirchhoff's laws, 37, 153, 263
Kirchhoff's current law (KCL), 32, 39
Kirchhoff's voltage law (KVL), 32, 38

Lagging current, 56, 77
 and phase angle, 64
Leading current, 60
 and phase angle, 64
LED clocks, 136
Light, 26
 speed of, 8, 27, 212, 288
Lightning, 6, 181
Limit, thermal and stability, 234.
  See also Ratings
 stability, 238
Line losses. *See* Losses
Line drop, 185. *See also* Voltage drop
Load, 127
 aggregate, 136, 146
Load duration curve (LDC), 138–139, 264–265
Load balancing
 among feeders, 159, 279
 among phases, 163

Load factor, 140, 144, 146
Load flow, 195. *See also* power flow
Load forecast(ing), 137–138, 267
Load growth and distributed generation, 275
Load-following plant, 204, 264–265
Load profile, 138
Load tap changer (LTC), 170, 185
Loop flow, 153
Loop system, 151–152
Lorentz equation, 24
Loss-of-load-expectation (LOLE), 230
Loss-of-load-probability (LOLP), 229
Losses
 and distributed generation, 274
 and load balancing, 159
 d.c. line, 49
 heat. *See* resistive
 line, 74, 201, 226, 263. *See also* resistive
 reactive, 74, 203, 208–209
 resistive, 15, 17, 67, 203, 209.
  See also line
Low-pass filter, 57

Magnetic field, 21, 23, 44, 46
 in armature, 94–95
 in inductor, 55
 in generator, 87, 90
 of rotor. *See* Rotor field
Magnetic flux. *See* Flux, magnetic
Magnetic moment, 22, 170
Magnetic poles, 21, 44
 and rotor windings, 97
Magnetization, 22
 curve, 46
Magnetomotive force (mmf), 47, 169
Maps, 284
Market(s), 266, 277, 292–294, 296
 and power flow, 227
Market power, 295
Matrix, 36, 40, 215, 222
Metals, 5
Mho, 10
Microturbines, 273
Microwave oven, 135–136
Mismatch, 218, 222
Monitors, TV and computer, 136
Monopoly, 137, 147, 232, 292

306 INDEX

Motor(s), 85, 131–134
 efficiency, 133–134
 energy consumption by, 131
 induction, 119, 122. See also Induction generator
 paper clip, 90–91
 single- and three-phase, 133
Multiphase. See three-phase

Negative sequence, 95
Network reduction, 35
Neutral, 140, 162, 164
Newton's method, 220–221
Node(s), 40, 197, 217, 289, 295
North American Electric Reliability Council (NERC), 147
Nuclear power, 264, 271–272

Oersted, Hans Christian, 22
Ohm, 9
Ohm's law, 8, 81, 223, 251
Oil, 172
One-line diagram, 149, 159, 166, 197–198
Operation
 challenges, 227, 282. See also System operator
 skill, 262
Optimal power flow (OPF), 196
Oscillation, 61
Outage, 158–159, 230, 250, 263
Outlet, 140–141

Parallel connection, 31, 33, 36
Paralleling of generators, 111, 262
Peaking plant, 264–265
Penalty factor, 264
Per-unit (p.u.) notation, 208
Period, 27, 50
Permeability, 63, 92
 constant, 45
Phase, 50
 conductor, 140
 in generator windings, 94
Phase angle, 199
 complex power, 84
 exponential notation, 82
 FACTS, 280–281
Phase shift
 between voltage and current, 69

 in capacitor, 60
 in inductor, 56
Phasor(s), 76, 80, 216
Phasor diagram, 77, 83
Phasors
 notation of, 78, 81
 operations with, 80–82
Philosophy, of electric service, 136
Photon(s), 28
Photovoltaics (PV), 126, 273, 294
Planning, 227, 267
Plasma, 6, 190
Polychlorinated biphenyls (PCBs), 172
Positive sequence, 95
Potential, 3, 4. See also Voltage
Power, 15, 136
 active, 70
 apparent, 68
 average, 68, 70
 complex, 84
 dissipated and transmitted, 67. See also Losses
 dissipation in load, 128
 instantaneous, 68
 standby, 135
 reactive. See Reactive power
 real, 70
 output from generator, 100
 transmitted on line, 67, 183, 237
 true, 70
Power angle, 110, 183, 204–205, 236–237, 239. See also Voltage angle
 and circulating current, 110
 difference between nodes, 240
 in generator, 108
Power factor (p.f.), 70–72, 167
 aggregate, 73
 and induction generator, 122–123
 and rotor field, 104
 of fluorescent lamps, 75
 of generator, 11, 123
 significance of, 73
 system, 115
Power flow, 195
 analysis, 154
 controlled by FACTS, 281
 decoupled, 221, 224–225
 direction of, 150
 equations, 217

nonlinearity of, 200
   solution of, 195, 200
Power pool, 265
Power quality, 118, 250–258
   and motors, 134
Power supplies, 135
Power system(s)
   behavior, 210
   design and evolution, 49. *See also* Grid
   interconnection of, 146–147
   operating state, 195
   structure of, 149
Precision, 286, 289
Predictability, 283
Price, 137, 145, 268, 293, 295–296
Prime mover, 86, 89
Protection, 150, 188–194, 262
   and distributed generation, 275
   zones, 192–194
Protons, 2
Public Utility Commission (PUC), 232
Pulse-width modulation (PWM), 125–126
Pumped hydroelectric power, 269–270

Radians, 51
   per second, 57
Radiation, 26
Radio, 28
   as example of load, 30
Ramp rates, 265–266
Rating(s), 295
   dynamic, 182
   emergency, 182
   generator, 116
   historical trend of, 116, 183
Reactance, 55
   capacitive, 58
   inductive, 57
Reactive capability curve, 116–117
Reactive power, 70, 72, 102
   allocation of, 112, 114
   and distributed generation, 274–275
   and generator, 102
   and motors, 132
   and voltage, 187
   balance of, 202, 204, 260
   conservation of, 112
   consumed or supplied, 71
   direction of flow, 208

from generator, 203
Recloser, 191–192
Redundancy, 150, 290, 292
Refrigerator, 137
Relay(s), 157, 189, 254, 262
Reliability, 229, 231–232, 285
Reluctance, 45, 169
Reserve(s), 144, 233
   spinning, 266
Reserve margin, 146, 229, 268
Resistance, 9, 175
   combined parallel and series, 33–35
   effect on power consumption, 16
   per unit length, 10
Resistivity, 9–10
Restoration, 159, 264
Rheostat, 130
Right-hand rule, 22–24, 55
Robustness, 290
Root-mean-square (rms), 53, 199
   derivation of, 54
Rotational frequency, 96
Rotor, 86, 92
   cylindrical, 97
   round, 97
   salient pole, 97
   squirrel-cage, 120
   wound, 120
Rotor current, 93
Rotor field, 92
   in induction machine, 121–123
   interaction with stator field, 90
Rotor winding, 92
Rubber bands, 212–213

Safety, 192, 285, 287
Sag, 253, 260
Scheduling coordinator, 264
Security, 229, 233
Self-excitation, 99
Series connection, 31–32
Service, 232
   valuation of, 232–233
Shedding of load, 159, 230, 263
Shock, 13–14
Siemens, 10
Significant figures, 74
Sine function, 51–52, 80
   argument of, 50

308 INDEX

Sine wave, 50
Sinusoid, 50, 88, 89, 256, 258
Situational awareness, 292
Slip, 118, 132
Solar power, 271. *See also* Photovoltaics
Solar thermal generation, 277
Solenoid, 55, 92
Solid-state technology, 124, 281
Source, current or voltage, 42
Speed, 286, 288
Spike, 252
Spin, 22
Square wave, 124–125
Stability, 109
   angle and voltage, 234
   as operational criterion, 288
   dynamic, 213, 235, 240, 244. *See also* transient
   steady-state, 213, 235
   transient, 233, 240, 248
   voltage, 249
Stability limit, 167, 183, 212, 280
Static VAR compensators (SVC), 186
Stator, 92
Stator field, 95
   decomposition of, 105
Steady-state condition, 211
Steam, 86
Steam generation, 271, 282, 287
Storage, 260, 268
   valuation of, 271, 277
Substations, 149
   distribution, 157–158
Subtransmission, 148
Sulfur hexafluoride (SF_6), 171–172
Superconducting magnetic energy storage (SMES), 270
Superconductivity, 6, 10, 213, 270
Superposition principle, 41, 73, 154
Supervisory Control and Data Acquisition (SCADA), 158, 278, 291–292
Surge impedance, 177
Surge impedance loading (SIL), 177, 184
Susceptance, 65–66, 215, 224
Swell, 252
Swing equation, 242
Switch(es), 157, 292
Switchgear, 190
Switching operations, 264

Synchronicity of generators, 108, 111
Synchronism, 211, 236
Synchronization, 111
Synchronous generator, 86, 92–95
Synchronous speed, 118
System operator, 227, 262, 266. *See also* Independent system operator

Taylor series, 219, 221
Telephone, 263, 278–279, 291
Tension, 3. *See also* Voltage
Terminal(s), 30
Territory(ies), 147, 263
Tesla, 23
Tesla coil, 178
Thermal limit, 182, 211, 213, 280. *See also* Ratings
Thermodynamics
   first law of, 27
   second law of, 203
Three-phase
   generator winding, 94
   transmission, 140, 160–163, 166
Thyristors, 167
Time
   and system frequency, 255
   as angle, 51
   as variable in a.c. circuits, 199
   conceptualization in synchronous grid, 211–212
   in operations, 260, 292
   in stability analysis, 245
   sensitivity of resources, 278
Time-of-use rates, 294
Torque, 100, 132, 160, 261
   for induction machine, 119, 122
   on generator rotor, 94
Total harmonic distortion (THD), 256–257
Transformer(s), 147, 157, 168
   and a.c., 49
   and harmonics, 256
   capacity limit, 171
   cooling, 171
   core, 169
   distribution, 150, 156
   primary and secondary side, 168
   tap(s), 141–142, 170
   turns ratio, 170

INDEX **309**

Transformer bank, 157, 173
Transistor(s), 45, 280
Transmission line(s)
  capacity, 295
  clearance, 180
  dimensions of, 177–178
  reactance of, 50
  sagging, 180
Transmission system, 148
Transparency, 288
Turbine, 86, 96
  and stability, 244

Uninterruptible power supply (UPS), 230, 231, 270
Unit commitment, 264
Utility(ies), 147, 232, 253, 258, 263

VAR compensation, 74.
  See also Reactive power
Variable speed drive (VSD), 134
Variables in power flow, 198–199, 206
Vector(s), 40, 64, 95, 222
  addition, 116
Veracity, 289
Voltage, 3
  and power consumption, 260
  and resistive loads, 129
  at generator terminal, 93, 97, 102
  at node, 216
  control in generator, 107
  difference between generators, 109
  excessively high or low, 251
  in power flow, 199
  line-to-line and line-to-ground, 181.
    See also phase-to-phase and
  phase-to-ground, 142, 165, 174–175, 257
  time derivative of, 60
  transmission, 17, 147
  tolerance, 118, 185, 251–252
Voltage angle, 204, 207, 217, 236
  in generator, 108
Voltage dip, 253
Voltage drop, 13, 67, 251
Voltage level, 260
Voltage profile, 184, 187, 266
Voltage regulator, 185, 187
Volt-ampere (VA), 70, 182.
  See also Apparent power
Volt-ampere reactive (VAR), 70.
  See also Reactive power
  supplied by generator, 103
Vulnerability, 147, 233, 276, 282, 291

Water
  flow analogy with electricity, 38, 39
  properties of, 3
Watt, 15
Watt-hours, 66
Wave, propagation of, 27
Waveform, 94, 257–258, 281
  of inverters, 124–125
Weber, 23
Westinghouse, George, 49
Wild a.c., 123
Wind power, 264, 271–272, 277
  and generator rating, 115
Work, 3–4, 89
Wye connection, 143, 164
  grounding of, 166